LABORATORY AUTOMATION USING THE IBM PC

STEPHEN C. GATES
IBM T. J. Watson Research Center

with JORDAN BECKER
IBM T. J. Watson Research Center

D1402840

PRENTICE HALL, ENGLEWOOD CLIFFS, NEW JERSEY 07632

Library of Congress Cataloging-in-Publication Data

GATES, STEPHEN C.
    Laboratory automation using the IBM PC / Stephen C. Gates
with Jordan Becker.

    p.   cm.
    Bibliography: p.
    Includes index.
    ISBN 0-13-519828-3
    1. Automatic control—Data processing. 2. Computer interfaces.
3. IBM Personal Computer. 4. Laboratories—Automation—Data
processing. I. Becker, Jordan. II. Title.
TJ223.M53G38 1989   88-32108
629.8′95—dc19   CIP

The author and publisher of this book have used their best efforts in preparing this book. These efforts include the development, research, and testing of the theories and programs to determine their effectiveness. The author and publisher make no warranty of any kind, expressed or implied, with regard to these programs or the documentation contained in this book. The author and publisher shall not be liable in any event for incidental or consequential damages in connection with, or arising out of, the furnishing, performance, or use of these programs.

Editorial/production supervision: *Debbie Young*
Cover design: *Diane Saxe*
Manufacturing buyer: *Mary Noonan*

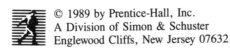

© 1989 by Prentice-Hall, Inc.
A Division of Simon & Schuster
Englewood Cliffs, New Jersey 07632

IBM is a registered trademark of International Business Machines Corporation.

Printed in the United States of America

10 9 8 7 6 5 4 3 2 1

ISBN 0-13-519828-3

ISBN 0-13-521295-2   {PBK.}

PRENTICE-HALL INTERNATIONAL (UK) LIMITED, *London*
PRENTICE-HALL OF AUSTRALIA PTY. LIMITED, *Sydney*
PRENTICE-HALL CANADA INC., *Toronto*
PRENTICE-HALL HISPANOAMERICANA, S.A., *Mexico*
PRENTICE-HALL OF INDIA PRIVATE LIMITED, *New Delhi*
PRENTICE-HALL OF JAPAN, INC., *Tokyo*
SIMON & SCHUSTER ASIA PTE. LTD., *Singapore*
EDITORA PRENTICE-HALL DO BRASIL, LTDA., *Rio de Janeiro*

CONTENTS

PREFACE

INTRODUCTION

Welcome to the world of laboratory automation! You are probably about to embark on a project in interfacing a scientific or industrial instrument to a computer. This book is an attempt to provide you with the tools to do that task and do it well.

The tools we provide in this book should allow you to:

- Attach a computer to an instrument.
- Control the instrument.
- Collect data from it.
- Analyze the data your computer has collected.
- Present the analyzed data in a fashion that makes it useful.

We have assumed in writing this book that you would like to learn both the hardware and software aspects of laboratory automation. We have tried to integrate these two equally important aspects into a single coherent text. In addition, we have tried to provide numerous examples to guide you -- because we have found that the best way to learn about laboratory automation is to see how others have done it.

HARDWARE AND SOFTWARE SUGGESTIONS

In order to take maximum advantage of this book, you will need to have an IBM PC, IBM PC/XT, IBM PC/AT, IBM PS/2 Models 25 or 30, or one of the many IBM PC-compatible computers. However, most of the examples of both hardware and software we describe in this text can be adapted to the PS/2 and even to other completely different microcomputer systems as well.

In order to use this text most efficiently, we recommend that your IBM PC have the following:

- Approximately 640K memory (i.e., random access memory), although the examples in this book can be worked with considerably less memory.

- A hard disk plus a floppy diskette (i.e., an XT or an AT), although a single floppy disk system is adequate for the material in this book. Most scientific users will prefer the PC/AT.

- A graphics adapter and monitor (color is nice but optional). For example, you might use either the IBM Color Graphics Adapter with a color monitor, or the IBM Extended Graphics Adapter with either the IBM Monochrome Display or the IBM Enhanced Graphics Display. The important criterion is that you be able to do on-screen graphics. Note that the IBM Monochrome Display can not display graphics when used with the IBM Monochrome Display Adapter. In our own labs most of our systems contain both the monochrome display adapter and a graphics adapter, so that we have one display for text and one for graphics.

- A math coprocessor chip (8087, 80287 or equivalent) is highly recommended.

- A printer and printer adapter.

- DOS 3.0 or later release.

- IBM BASICA or other BASIC interpreter. All of the programs in this book are written in BASIC because that is the one language that comes with every PC. *For most users we recommend a more modern version of BASIC* (such as Microsoft QuickBASIC) that has support for the math coprocessor and is more structured. You may have some other language you prefer, and we encourage you to convert the programs to that language. *We have tried to write programs in structured BASIC so that they can be easily converted to other languages*. Common languages in the laboratory other than BASIC are C, Pascal, Fortran, and Forth, and we encourage you to use whichever language seems most familiar or appropriate.

- A BASIC compiler compatible with the BASIC interpreter you are using. For example, the IBM BASIC Compiler, Version 2.0 or later is designed to be compatible with BASICA. QuickBASIC can also be used in the compiled mode, and is again our preference over the older IBM Basic Compiler for most applications.

- An analog and digital I/O interface board. Most analog and digital I/O boards contain an analog-to-digital converter, a digital-to-analog converter, a digital I/O section, and a timer. You will probably also need to purchase cables and a connector panel to accompany the board. You may wish to read Chapter 3 before purchasing the board in order to understand the many options available for the analog-to-digital converter portion of the board.

- A serial port (e.g., IBM's Asynchronous Communications Adapter). Many of the multifunction memory expansion boards currently available include one or more serial ports, and the IBM PC/XT and PC/AT both already include a serial port in their basic configurations.

- A board implementing the General Purpose Interface Bus (IEEE-488) protocol. Because of the complexity of this board, we also suggest purchasing the commercial software from the same manufacturer. (For example, we recommend using the National Instruments General Purpose Interface Bus software when using the IBM GPIB board.)

We have used data acquisition and control boards sold by Data Translation, National Instruments, and IBM as examples in the text, but those by other manufacturers would serve as well. However, please note that most of the software developed in this book will need to be modified before it can be used on boards of other manufacturers.

DESCRIPTION OF SCOPE OF THIS TEXT

The organization of this text, as previously mentioned, is to integrate the hardware and software aspects of laboratory automation. In the first part of the text we discuss the hardware and software aspects of laboratory data collection and instrument control. Thus, for example, Chapter 2 discusses both the theory and applications of digital-to-analog converters; these are often used to synthesize analog signals that control devices such as recorders and oscilloscopes. Chapter 3 then discusses the critical topic of analog-to-digital (A/D) converters and software techniques for using them. It also includes sections on connecting the A/D converter to instruments, and preamplifiers for the A/D.

This leads naturally into a discussion of noise reduction techniques, because analog signals are particularly prone to noise problems. Hence, Chapter 4 presents both hardware and software techniques for reducing noise in data collected by the A/D.

Chapter 5 discusses digital input and output techniques that can be used to sense switches, close relays, and read digital output from some types of devices. Chapter 6 describes the use of the IEEE-488 protocol that is being used increasingly in the laboratory for transferring data to and from instruments at very high speeds. Chapter 7 discusses serial interfaces; these are used both to communicate with other computers and to interact with devices such as printers, plotters and some laboratory instruments. Because data collection often must occur at precise intervals, Chapter 8 discusses the use of timers and counters. The first part of the text closes with Chapter 9, which suggests ways to combine the types of hardware and software described in the earlier chapters.

Chapter 10 begins the second part of the text, in which we describe some useful techniques for analyzing laboratory data. Chapter 10 describes software techniques for making the data collected in the laboratory more useful, emphasizing practical, "user-friendly" programs. Chapter 11 contains a discussion of how to analyze data that contains peaks, with an emphasis on chromatography data analysis. In the last chapter in this section, Chapter 12, we introduce you to image analysis techniques.

In the third part of this book we describe some techniques that can be employed to increase the power and efficiency of laboratory computer usage, but that require a more sophisticated application of both software and hardware. Specifically, Chapter 13 describes the use of local area networks to speed the flow of data between computers. Chapter 14 contains the details of using assembly language for applications that require higher speeds than those obtainable with high-level languages such as BASIC, and Chapter 15 describes interrupts and direct memory access as methods of increasing the efficiency of the computer in the laboratory.

Although it is at the very back of the book, the Scientific Routines Diskette is one of the most important parts of this book. This includes all of the programs listed in the text. This software is described in detail in Appendix B. Appendix B also contains information on several routines included on the diskette that utilize the IBM Data Acquisition and Control Adapter; these routines are similar to programs listed in the textbook for the DT2801. All of the software included on the diskette is written specifically for this book, but it is based on the authors' experience in writing software for their own applications. We have included this diskette to make it easy to try out the software described in the book, and to modify it for your own applications.

Hence, this book is about a computer that has become an industry standard, and about a computer with which we have a great deal of personal experience. However, it is also in a more general sense a book about using personal computers in the laboratory. No longer, for example, do most of us care about building our own laboratory data acquisition boards; those days are gone. No longer do we worry about coding everything in machine or assembly language to save memory space; those days are also gone. The personal computer has made it possible to concentrate more on the issues that really count--like how can we get the data out of our instruments most efficiently, how can we ensure the quality of that data, how can we present summaries of our data most effectively, and how we can process the data to obtain the needed scientific or engineering information.

Thus, we have tried to give you all of the help you need to automate your own laboratory. No particular knowledge of electronics is required. If you follow the guidelines in this text, you should be able to develop an interface for almost any instrument in a very short time and with a minimum of difficulty!

DISCLAIMERS

Although we have made every effort to locate any errors in this textbook and to test the programs included on the diskette, we remind the reader that neither we nor IBM nor Prentice Hall warranties any information or programs contained in this book, except for the replacement of the diskette if it is faulty. In addition, mention of specific vendors of hardware or software should not be construed as an endorsement of those vendors or their products.

ACKNOWLEDGMENTS

This is unabashedly a book about using the IBM PC in the laboratory. It summarizes our years of experience with what has become the *de facto* standard in the world of laboratory computers. It grew out of a course that I taught when I was a biochemistry professor, but reflects our combined years of experience with the IBM

PC in our own research environment. We have each shared our experiences; Jordan Becker, by writing almost all of Chapter 4 and some of Chapter 3; I, the remainder.

However, the book is also the result of interactions with literally hundreds of other scientists who over the years have shared their expertise on computers and scientific instruments with us, and who not infrequently generously shared programs they had written for use in their own laboratories. We owe all of these women and men a considerable debt of gratitude. In particular, we appreciate the generosity of Gerard Kopcsay and Norman Brenner here at IBM sharing with us the BASIC program for computing FFTs which forms the basis of Program 4.2.

We also would like to acknowledge the individuals at other companies who have helped us along the way. Almost from the day of the PC's release, dozens of smaller companies have developed literally hundreds of boards and other products to be used in the laboratory with the IBM PC. It is largely to these companies that we owe the overwhelming use of the PC in the laboratory. We particularly appreciate the efforts of individuals at these companies to answer technical questions when we called.

In putting this book together, we have, of course, used IBM PCs to prepare the material. We edited the book at our PCs, and then loaded it via a local area network connection to one of the IBM 3090 mainframe computers for printing. We prepared rough sketches of some of the figures at our PCs. We made changes suggested by the Prentice Hall editorial staff at our PCs. We controlled the printing of the final, camera-ready copy from our PCs. Hence, we take full responsibility for any errors that may remain in the text.

In preparing the book, however, we got much needed technical help from several individuals. In particular, Joan Dunkin here at IBM spent a considerable amount of time writing a text editor (SCRIPT) profile that corresponds to the Prentice Hall standards. And Tim Bozik and his team at Prentice Hall spent much time guiding us through the intricacies of publishing the book. We particularly appreciate Tim's willingness to accommodate both the inclusion of the diskette with this text, and a variety of non-conventional methods of producing the copy for the text, and Debbie Young's help during the book's production.

Finally, I would like to express my appreciation to my wife, Rita O'Donnell, for her patience and support during the lengthy process of writing this book.

To all of these people we give our sincere thanks.

S.C.G.

PART ONE

DATA COLLECTION AND INSTRUMENT CONTROL

1

INTRODUCTION

1.1 WHAT IS LABORATORY AUTOMATION?

What exactly is laboratory automation? In most modern applications, laboratory automation involves the use of computers or computer-based systems to control one or more instruments, collect data from those instruments, and process the data to make it useful to the scientist or other user. In a less formal sense it involves hooking up a computer to an instrument and writing some software. Thus, laboratory automation involves most of the computer-based tasks routinely performed by scientists and engineers.

The definition of laboratory automation also can be expanded to include the area commonly referred to as laboratory robotics, in which a computer-controlled device is responsible for some or all of the sample handling during the experiment. It can even include many kinds of large-scale applications in the factory or laboratory such as process control and laboratory information management. However, we shall not discuss these areas in this book.

Laboratory automation can be applied in areas from physics to physiology to environmental control. The details of laboratory automation vary tremendously, however. Laboratory automation can be as simple as plugging a data acquisition board into your computer and using commercial software to collect and process the data. Or it can be as complicated as building a room full of custom control and data acquisition hardware and writing a complete operating system in assembly language.

Most modern automation projects, fortunately, fall much closer to the first case than the second.

Typically, laboratory automation includes the following steps:

- Purchasing one or more data acquisition and control boards.
- Electrically connecting the boards to your instrument.
- Writing data acquisition and control programs.
- Writing data analysis software.

These steps are, of course, very general. Perhaps you can obtain a better feeling for laboratory automation by examining a few illustrative examples.

1.1.1 Example 1: Automating a Polarograph

Sometimes, automating an instrument can be quite easy. For example, as part of a class in laboratory automation, one of the students chose to interface a polarograph used in an analytical chemistry class to an IBM PC. He was required to complete the project in one week, so he decided not to attempt computer control of the instrument. Thus, the steps in his project were the following:

- He used a commercially-available data acquisition board; it contained a 12-bit analog-to-digital converter that could be used to acquire signals in the -10V to +10V range.
- He connected the data acquisition board directly to the strip-chart recorder outputs of the polarograph. This process was straightforward because the polarograph already had a commercial signal conditioner that amplified and filtered the signal to provide a 0 to +10 V output.
- Data collection was accomplished using a public domain assembly-language data collection routine called from a program written in BASIC. (See Chapter 14 for similar programs.)
- Thus, the major task faced by the student was to analyze the data collected. This involved two different methods for calculating the concentration of the substance being analyzed: finding the inflection point in a curve that had a sawtooth waveform superimposed on it, and integrating a curve. The results were plotted on a screen, printer or plotter and a permanent set of information typed out for the user on a printer.

This program was written in BASIC and subsequently converted to compiled BASIC to increase its speed. With only a few minor modifications (like adding the the fight song of the analytical chemistry faculty member's alma mater), it is still routinely used by students taking the analytical chemistry class.

1.1.2 Example 2: Automating a Time-of-Flight Mass Spectrometer

Often it is possible to use pre-existing hardware or software components to speed the automation process. An example of this occurred during automation of a time-of-flight mass spectrometer used to measure laser-produced reaction products. In this case, the automation process could be viewed as the following:

- For data acquisition, a system was purchased that would gather data from the time-of-flight mass spectrometer. This system required the use of an IEEE-488 interface at very high speeds, so an IBM General Purpose Interface Bus board was installed in an IBM PC/AT. The electrical connection between the experimental apparatus and the computer was accomplished very simply with an IEEE-488 interface cable.

- The physicists running the experiment needed to be able to see the results of the experiment as the data were collected, but the data rate was very high. Hence, software was written in BASIC to manage the data acquisition, but the real-time data processing -- the actual data acquisition, sorting the data (putting it into "bins"), and displaying the data as soon as it was processed -- was done in assembly language.

- All of the post-run data processing was done either on the IBM PC using general-purpose data analysis software or on a mainframe computer to which the IBM PC was connected.

Thus, in this case, almost all of the time required for automating the experiment (roughly three weeks) was spent in writing data acquisition software. Very little time had to be spent on doing the data analysis, because pre-existing programs were used.

1.1.3 Example 3: Automating a Scanning Tunneling Microscope

Sometimes, however, a project can require months or even years of development time. An example of this is the automation of scanning tunneling microscopes, devices for imaging atomic details on surfaces. In this case, we desired to provide a completely general package that could be used in a wide variety of research efforts that used some type of high resolution microscopy. The steps included:

- Data acquisition boards were purchased that would acquire data at rates of 80 to 100 kHz. In addition, a 1024 by 1024 pixel high-resolution graphics board to display images from the microscope was purchased along with a suitable monitor. Control of the instrument required four digital-to-analog converters.

- Because the scanning tunneling microscopes were a newly-invented type of instrument, no electronics existed for them. For this reason, custom electronics were constructed to amplify and condition the signals from the instruments.

The analog-to-digital and digital-to-analog converter signals were then passed through the electronics on their way to or from the microscope.

■   The data acquisition and control software program was written in compiled BASIC (and later converted to C when the program became quite large), but much of the acquisition and display was done using assembly language routines. This included controlling the sample positioning, acquiring the data from as many as 50 different signal sources in the instrument, providing real-time background correction, and display of images as they were acquired. As we will discuss in Chapter 14, writing of assembly-language routines was very time-consuming, but resulted in a program that could give very high-speed performance.

■   The data were then displayed as either graphs or images with color added. Several types of post-run data processing, including filtering, contrast enhancement, and zoom, pan and scroll functions, were implemented, usually in assembly language.

■   Finally, data were sent to a large mainframe computer via a local area network. Several programs that operated concurrently on the PC and the mainframe were used to perform image analysis functions and to make the images appear three-dimensional or even to animate them.

Of course these examples illustrate only three of the many automation projects we have undertaken. It may appear that your own applications are quite different from these. However, rest assured that the principles involved in automating almost any device are the same and that most of what you need to know is discussed somewhere in this book.

1.2 INSIDE VIEW OF THE IBM PC

Before examining the details of laboratory automation, we probably should look at the components of the computer typically used in the laboratory. We shall concern ourselves only with the IBM PC but the components of most laboratory microcomputers are similar.

As shown in Figure 1.1, the heart of the computer is the **central processing unit** (**CPU**). In the IBM PC and IBM PC/XT the CPU is an Intel 8088, whereas in the IBM PC/AT it is an Intel 80286. It is this unit that controls the rest of the computer and also performs many of the mathematical operations, such as addition and subtraction. The 8088 and 80286 are 16-bit processors; that is, the normal size of an integer in the computer is 16 bits (see Appendix A for a discussion of bits, bytes, and binary numbers). Most internal operations in the IBM PC are thus designed for use with 16-bit numbers.

The 80286 differs from the 8088 in two ways: it is faster and it has a larger bus size. The **bus** is the electrical pathway along which data and commands are sent to the rest of the computer and to the external devices connected to the computer. In general, a larger bus size means that the computer can transfer more data in a single

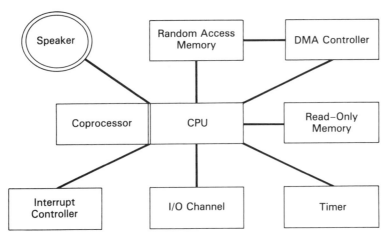

Figure   1.1.   Components of the IBM PC Microcomputer. These components form the basic system board of the computer. Lines are included to show some of the major electrical connections among the components.

operation (i.e., is faster) and can address more memory. The PC/AT is faster than the PC (by a factor of 2 or more) because of its higher speed processor and wider bus, and it can address 16 megabytes of information compared to the 1 megabyte of the IBM PC and the PC/XT. (The AT also has other features not available on the PC or PC/XT; these features are largely related to operating in a multi-user or multi-tasking environment.) All of the descriptions in this book will apply equally to the PC or XT or AT.

The **system board** or main board of the IBM PC series also contains a socket normally used for a **math coprocessor**. A coprocessor is a special-purpose processor designed to perform certain tasks much more quickly and efficiently than the CPU. Because the coprocessor can execute instructions concurrently with the CPU, additional time savings are gained.   Currently, the coprocessor used almost universally for scientific applications is the Intel 8087 (80287 if used with the 80286 PC/AT) math coprocessor. This coprocessor is used to speed floating-point calculations because the 8088 or 80286 CPU performs such calculations relatively slowly.

In addition, the system board contains space for **random access memory (RAM)** and **read-only memory (ROM)**. RAM is used to store the programs and data while the computer is running. Whenever the computer is turned off, all information in RAM is lost. When users say that their computers have 256K of memory, they are referring to RAM. Some of the RAM may be on another board than the system board, however, because the system board has a limited capacity for RAM.

In contrast to RAM, ROM is used to store information permanently. In the IBM PC it is used to store **BIOS (Basic Input/Output System)**.

BIOS is a collection of data tables and machine language subroutines that perform hardware dependent functions. BIOS routines handle such tasks as operation of the disk or diskette, character generation for the display, routine keyboard input, time-of-day and date, error checking, and diagnostics. Each machine (PC vs

XT vs AT) uses a different BIOS. As a result new hardware (e.g., higher capacity disks) can be introduced and still allow existing programs to function properly with the new hardware. However, only the most sophisticated users need to know about the details of the BIOS routines, because languages such as BASIC automatically use them without even asking the user.

BIOS also distinguishes IBM PCs from compatible PCs, which typically use a different BIOS. As a result, any incompatibilities associated with IBM PC-compatible computers, particularly older compatibles, often have been associated with the different BIOS used.

Also on the system board are a small speaker, a connection to the keyboard, a timer-counter for timing events, an interrupt-handling device and a direct-memory access device, all of which are important for the proper functioning of the IBM PC, but about which the average user needs to know very little. More important to the user is the **I/O channel**, which is often referred to as the "plug-in slots," or just the **slots** of a computer. It is into these slots that any additional boards are inserted, because the slots contain all of the necessary electrical connections to the CPU and other parts of the computer. Finally, the system board contains a power supply for powering all of the components on the system board and the I/O channel. Many of these components are visible in Figure 1.2.

An amazing variety of devices can be connected to the IBM PC via the slots. Among those that are commonly used in the laboratory are boards for controlling a disk or diskette, additional RAM, boards for running various display devices such as the IBM Monochrome Display Adapter and the IBM Enhanced Graphics Adapter, boards for sending information to a printer or to another computer via a modem and telephone wires, and finally, devices for communicating with laboratory equipment (such as the Data Translation 2801 Single Board Analog and Digital I/O System and the IBM General Purpose Interface Bus adapters discussed later in this book). In some cases there can be more than one of a given type of board in the IBM PC. The maximum number of slots available for these peripheral boards varies with the type of IBM PC (typically from five to eight slots). Additional slots are available if an **expansion chassis** is purchased. The expansion chassis contains not only additional slots but a separate power supply as well.

Thus, a typical system used in our laboratory might be an IBM PC/AT with a system board, an additional memory board, boards to control the diskette and hard disk, a serial communications and parallel printer adapter, a color display adapter, and one or more boards for data acquisition and control. Many of the systems in use in the authors' laboratories also contain a monochrome display and a board for communicating with a mainframe computer via a local area network. Almost all of our systems use the 8087 (or 80287) math coprocessor. We prefer the PC/AT to the PC or PC/XT for most laboratory applications because of the higher computing speed, the higher capacity hard disk and the extra slots in the AT; however, many applications can run perfectly acceptably on the original PC or the PC/XT.

Figure 1.3 shows the variety of devices that might be attached in a typical laboratory system. Notice that, with one exception, all of the devices attached to the

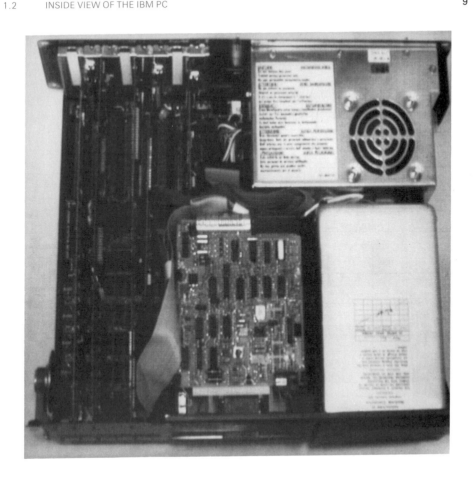

Figure 1.2. Inside the IBM PC. This image of the IBM PC/XT (see Chapter 12 for details of how such images are obtained) shows some of the major features of the IBM PC. The most visible items are the power supply (upper right), the hard disk drive (lower right), and the diskette drive (lower middle).   The speaker is at the extreme lower left. Running from top to bottom on the left side are the boards occupying the slots. Between the power supply and the right-most of the slots are the CPU and the math coprocessor, although they are too dark to be seen in this image. All of these components plug into the system board. The RAM, ROM, and other components also plug into the system board.

computer are connected ultimately to the CPU (the one exception is direct memory access devices, which are discussed in Chapter 15).

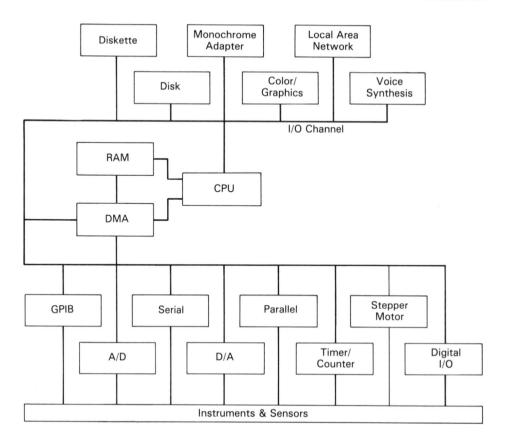

Figure   1.3.   Laboratory Configuration of the IBM PC.   Many types of devices can be connected to the IBM PC to form a laboratory system. In the upper part of the figure are typical peripheral devices that might be used for non-laboratory applications as well; in the bottom portion of the figure are some of the devices that might be used for laboratory data acquisition and control. The number of both kinds of devices in a given system is limited by the number of slots in the system unit and any expansion chassis. However, most systems would have only a few of the devices shown.

2

DIGITAL TO ANALOG CONVERTERS

All of us have become accustomed to seeing computers and computer-controlled devices in our laboratories. We tend to treat the computer as a black box that gets data, controls the instrument, and produces large amounts of information for us to digest. How exactly does the computer do these things? The first section of this book is an attempt to help you answer that question.

Recall that modern computers are essentially **binary** devices. That is, all data and instructions are represented inside the computer as 0s and 1s (or in more direct physical terms, as two different levels of electrical signals). However, most of the devices that scientists and engineers use to measure and control the world around them are **analog** devices; that is, they produce or require a continuous range of signals such as voltages. How then, do we get a binary device (the computer) to communicate with an analog device? In this and the following chapter, we shall see two devices that essentially convert signals from the binary domain to or from the analog domain.

Thus, the first device that we are going to examine is the **digital to analog converter**, usually abbreviated **DAC**, **dac**, or **D/A**. It is a device designed to convert the binary signals in a computer to analog signals understandable by a wide range of scientific and engineering devices. (Hence, it might more accurately be called a binary to analog converter. When humans (as opposed to other computers) deal with computers, however, we usually do so using decimal numbers, so we refer to computers as digital devices.) In simple terms, the D/A receives a binary value from the CPU and converts it to an analog voltage (or current) that can be used or displayed by some external device.

Before we discuss the theory of operation of the D/A, it may help you if we give some examples of where a D/A might be used. Many of the applications of the D/A involve some sort of positioning. For example, we might use the D/A to position a pen on a standard laboratory recorder -- or use two D/As to position the pen on an X-Y recorder. Equivalently, we might use two D/As to position the light beam on an oscilloscope (one each for the X and Y directions). Or we might use D/As to move a device with piezoelectric positioners.

Other applications of the D/A are possible. For example, we sometimes use the D/A to provide an offset voltage when a voltage might otherwise go out of a desired range. It can be used to provide small voltages in a physiology experiment or high voltages or currents in scanning instruments if the output is suitably buffered. Or, finally, we might use it in any application where we need to produce a particular waveform (e.g., a sine wave).

Digital to analog converters are available for the IBM PC in a number of forms. The most common form is as one of the components on a general purpose laboratory board, which plugs into one of the slots of the IBM PC and which typically contains an analog to digital converter, one or more D/As, and sections for digital input/output and counting/timing. However, there are also boards that contain multiple D/As, and specialized D/A boards that can be preprogrammed with a specific desired waveform (these latter are usually called digital waveform generators or waveform synthesizers).

Actually, the most common applications of D/As are those in which the D/A is built into an instrument. Many modern instruments use internal D/As to provide needed voltages. These internal D/As are then often externally controlled using some type of digital technique such as RS-232 or GPIB (discussed in the following chapters).

Before we discuss the operation of D/As, it is useful to point out that two steps were mentioned above: receipt of digital data from the CPU and conversion to an analog output. Generally, the digital to analog conversion process is similar for most computers; however, the transmission of the data from the computer to the D/A is highly dependent upon the type of computer used. Thus the commercial product you purchase includes not only a D/A but also the associated electronics to receive the digital signal from your computer. More specifically, the D/A must be connected to the computer's bus (see page 6). The number, arrangement, and signal levels of these bus lines vary widely. Hence, we will concentrate on the digital to analog conversion process only; the user interested in the specifics of connecting to the IBM bus (or I/O Channel) is referred to other books available on this subject. We also recommend that you review binary arithmetic (Appendix A) at this point if you have not done so already.

2.1 THEORY OF RESISTOR-LADDER TYPE D/A'S

In order to understand the selection, operation, and programming of digital to analog converters, it is useful to first understand some of the theory underlying D/A design. There are many types of D/As with somewhat different modes of operation; these are of interest primarily to engineers designing D/As into electronic equipment. We shall look at two of the designs to give us some idea of how most of the D/A systems available for the IBM PC operate.

One type of D/A of interest primarily because of its simplicity uses **binary-weighted resistors**. A version of such a system is shown in Figure 2.1.

In the 4-bit version shown, the binary code from the computer is used to control 4 switches. The switches control the flow of current from the reference source through a series of precision resistors. In this type of D/A, the resistors are each twice the resistance of the previous resistor; hence the name for the device. The output currents from the resistors are thus related by powers of two; i.e., the output of each successive resistor (starting at the left) is half that of the preceding resistor. The outputs from the resistors are all added by the summing amplifier, which thus provides an output voltage that is *proportional to the binary number input.*

For example, suppose that we input the number 1101 into this D/A. Switches 3, 2, and 0 are closed, and switch 1 is open. If the current from switch S_1 is 1 mA, for example, then the other currents will be $S_2 = 0.5$ mA, $S_3 = 0$ mA, and $S_4 = 0.0125$ mA. The output will then be a voltage proportional to the sum of these currents. The possible outputs from this D/A are graphed in Figure 2.2.

A major disadvantage of the binary-weighted resistor type of D/A is the difficulty in extending this scheme beyond approximately 4 bits. In typical integrated circuits it is difficult to have an extreme range of resistance values. The resistances in the binary-weighted resistor D/A range over a range of $1:2^{(N-1)}$, where N is the number of bits. Hence, for the common 12 or 16 bit D/As, the range of resistances would be too large to be practical.

A D/A that is more complex electronically, but that can achieve high resolution in a practical manner is the **R-2R network D/A**. In this type of D/A, a set of resistors is used in a resistance ladder network. An example of one such R-2R network is shown in Figure 2.3

In the R-2R network D/A, we again have a series of switches connected to a reference voltage and a group of resistors. However, by increasing the number of resistors above that used in the simple binary-weighted scheme and arranging them cleverly, we can use only two sizes of resistors, where one of the sizes of resistors has a resistance of twice the other (hence, R-2R). Again, the output is related to the size of the binary input because the resistance ladder arrangement gives us a voltage from each bit input that is half the voltage output of the bit above it. Thus, the net effect is essentially the same as that achieved by the simpler binary-weighted D/A.

Regardless of the type of D/A, additional circuitry is required beyond that shown so far. In particular, we must have a method of changing the value of D/A; this is usually done by having the CPU send a **chip select** signal to the D/A, which in turn causes the digital input from the CPU to be latched by a series of latches. The

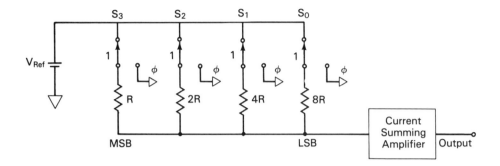

Figure   2.1.   Binary Weighted Resistor D/A.   A very simple D/A might contain a refer-
          ence voltage source, a set of binary-weighted precision resistors, and a set of
          digital switches. The output of such a system is related to the value of the
          binary input. The input to the digital switches is supplied by the CPU.

latched input is then switched into the D/A circuitry, where it is converted to the
desired analog voltage. The analog output can usually be adjusted by changing the
gain and offset of the summing amplifier. It is a good idea to occasionally check the
output of the D/A to be sure it matches the nominal (desired) output; on most
common general-purpose data acquisition boards, the gain and offset of the D/A are
therefore accessible at the top edge of the card so they can be adjusted without
turning off the computer. Figure 2.4 shows the general arrangement of a complete,
practical D/A.

2.2 PRACTICAL CONSIDERATIONS IN SELECTING D/A'S

When we use a D/A, there are several other matters we must consider. First,
although we have shown the output of the D/A as being linear over its entire range,
this may not always be the case, particularly for very cheap D/As. Hence, you
should always examine the specification sheet of the D/A to make sure that it is in
fact linear; in critical applications, you should test it yourself.

2.2.1 Speed

Another consideration is the speed of the D/A. Common commercial D/As are very
high speed; maximum rates of 40 kHz or more are typical. For applications that
require even higher speeds, you may need to purchase special-purpose systems. A
good rule of thumb is to purchase a D/A that will go considerably faster than the
fastest application you have planned.

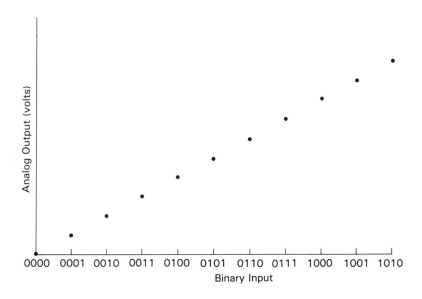

Figure   2.2.   Output from Binary Weighted Resistor D/A. The output from the D/A is a function of the binary input. However, the linear relationship shown here is for an "ideal" D/A; real D/As may vary significantly from this ideal.

A question related to the speed of the D/A is the availability of an **on-board memory buffer** for the output data. As we suggested at the beginning of this chapter, many applications require some type of waveform output from the D/A. If your application is very high speed (say above 10 kHz), a very significant amount of the computer's time may be spent in sending the appropriate values to the D/A. Some types of D/A boards, typically referred to as waveform synthesizers, allow you to preload the board's memory with the desired values, so that the computer or some other device only need trigger the board once to produce the entire waveform at some predetermined rate. This leaves the computer free to perform other tasks, such as data collection and analysis, while the waveform generator performs in a largely automatic fashion. Some boards (e.g., video display boards) use even higher speed D/As, as fast as several MHz, for special purposes.

2.2.2 Resolution

Another very important issue is the **resolution** of the D/A. This is because the precision with which you can output a value on the D/A is limited by the value output when the least significant bit is true. For example, if we use a 4-bit D/A that has a 10 volt full-scale value, then the least significant bit represents 2^{-4} times the full

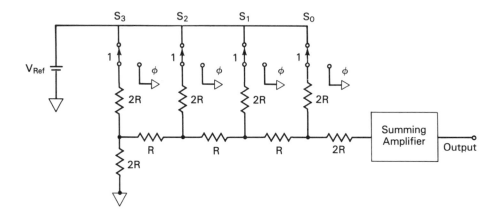

Figure 2.3. R-2R Network D/A. The output of this R-2R resistance ladder network D/A is proportional to the binary input. This is achieved by a series of resistors that provide a voltage output that varies by a factor of two at each bit. In this particular case, the output is sent to the noninverting terminal of an operational amplifier.

scale value, or 0.625 V. Thus, the possible output values are 0 V, 0.625 V, 1.25 V, etc. Then if we wish to output 0.92 V, for example, we cannot do so; in fact, the closest we can output is 0.625 V. The resolution of a D/A is therefore typically considered to be 1/2 of the value of the least significant bit; i.e., the difference between the desired value and the value we output will typically not be more than this value.

Most commercially available D/As for the IBM PC have resolutions of 12 bits. Hence, the resolution of a 12-bit, 10 V D/A will be one half of $10/(2^{12})$, or 0.0012 V. D/As for scientific use should only rarely be less than 12 bits resolution. Sixteen-bit D/As are often used in cases where highly precise movements are required, and even higher resolution D/As are available for special purposes. Typically, the higher the resolution, the higher the cost, however.

2.2.3 Other Considerations

Many of the commercially available D/As have selectable output ranges. On some D/As, this is selected using jumpers while on other systems the range is selectable in software. It is usually possible to select unipolar or bipolar output ranges, and often several different voltage ranges are allowed. For example, it may be possible to change from a -10 V to + 10 V range to a 0 to 5 V range by simply changing jumpers. If you need ranges outside of these, it is possible to amplify the signal; however, an external amplifier may change the practical speed or other characteristics of the D/A.

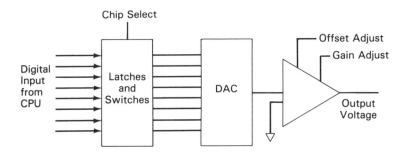

Figure   2.4.   Practical D/A.   A typical D/A contains the basic D/A plus a set of latches and associated control circuitry.   Typically, when the chip select line(s) are activated, the digital input is latched and switched across the D/A.   The output from the D/A is summed by the output amplifier, which usually has both gain and offset adjustments accessible to the user.

A more subtle, but still very important, characteristic of D/As is the exact shape of the waveform produced.   This may be particularly important in high-speed applications.   The D/A output does not jump instantaneously from value to value, but has some characteristic rise and settle times.   It is not uncommon to find that at very high speeds the D/A may exhibit some spiking during the settling period.   It may therefore prove useful to connect an oscilloscope to your D/A to examine its output at the speeds at which it is to be used.   Very often, such effects are totally harmless, but may occasionally prove troublesome when the D/A is used to drive a very high speed device.

2.3 PROGRAMMING THE D/A

Although it is helpful to understand the theory of operation of the D/A and the various types of D/As available, you probably would also like to see some actual examples.   Thus, the following examples are included to help you see some of the possible applications of the D/A in the laboratory.   We have listed programs written in IBM's BASICA language because this language is purchased by every user of the IBM PC as part of the DOS software package.   Other languages may be more efficient or useful to the reader; hence you are encouraged to convert the examples to the language of your choice.   The examples shown here and elsewhere in the text assume the use of the Data Translation DT2801, which is a typical general-purpose data acquisition board; this software is included in the Scientific Routines Diskette attached to this text (see Appendix B).

2.3.1 Using Ports and Registers

Before writing our first program, we need to discuss two basic methods that the IBM PC uses to communicate with external devices, namely memory and I/O ports.

Some boards are designed to use **memory-mapped I/O**; that is, they respond whenever a certain address in memory is referenced. For example, this is the manner in which the monochrome and color/graphics boards work; data placed in certain memory locations are automatically displayed on the appropriate monitor by the monochrome or graphics adapter, respectively.

However, many of the D/A boards, including the DT2801, use **I/O ports**, which are simply non-memory locations that can nonetheless be addressed by the IBM PC's central processor. (Electronically, these are distinguished by the fact that an address placed on the address bus of the IBM PC is assumed to be a memory address if the "Memory Write" line is low, whereas it refers to the I/O port if the "I/O Write" line is low.) The advantage of ports over memory is that when ports are used, no memory locations have to be reserved for use by that device. Typically, memory addressing is used only when the data sent to the external device have to be stored or continuously available to the device. Thus, for example, the IBM color/graphics monitor adapter uses memory addressing for the data to be displayed on the screen and port addressing for control information sent to the adapter.

Most data acquisition boards therefore use port I/O for D/As. The method by used IBM BASICA to send data to a port is very simple. The BASICA instruction is:

$$\text{OUT port,datum}$$

where **port** is the integer port number (in the range 0 to 65535) and **datum** is the datum to be transmitted. The datum is only one byte and so must be an integer between 0 and 255.

The BASICA instruction to read data from a port is equally simple, namely:

$$\text{datum} = \text{INP(port)}$$

where **port** is again the integer port number and **datum** is the 8-bit value read from the port.

One rather confusing point should be made clear. The INP and OUT instructions in BASICA read or write only one byte of data. How then can we read or write a 16-bit word, which is the usual size of data on the data acquisition boards for the IBM PC? The answer is that we must do two reads or two writes, one for each byte of the datum. The manner in which this is done varies from D/A board to D/A board. As an example, for the DT2801 there are two simple rules:

- We must read or write the datum in the order LOW byte then HIGH byte.

- Both bytes of the datum are read from or written to the same 8-bit port.

Thus, for example, we can send a 16-byte datum (VALUE%) to a port (labeled PORT%) using the following steps:

1.   Compute the low and high bytes of the datum to be sent:

LOW%=VALUE% AND &HFF
HIGH%=((VALUE% AND &HFF00) / 256) AND &HFF

where &H indicates a hexadecimal number in BASICA. Notice that we use the logical AND instruction to mask out all but the desired bits. Thus, ANDing with hex FF allows us to set the high byte to zero, and ANDing with hex FF00 sets the lower byte to zero. Dividing by 256 in effect moves the upper byte to the lower byte and then we mask out the upper bits again with another AND instruction.

2.   Send the data to the port corresponding to the desired low and high address, respectively.

OUT PORT%,LOW%
OUT PORT%,HIGH%

A second common system (used by many boards, but not by the DT2801) is to utilize *16-bit* port addresses when sending data to ports. This is done using an 8-bit OUT instruction as follows:

OUT PORT%,LOW%
OUT PORT%+1,HIGH%

where PORT% is an even number (divisible by 2). Note that the information is sent in the order low byte, then high byte; the high byte is sent to a port address one greater than that to which the low byte was sent. If you are using some language other than BASICA, you may find that 16-bit port I/O is allowed; in this case, you use the even address as the address for the instruction.

The information written to the ports corresponding to the D/A is rapidly transferred to **registers** on the D/A board (registers are simply devices designed for rapid transfer and storage of information). Generally, each register can be addressed at a different port address. For the D/A on the DT2801 card, the registers of interest are shown in Table 2.1.

More complete information can be obtained from the **User Manual for DT2801 Series**. In particular, you should consult this manual if you have any of the options for this board, or a higher performance version of the DT2801 series.

Table   2.1.   Registers on the DT2801 used by the D/A

Register[a]	Name	Address[b]	Meaning of Bits
1	Command	02ED	bits 7-4 should be 0000
			bits 3-0 = 1111 for STOP
			bits 3-0 = 0001 for CLEAR ERROR
			bits 3-0 = 1000 for WRITE D/A
			IMMEDIATE
1	D/A Status	02ED	bit 1=datum full
			bit 2=previous command finished
			bit 3=previous byte was command
			bit 7=composite error
0	D/A Channel	02EC	bits 1-0=channel
			(00 or 01; 10=both channels)
0	D/A Datum	02EC	low or high byte of datum

[a] Only those registers and bit meanings are shown that are used in programming the
    D/A in the WRITE D/A IMMEDIATE mode.
[b] The addresses shown are for the board at its factory-default location. You may
    select other base addresses using jumpers on the DT2801. Values are in
    hexadecimal.

2.3.2 Program to Output One D/A Voltage

We are finally ready to write our first program! This is a very simple program that is
designed to **output a single value to the D/A.** In Program 2.1 we show how to accomplish this with the Data Translation DT2801 board using the information in
Table 2.1.

There are several methods of using the D/A on the DT2801. The one we will
use first is the simplest, namely the WRITE D/A IMMEDIATE command. The
sequence of steps we shall follow is:

1.   Get parameters from the user (lines 180-230).
2.   Set up the board for using the D/A (lines 260-300).
3.   Send the WRITE D/A IMMEDIATE command (lines 330-340).
4.   Select the appropriate channel of the D/A (lines 350-360).
5.   Send the datum, low-byte first (lines 370-420).
6.   Check for errors (lines 430-450).

```
10 REM PROGRAM ONE_D/A
20 REM LABORATORY AUTOMATION USING THE IBM PC
30 REM Program to output one datum on a D/A using the DT2801
40 REM ***************************************************************
50 REM Define all registers, etc.
60 BASEADD%=&H2EC                          'base address of board
70 COMMAND%=BASEADD%+1                     'command register
80 STATUS%=BASEADD%+1                      'status register
90 DATUM%=BASEADD%                         'data register
100 REM ******************************
110 REM              Main program
120   GOSUB 170                            'get user parameters
130   GOSUB 260                            'set up board
140   GOSUB 330                            'output on D/A
150 END
160 REM ******************************
170 REM Get parameters from user
180    SCREEN 0:CLS:VALUE%=-1
190    INPUT "What channel do you wish to use (0 or 1)? ",CHAND2A%
200    WHILE ( VALUE% < 0 OR VALUE% > 4095)
210       INPUT "What value do you wish to output (0-4095) ",VALUE%
220    WEND
230 RETURN
240 REM ******************************
250 REM Set up board for correct operation
260    OUT COMMAND%,&HF                    'stop board
270    TEMP%=INP(DATUM%)                   'clear data register
280    GOSUB 520                           'ready to send command?
290    OUT COMMAND%,&H1                    'clear errors
300 RETURN
310 REM ******************************
320 REM Output to D/A by selecting the D/A, setting the channel, sending value
330    GOSUB 520
340    OUT COMMAND%,&H8                    'give command WRITE D/A IMMEDIATE
350    GOSUB 490                           'ready to set register?
360    OUT DATUM%,CHAND2A%                 'select the channel
370    LOW%=VALUE% AND &HFF                'get low byte
380    HIGH%=(VALUE% AND &HF00)/256        'get high byte
390    GOSUB 490
400    OUT DATUM%,LOW%                     'output the datum, low byte first
410    GOSUB 490
420    OUT DATUM%,HIGH%
430    GOSUB 520
440    ERRORCHECK%=INP(STATUS%)            'check for D/A error
450    IF (ERRORCHECK% AND &H80) THEN PRINT "Error writing to D/A"
460 RETURN
470 REM ******************************
480 REM check to see if ready to set a register
490    WAIT STATUS%,&H2,&H2                'wait for bit 1 to be reset
500 RETURN
510 REM check to see if ready to send another command
520    WAIT STATUS%,&H2,&H2                'wait for bit 1 to be reset
530    WAIT STATUS%,&H4                    'wait for bit 2 to be set
540 RETURN
```

Program   2.1.   Output One D/A Voltage.

Before examining the steps of the program in detail, we should note a characteristic of programs for the DT2801: they consist essentially of a series of INP and
OUT commands. Each of these commands tests or controls, respectively, the operations of a specific register on the DT2801. The registers then determine the functioning of the D/A. However, before each instruction to the DT2801 (using the

OUT command), we also need to ensure that the previous instruction on the DT2801 has finished executing (because the PC instruction cycle, particularly for the PC/AT, may be faster than the instruction cycle on the DT2801). To do this, we use the BASICA WAIT command, which is of the form:

WAIT register,bits,bits0

where **register** is the port or register to be tested and **bits** is a number representing the set of bits to be tested. If **bits0** is specified, it indicates those bits that must be zeros before proceeding to the next instruction; otherwise, we wait for the bit to be a 1. In essence the WAIT instruction is equivalent to INPutting a status byte and testing individual bits of the status byte.

We use the WAIT command with the DT2801 under two slightly different circumstances. First, we must wait after each OUT instruction to the board to ensure that the register received the information; this is done in the subroutine in line 490, where we wait for bit 1 of the Status Register to be reset (become zero). Second, after each command (i.e., after instructions sent to the DT2801 Command Register), we test to see if the command has completed; this is performed in the subroutine in lines 520-530, where we first test to see if the register received the information (line 520), and then test to see if the command has completed (line 530) by waiting for bit 2 of the Status Register to be set (become one). (In this case, these are probably redundant checks, but many commands to the DT2801 may take seconds or even minutes to complete, so we need a method of testing for completion of the command.)

You should also note that the DT2801 expects instructions to be issued in a very specific sequence; in this case, we must issue the WRITE D/A IMMEDIATE command, then select the D/A channel, and then write the data, in precisely that order. Because the sequence is specific, the same register can be used to perform multiple functions; e.g., in this case, the same physical register is used for the channel selection and for both bytes of data.

Thus, the first step in operating the DT2801, as shown in Program 2.1, is to clear the DT2801 of any previous commands. This is done with a HALT command to the Command Register (line 260). The Data Register is then read once (line 270) to clear any remaining data in that register after the board has been halted.

Then we issue a CLEAR ERROR command (line 290). To do this, we first check to see that there is no previous command in the Command Register and that the board has finished executing the previous command by a call to the appropriate subroutine (line 280), which causes the program to wait until the board is ready to issue another command.

The combination of halting the board, reading the Data Register, and issuing a CLEAR ERROR command puts the board in a known, well-defined state, ready to accept further commands. The equivalent of lines 260-290 thus should be included in all programs for the DT2801.

We are now ready to issue a series of commands to put the D/A in the WRITE IMMEDIATE mode. To do so, we issue the WRITE D/A IMMEDIATE command (line 340) and then the channel number (line 360) followed by the low and high bytes of the datum (lines 370-420); recall, they must be in precisely this order. Finally, we check whether an error occurred during the use of the D/A by checking bit 7 of the Status Register. (Notice that we again use the AND operation to mask out all but the desired bit.) Preceding each of these steps, we check to make sure the register is ready to receive more input, and before commands to make certain the previous command has finished. To run the program, we type:

BASICA path\PROG21

For example, if the program is on a hard disk (drive C) in the PROGRAM subdirectory, we type:

BASICA C:\PROGRAM\PROG21

The program outputs a value to the D/A; by monitoring the appropriate channel with a voltmeter, we can watch the voltage change.

We've finished our first program!

2.3.3 Program to Output Ramp Voltage from D/A

Now before we celebrate too vigorously, let's make the program a bit more useful. First, instead of sending a single value, why don't we send a range of values? That is very easy to do and is shown in Program 2.2.

This program is very similar to Program 2.1, except that we are sending a series of values to the D/A. These values are constrained by the program to stay within the allowed values of the D/A, namely 0 to 4095 (recall that a 12-bit D/A can output values between 0 and 2^{12}). Notice that the major alteration in the program is that we have added a FOR...NEXT loop in lines 140-200. All of the subroutines have remained the same as in Program 2.1. This is one reason why we consistently recommend using subroutines: it allows you to move sections of code from one program to another more easily. Subroutines often also make it easier to find errors in your program.

The observant reader will also notice that Program 2.2 is relatively inefficient, because the WRITE D/A IMMEDIATE command is really designed for a single triggering of the D/A. The DT2801 has a more sophisticated pair of commands, SET D/A PARAMETERS and WRITE D/A, that can be used to perform the function of Program 2.2 more efficiently. However, it involves setting a clock, so we will wait until Chapter 9 to take advantage of this feature.

2.3.4 Program to Output Waveform on Oscilloscope

As a final modification of our previous programs, we show a program to output the same waveform repetitively using two D/As simultaneously. For example, we might use this program to display a waveform on an oscilloscope, or to position some device

```
10 REM PROGRAM RANGE_D/A
20 REM LABORATORY AUTOMATION USING THE IBM PC
30 REM Program to output a range of values on a D/A using the DT2801
40 REM ****************************************************************
50 REM Define all registers
60 BASEADD%=&H2EC                              'base address of board
70 COMMAND%=BASEADD%+1                         'command register
80 STATUS%=BASEADD%+1                          'status register
90 DATUM%=BASEADD%                             'data register
100 REM *******************************
110 REM                Main program
120    GOSUB 240                               'get parameters from user
130    GOSUB 310                               'set up board
140    FOR I%=1 TO NUMSTEPS%                   'main loop
150        LOCATE 10,1:PRINT VALUE%
160        GOSUB 390                           'output current value on D/A
170        VALUE%=VALUE%+STEPSIZE%             'calculate new value
180        IF VALUE% < 0 THEN VALUE%=0         'check if value is within range
190        IF VALUE% > 4095 THEN VALUE%=4095
200    NEXT I%
210 END
220 REM *******************************
230 REM Get parameters from user
240    CLS:INPUT "What channel do you wish to use (0 or 1)? ",CHAND2A%
250    INPUT "What initial value do you wish to output (0-4095) ",VALUE%
260    INPUT "What step size do you wish to use? (0-4095) ",STEPSIZE%
270    INPUT "How many steps do you wish to use? (0-4095) ",NUMSTEPS%
280 RETURN
290 REM *******************************
300 REM Set up board
310    OUT COMMAND%,&HF                        'stop board
320    TEMP%=INP(DATUM%)                       'clear data register
330    REM wait for command to finish, then clear errors
340    GOSUB 580
350    OUT COMMAND%,&H1
360 RETURN
370 REM *******************************
380 REM Subroutine to output VALUE% on D/A
390    GOSUB 580
400    OUT COMMAND%,&H8                        'give command WRITE D/A IMMEDIATE
410    GOSUB 550
420    OUT DATUM%,CHAND2A%                     'select the channel
430    LOW%=VALUE% AND &HFF                    'get low byte
440    HIGH%=(VALUE% AND &HF00)/256            'get high byte (upper 4 bits zero)
450    GOSUB 550
460    OUT DATUM%,LOW%                         'output the datum, low byte first
470    GOSUB 550
480    OUT DATUM%,HIGH%
490    GOSUB 580
500    ERRORCHECK%=INP(STATUS%)
510    IF (ERRORCHECK% AND &H80) THEN PRINT "Error writing to D/A"
520 RETURN
530 REM *******************************
540 REM check to see if ready to set a register
550    WAIT STATUS%,&H2,&H2                    'wait for bit 1 to be reset
560 RETURN
570 REM check to see if ready to send another command
580    WAIT STATUS%,&H2,&H2                    'wait for bit 1 to be reset
590    WAIT STATUS%,&H4                        'wait for bit 2 to be set
600 RETURN
```

Program   2.2.   Output a Range of D/A Voltages.

in a repetitive fashion. Program 2.3 shows how we might do this. Again, the modification is simple, namely to read the data pairs using a READ statement.   Notice that we can still use the same subroutine to output the data from the D/A, and that much of the rest of the program is the same as the previous two programs. We can output data on both D/As of the DT2801 simultaneously by using a channel number of 2.

Actually, the D/A is most often used in concert with an A/D or other devices. In Chapter 9, we shall see several examples of this. However, first we need to learn how to use these other types of devices.

```
10 REM PROGRAM DUAL_D/A
20 REM LABORATORY AUTOMATION USING THE IBM PC
30 REM Program to output a range of values on both D/As of the DT2801
40 REM ******************************************************************
50 DIM X%(1000),Y%(1000)                'set up arrays to hold output data
60 REM Define all registers, etc.
70 BASEADD%=&H2EC                       'base address of board
80 COMMAND%=BASEADD%+1                  'command register
90 STATUS%=BASEADD%+1                   'status register
100 DATUM%=BASEADD%                     'data register
110 CHAND2A%=2                          'output on both channels together
120 REM *********************************
130 REM Main program
140   GOSUB 250                         'get user parameters
150   GOSUB 330                         'set up board
160   FOR I%=1 TO NUMLOOP%
170       LOCATE 10,1:PRINT "Iteration number ";I%
180       FOR J%=1 TO NUMPOINTS%
190           GOSUB 410                 'output current values on D/A
200       NEXT J%
210   NEXT I%
220 END
230 REM *********************************
240 REM Get parameters from user
250   CLS:INPUT "How many pairs of data do you wish to output? ",NUMPOINTS%
260   FOR I%=1 TO NUMPOINTS%
270       INPUT "Enter an X value and a Y value, <X,Y> ",X%(I%),Y%(I%)
280   NEXT I%
290   INPUT "How many times would you like this waveform output?",NUMLOOP%
300 RETURN
310 REM *********************************
320 REM Set up board
330   OUT COMMAND%,&HF                  'stop board
340   TEMP%=INP(DATUM%)                 'clear data register
350   REM wait for command to finish, then clear errors
360   GOSUB 590
370   OUT COMMAND%,&H1
380 RETURN
390 REM *********************************
400 REM Output VALUE% on D/A
410   GOSUB 590
420   OUT COMMAND%,&H8                  'give command WRITE D/A IMMEDIATE
430   GOSUB 560
440   OUT DATUM%,CHAND2A%               'select the channel
450   LOCATE 20,1:PRINT X%(J%),Y%(J%)
460   VALUE%=X%(J%)                     'get X value
470   GOSUB 640                         'get low, high bytes of it and output
480   VALUE%=Y%(J%)                     'get Y value
490   GOSUB 640                         'output it
500   GOSUB 590
```

Program 2.3 (Part 1 of 2).   Waveform Output with D/As.

```
510    ERRORCHECK%=INP(STATUS%)              'check for error
520    IF (ERRORCHECK% AND &H80) THEN PRINT "ERROR WRITING TO D/A"
530 RETURN
540 REM *********************************
550 REM Check to see if ready to set a register
560    WAIT STATUS%,&H2,&H2                  'wait for bit 1 to be reset
570 RETURN
580 REM Check to see if ready to send another command
590    WAIT STATUS%,&H2,&H2                  'wait for bit 1 to be reset
600    WAIT STATUS%,&H4                      'wait for bit 2 to be set
610 RETURN
620 REM *********************************
630 REM Output low and high byte of datum
640    LOW%=VALUE% AND &HFF                  'get low byte
650    HIGH%=(VALUE% AND &HF00)/256          'get high byte (upper 4 bits = 0)
660    GOSUB 560
670    OUT DATUM%,LOW%                       'output the datum, low byte first
680    GOSUB 560
690    OUT DATUM%,HIGH%
700 RETURN
```

Program   2.3 (Part 2 of 2).   Waveform Output with D/As.

EXERCISES

1.   Modify Program 2.2 to input from the user the desired VOLTAGES rather than the desired binary range. This is always helpful--to allow the user to enter data in scientific or engineering units rather than in computer units. Be sure to ascertain whether the user has selected unipolar or bipolar outputs.

2.   Write a program to input a table of desired voltages from a disk, and then repetitively output the tabulated voltages on the D/A. (For those who wish to be more fancy, allow the user to shell to a text editor to create the table as part of your program.)

3

ANALOG TO DIGITAL CONVERTERS

3.1 THEORY

Probably the single most useful device for getting a computer to communicate with a laboratory instrument is the **analog to digital converter** (usually abbreviated as A/D or ADC). Most instruments, as well as many kinds of machinery, produce an analog output; that is, an electrical signal that can vary continuously over some range. The A/D simply converts this analog electrical signal, most often a voltage, into a discrete (digitized) form and then transmits it in a form that can be interpreted by the computer. This process is illustrated in Figure 3.1.

An A/D is built into the electronic circuitry within many "digital" instruments, including digital multi-meters. In such cases the user obviously does not need to attach an external A/D; instead, the user may simply obtain the digitized data using devices such as the General Purpose Interface Bus, a serial port, or digital I/O, as discussed in subsequent chapters.

However, in many applications, especially with older instruments, you need to attach an A/D, either to the instrument or to the computer, to record analog signals. As we mentioned in Chapter 2, in most analog data acquisition systems available for the IBM PC, the A/D is available as part of a multi-function interfacing board that includes other frequently used devices such as a timer, digital-to-analog converter,

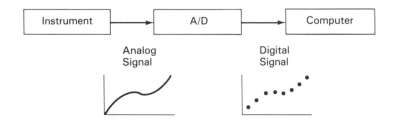

Figure   3.1.   Digitization using an A/D.   The basic function of the analog-to-digital converter is to take a continuous (analog) signal produced by an instrument or
sensor and convert it to a discrete (digitized) signal. This digitized signal is
then transmitted in binary form to the computer.

and digital input/output. We will see in Chapter 9 how all of these functions can be
used together in a general lab automation system.

3.1.1 Successive Approximation A/Ds

By far the most popular type of A/D for the IBM PC (and other computers, for that
matter) is the **successive approximation A/D**. It has the advantage of being relatively
cheap and can operate over a wide range of sampling rates that includes those used in
all but the most demanding applications.

As the name suggests, the successive approximation A/D arrives at a specific
digital output by a series of approximations. The output of a successive approximation A/D is illustrated in Figure 3.2. Note, however, that the output is not sent
to the computer until time T_1, when the output is as close as possible to the analog
input.

There are several different ways in which successive approximation A/Ds can
be designed. The basic principles for all of them are similar, however. If, for
example, we have a 4-bit A/D with a binary output, we can describe each bit as
representing a voltage half that of the preceding (more significant) bit. Thus, if we
assume it is a 0 to 10 volt A/D, one binary representation would be:

$$1000 \quad \text{is} \quad 5.333 \text{ V}$$
$$0100 \quad \text{is} \quad 2.667 \text{ V}$$
$$0010 \quad \text{is} \quad 1.333 \text{ V}$$
$$0001 \quad \text{is} \quad 0.667 \text{ V}$$
$$0000 \quad \text{is} \quad 0.000 \text{ V}$$

This representation was designed so that a 10-volt input (i.e., full scale) could
be represented by a binary value of 1111 and a 0V input could be represented as

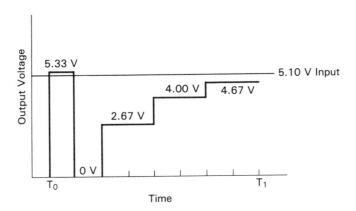

Figure   3.2.   Output of Successive Approximation A/D.   A 4-bit A/D is used to measure a 5.10V input signal. At time T_0 the A/D receives a signal to begin the conversion process. A voltage (5.33 V in this case) is compared to the 5.10 V input signal. Because the input is less than 5.33 V, the trial output is reset to zero and a voltage half the original value is tried. Because this trial voltage (2.67 V) is less than the 5.33 V input voltage, it is retained and all further voltages added to it. This process continues until all four of the bits of the A/D are determined, i.e. time T_1. At this time, the computer is notified that the conversion process is complete and the datum can be read from the A/D.

0000. It should be pointed out that this encoding scheme is not unique; other schemes can be used as well.

The method by which the A/D converts analog input into a digital (binary) output is to perform a series of comparisons between the analog datum and a series of reference voltages. Whenever the reference voltage is less than or equal to the input voltage, a binary 1 is output; if the reference voltage is too large, then a 0 is output.

With our example of a 4-bit A/D, suppose we wish to measure a 5.100 V input. We can imagine the process of converting this 5.100 V input to a binary output as proceeding in the following fashion, starting at the left-most (most significant) bit:

1.   Try   1000   which is 5.333 V.
      This is larger than our input signal, so reset the most-significant bit to produce
                                         0000   which is 0.000 V.

2.   Try   0100   which is 2.667 V.
      This is smaller than our input signal, so we keep the second-most-significant bit set (to 1).

3.   Try   0110   which is 2.667 + 1.333 = 4.000 V.
      This is smaller than our input signal, so we keep the third-most-significant bit set.

4.   Try   0111   which is 4.000 + 0.667 = 4.667 V.
      This is smaller than our 5.100 V input signal, so keep the last bit set.
5.   Report the binary representation of the 5.100 V signal as 0111.

Now of course 4.667 V is not very close to 5.100 V. This is because we have
used only a 4-bit A/D. The precision of this A/D is only 1 part in 2^4, or one part in
16. For a 10 V input this would be a precision of ± 625 mV at best. In a fashion
similar to that described for D/As, we can define the **resolution** of the A/D as one-
half of the smallest voltage step it can detect. In this case, the 4-bit 10 V A/D has a
resolution of 312 mV.

With a higher-resolution A/D, the results will be much closer to the input data.
With a 12-bit A/D, for example, we can have precision of one bit in $2^{12} = 4096$. For
a 10 V input this would be a precision of ± 2.44 mV at best. This error is often called
A/D quantization noise. However, each additional bit measured takes an additional
amount of time, so that a 12-bit A/D is much slower than a 4-bit A/D of the same
type.

The manner in which this process is achieved in hardware is illustrated in
Figure 3.3, which is a simplified diagram of the successive approximation A/D. The
principal components of the A/D are a comparator, a clock, a register, and a digital
to analog converter (D/A) (see Chapter 2 for a discussion of D/As). Upon receipt of
a command from the computer to "start a conversion," the register is set to all zeros.
Each time the clock is incremented (i.e., whenever the clock "ticks"), a new value is
placed in the register. Starting at the most significant (leftmost) bit, a 1 is temporarily placed in the register. In our 4-bit example, this causes the D/A to output a
voltage corresponding to the value 1000, i.e., 5.333V. The comparator is used to test
whether this output is larger than the input signal. If it is not, then the 1 is left in the
register; if it is, then the bit is reset to 0. This process continues for each bit of the
register, until appropriate values have been selected for each bit. When finished, the
A/D generates a signal indicating that the conversion process is finished; the register
can then be read by the computer to obtain the binary output.

Finally, three technical points about the successive approximation A/D should
be mentioned. First, the successive approximation process obviously will not work if
the input voltage changes while a conversion is occurring. Hence, a sample and hold
circuit is normally used to hold the input voltage fixed during the A/D conversion
process. Second, in our example we have used a scheme where each voltage is half of
the preceding voltage. This is not the only possible relationship; for example, binary-
coded-decimal (BCD) outputs are generated by some A/Ds, particularly those used
to drive seven-segment LED digital displays rather than computers. Third, in the
example just described the reported voltage is always less than or equal to the input
voltage. To make the errors symmetric about the true value, a voltage equal to one-
half the value of the least significant bit (e.g., to 0.333 V in our 4-bit example) is
usually added to the internally generated voltage before each comparison occurs.

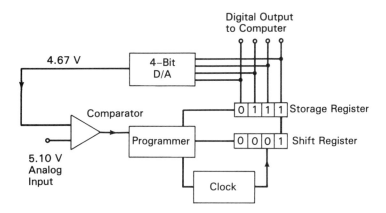

Figure   3.3.   Simplified Successive Approximation A/D.   The components of a hypothetical 4-bit successive approximation A/D might include a comparator for comparing the input and output of the A/D; a programmer that generates all of the control logic; a clock for controlling the rate at which the various bits of the A/D output are changed; two registers that generate the binary output; and a 4-bit D/A that converts the binary output back into a voltage for comparison with the input voltage. The A/D is shown in the state it would be at time T_1 in Figure 3.2.

3.1.2 Integrating A/Ds

The other major type of A/D available for the IBM PC and most microcomputer systems is the integrating A/D. As the name implies, it has the advantage of integrating the input voltage over a period of time, so that fluctuations resulting from random noise contained in the signal are averaged together and thereby largely eliminated.

The mode of operation of an integrating A/D can be illustrated using the dual-slope A/D, a simplified diagram of which is shown in Figure 3.4. In its simplest form the dual-slope A/D has two phases of operation. During the integration phase the input voltage stores charge on a capacitor. During the reference phase the capacitor is switched from the input voltage and connected to a fixed reference voltage. This voltage is opposite in sign to the input voltage and hence discharges the charge stored on the capacitor. The time needed to discharge the capacitor with a fixed voltage is directly related to the charge on the device and hence to the input voltage. The time required to discharge the capacitor is then used to determine the binary output of the A/D. This process is illustrated in Figure 3.5.

The name dual-slope A/D simply refers to the shape of the voltage curve, as shown in Figure 3.5. However, integrating A/Ds in common use may vary consider-

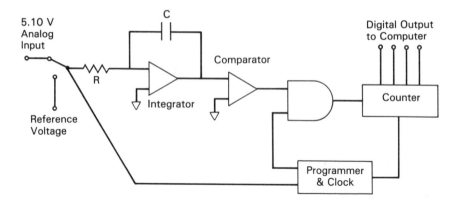

Figure   3.4.   Simplified Dual-Slope A/D.   Upon receipt of a command to initiate an A/D conversion, the charge on capacitor C is brought to zero, and the analog input voltage is connected to the integrator. As soon as the output of the integrator rises above zero, the comparator output goes high, starting the counter. Upon counting to a specified limit (that ultimately determines the noise reduction), the input to the integrator is switched to a reference voltage. This voltage, which is fixed and opposite in sign to that of the analog input, brings the voltage of the integrator back to zero, at which point the comparator output goes low and switches off the counter. The counter output is thus proportional to the time taken to bring the integrator output back to zero, which in turn is proportional to the input voltage. The accuracy of this type of A/D can be quite good, because both the analog input and reference voltages are integrated by the same circuitry.

ably from the dual-slope A/D described. They often, for example, include circuitry for automatic zeroing and may also be "quad slope" instead of dual-slope.

The major disadvantage of the integrating A/D is that it is slow: typical conversion rates are only 1 to 10 Hz (samples/second). Hence, the major use of this type of A/D has been in digital voltmeters and other instruments where the signal level changes rather slowly. In addition, proper digital signal averaging or digital filtering techniques can achieve similar averaging effects with a successive approximation A/D. Such techniques require more processor time from the computer's CPU. However, they allow the use of the successive approximation A/D in almost any situation where an integrating A/D might be used. For this reason the integrating A/D has not been widely used with the plug-in adapter cards for the IBM PC.

It is useful to remember that the output from some instruments, such as digital voltmeters and pH meters, may be produced by an A/D. Similarly, the typical laboratory analog X-Y recorder performs some integrating of the signal because of the relatively slow pen response. Hence, one of the common surprises that awaits the

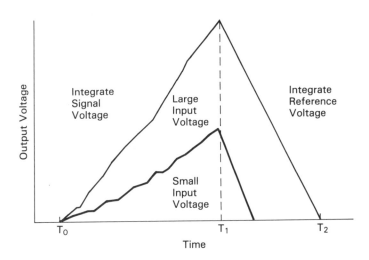

Figure   3.5.   Operation of Dual-Slope A/D. At time T_0 the A/D begins integrating the analog input signal. At a fixed time interval later $(T_1 - T_0)$ the input to the integrator is switched from the analog input voltage to a fixed reference voltage. Depending upon the amplitude of the signal at time T_1, the voltage reaches zero at some later time, T_2. The time interval $(T_2 - T_1)$ is proportional to the amplitude of the input voltage.

novice user of the laboratory computer is to find that when data are collected with a non-integrating A/D there is very often a large amount of "noise" in the signal that did not appear to be present when the data were displayed on a recorder or digital meter. This noise is not generated by the computer system but is simply averaged out by the integrating A/D or recorder.

3.2 PRACTICAL CONSIDERATIONS

There are several important practical options that you will need to consider when you purchase your A/D. We should point out in advance of your reading this section that these options can be quite confusing. It may help you to know that in our own laboratories the most common options used are probably the following: internal A/D, 20 to 40 kHz maximum rate, 12 bit resolution, external amplification, 4 or more channels, and polling mode of operation. For PC/AT applications, we are more likely to use 100 kHz, DMA, 4 channel boards. Furthermore, many of our more recent applications require 16-bit precision.

3.2.1 Internal or External A/D

First, you must decide whether to purchase a board designed specifically for the IBM PC or to use a generic system that does the A/D conversions in a separate box and then sends the binary data to the IBM PC via a serial port or other communications method such as the GPIB (IEEE-488) interface. Because most of their customers use an A/D with a specific computer, the majority of the manufacturers have chosen the first approach, namely to develop boards specifically for the IBM PC. Such systems are generally faster or cheaper than the external, generic systems. However, having the A/D in an external box may provide increased isolation from electrical noise and may also make it more convenient to add other signal conditioning equipment to the system.

3.2.2 Sampling Rates

The maximum speed at which the A/D can operate may also be an important consideration for your application. We can generally divide experiments into three categories: slow, fast and ultra-fast. Slow experiments are those that involve sampling rates less than 1 to 10 Hz, e.g., measurements of weather conditions or plant growth rates and some types of chromatography. Fast experiments are those in the 1 Hz to 100 kHz range, which includes most laboratory experiments. Ultra-fast experiments are those requiring rates above 100 kHz, such as measuring flash photolysis or nuclear chemical reactions.

For the slow experiments, particularly if the measurements include much high-frequency noise, integrating A/Ds can be very useful. You can also use a standard non-integrating type A/D for these applications, but you may need to perform additional filtering of the signal, as described in the next chapter.

For fast experiments, you normally will use one of the general-purpose successive approximation A/Ds. Most commercially-available boards can achieve rates of at least 20 kHz, and rates over 100 kHz are achievable on a few boards. Filtering of the data, if required, can be performed as described in the next chapter.

Ultra-fast experiments require special hardware. This is because for data rates significantly above 100 kHz, the user is limited by the execution speed of the IBM PC itself, and hence, the A/D system must have its own data processing and local storage facilities. Such systems are often described as digital oscilloscopes or frame grabbers, depending upon the application. By using other types of A/Ds, they can achieve rates as high as a few MHz. At those rates, data can be collected only over very short time intervals before the memory of the system (or even the IBM PC) is completely filled. Frame grabbers have the capability of storing data on video tape but typically use only 8-bit A/Ds. In addition, special-purpose boards are available for other types of ultra-fast experiments.

3.2.3 Resolution

As mentioned earlier, the resolution of the A/D is also an important consideration. The resolution of most commercial A/Ds is currently is between 8 and 16 bits. Typical manufacturing floor data acquisition systems use 8 bit A/D converters. For scientific applications, the 12-bit A/Ds are currently the most popular; A/Ds with 14 or 16 bits resolution are readily available, but the higher cost and slower speed of these units has made them considerably less popular. Recall also that the higher the resolution, the lower the sampling rate, given the same type of A/D. Hence, the maximum data conversion rate of the A/D may be the determining factor for your application. In addition, with the advent of true 16- and 32-bit processors and progress in integrated circuit fabrication technology, there may well be an increased interest in the higher resolution A/Ds.

3.2.4 Gain

The input required by most commercially-available A/Ds is in the -10 V to +10 V region. The analog output from many instruments is at some other range. An amplifier is thus often required. In order to take full advantage of the resolution of the A/D, the amplifier should be designed to provide outputs over the entire range of the A/D. For example, a signal from an instrument that varies over the range of 0 to 10 mV should be amplified by a factor of 1000 to provide a 0 V to 10 V output.

There are two basic methods for amplifying the signal from an instrument before the A/D conversion occurs: using an external amplifier or buying an A/D with a built-in amplifier. External amplifiers are discussed later in this chapter (see page 40). It is useful to note here, however, that the primary advantage of an external amplifier is that it can be placed inside the instrument, or at least very close to it. This reduces the distance traveled by the unamplified signal, which in turn reduces the amount of noise added to the signal. Hence, serious consideration should be given to the use of external amplifiers, particularly in environments where electrical and magnetic noise is significant or where the required gain is 100 or more.

The second alternative, that of using an amplifier built into the A/D, has three advantages: it is convenient, it allows the amplification to be changed by a software command, and it avoids possible grounding and impedance matching problems that are sometimes encountered with external amplifiers. Many of the commercially-available A/D boards give the user the option of an on-board amplifier. In some cases the gain of the amplifier is fixed; in others it can be modified by changing the hardware on the A/D board (e.g., changing the size of a resistor). Probably the most desirable alternative for many applications is to use an A/D that makes it possible to change the gain in software between data conversions; this is referred to as programmable gain. However, using programmable gain imposes an additional burden on the software. It may slow down data collection because of the need to check the size of the input signal and to reset the gain when appropriate. Its chief advantage is the added flexibility it gives the user in selecting the gain as the experiment proceeds.

3.2.5 Number of Channels

Another consideration for the user is the number of analog signal lines, or **channels**, to be monitored by the A/D. The number of input channels varies considerably, from as few as one to as many as 256 or more. Most of the multifunction laboratory boards are built with 4 to 16 analog input channels; even 4 channels is certainly sufficient for most applications. However, many manufacturers offer options for more channels. Because these are typically multiplexed (i.e, there may be 256 channels but there is only one A/D), the rate quoted for the A/D is the maximum rate for all channels combined, rather than for each channel. This means that a 25 kHz A/D with 10 channels can take data from all 10 channels at a maximum of 2.5 kHz per channel; in practice, the additional overhead of switching channels may decrease the maximum even below this rate. Of course, if only one channel of the 10 is monitored then the maximum rate is still 25 kHz in this example.

3.2.6 Method of Operation

Probably the major feature differentiating the A/D boards developed by different manufacturers is the mode of operation of the boards. Three different modes are commonly available: software polling, vector interrupts, and direct memory access. Many boards allow the choice of more than one of these modes.

The simplest and most commonly-used mode is **software polling**. In this mode the data conversion and storage process is done in the following steps:

1. The A/D is programmed to initiate an A/D conversion. This can be done either by the software or in the hardware.
2. The program then waits until the A/D conversion is completed. Typically, while it is waiting no other processing occurs.
3. When the A/D conversion is completed, the datum is stored under program control.

Note that considerable time may be lost during step 2, during which the computer waits for the datum to become available. For example, if the A/D is running at 100 Hz, as much as 99.9% of the time the computer is doing nothing except waiting. However, the amount of time spent waiting decreases as the rate of data collection increases. At rates near 100 kHz the computer is collecting data essentially without waiting.

The second alternative is to use **interrupts**. Interrupts will be discussed in detail in Chapter 15, but without knowing what interrupts are, we can still look at the process as consisting of the following steps:

1. The A/D is set up so that a timer or some other device triggers the A/D automatically at the appropriate intervals.

2. The program proceeds to other useful activities; these activities may even be unrelated to data collection.

3. When the A/D finishes converting each analog datum to binary, it interrupts the program and causes control to be passed to a special **interrupt service routine** that stores the datum or otherwise processes it. Control is then returned to the program that was interrupted.

Notice that when interrupts are used there is no waiting for the datum. It turns out, however, that there is considerable overhead involved in using interrupts, so that the maximum rate for which interrupts are usable is in the 1 to 10 kHz region on an IBM PC. This is due to the software overhead required to save the current task and switch to the interrupt service routine. In practice, interrupts are most appropriate at rates of less than 1 kHz. For this reason most boards that support interrupts also support polling.

The third alternative is **direct memory access** (DMA), which is also fully discussed in Chapter 15. In this mode the following steps occur:

1. The A/D is set up so that a timer or some other device triggers the A/D automatically at the appropriate intervals.

2. The program then proceeds to other useful activities; these activities may even be unrelated to data collection.

3. Whenever the A/D finishes converting an analog datum to binary, a processor on the A/D board directs the storage of the datum into a pre-selected area of memory via a direct memory access channel. Using this channel is a high-speed method of transferring data to and from the main computer memory without using the central processor of the IBM PC. Hence, the execution of the program is not disturbed by the data storage process.

The DMA method is potentially the fastest and switches most of the burden of data collection from the IBM PC's CPU to the processor on the data collection board. With the IBM PC rates of up to 200 kHz or more should theoretically be possible with this method; in fact, commercially available A/Ds using DMA often run at much less than this rate.

The choice among these three methods is not at all obvious and may depend upon both the application and the degree of custom programming needed. Here, an important consideration is whether you have an IBM PC, PC/XT, or PC/AT. On the PC or PC/XT, the practical upper limits are approximately 1 kHz for interrupts and 40 kHz for polling. DMA on the PC and PC/XT can reach rates of 40 kHz or more. However, the number of DMA channels on the PC and PC/XT is quite limited and other devices (disks, communications boards, GPIB, etc.) often occupy all of the DMA channels. Hence, DMA is probably not the method of choice for most PC and PC/XT applications.

On the PC/AT the situation is somewhat different, however. The AT's CPU runs faster, so the maximum rates for interrupts and polling are at least double those of the PC or PC/XT (depending upon whether the card uses an 8 or 16-bit data transfer). The PC/AT has additional DMA channels compared to the PC or PC/XT, so there are almost always enough DMA channels available for the A/D board to use one. DMA A/D boards are now available that achieve data rates in excess of 100 kHz. Hence, on the PC/AT's now in our laboratory, we usually use A/Ds that utilize DMA. You may wish to read Chapter 15 for a further discussion of these types of boards.

Another, and perhaps not-so-obvious factor determining the choice among the three methods is the degree to which custom programming is required. Typically, the software for both the DMA and interrupt driven data collection methods assumes that the data will be stored as they are collected. Hence, users who wish to implement such tasks as real-time averaging of data or real-time plotting may have to develop their own software. The DMA method essentially removes any control by the user over the real-time processing by giving complete control over data collection and storage to the processor on the A/D board. Thus it may be unsuitable for some real-time processing tasks, particularly those requiring synchronous events on several control or data acquisition boards. Interrupts allow the sophisticated user to program his or her own interrupt service routine, but this, as we shall see in Chapter 15, is not an easy task for the amateur. Clearly, real-time processing is most easily done with the polling method, and so we shall discuss this method in all of the examples in the remainder of the chapter.

3.3 CONNECTING THE A/D TO AN INSTRUMENT

Although we have discussed many of the fundamentals of A/D use and design, there are many other factors that you will need to consider before actually using the A/D. Some of the factors discussed below that you will need to understand include: how to connect the A/D to an instrument; amplifying the signal; and using the A/D to make resistance measurements. Connecting analog instruments and sensors such as thermocouples, strain gauges, operational amplifiers, manometers, and flow transducers requires careful consideration of the cabling and analog signal communications design. Inevitably, you will have to decide which trade-offs to make among bandwidth, noise susceptibility, and cost.

3.3.1 Distance

The first factor you will need to consider is the distance between an analog sensor or instrument and the computer or A/D. There are two reasons for considering the distance of an analog signal connection to the IBM PC. The first reason is that all conducting cables have some finite resistance associated with them. The longer the cable is, the greater the resistance. Because some finite current is to flow through the cable, a voltage drop equal to

$$V_{drop} = I_{cable} R_{cable}$$

will exist between the ground side of the analog instrument and the A/D analog ground. Thus, the voltage measured at the A/D will be less than the voltage at the sensor or instrument. Worse yet, if the current flowing in the cable changes periodically, then the voltage between the instrument ground and the A/D ground will vary, causing noise to be introduced during A/D conversions. This condition is known as a ground loop. Ground loops can often be minimized by using differential input connections to the A/D (see Chapter 4) and by choosing sensors that do not generate large currents.

The second reason for considering the length of an analog sensor connection is that interference or noise may be picked up on the cable. This is especially true for flat cable connections or unshielded conductors. A long conductor will act as an antenna, picking up low frequency noise such as 60 Hz noise from fluorescent lights and transformers. High frequency noise from switching power supplies can also be picked up by long cables; 10 mV of noise on a long unshielded conductor connecting a sensor to the A/D is not uncommon.

Hence, the simplest advice is to keep cable lengths between the instrument and the A/D as short as possible. Once the signal is converted to digital form by the A/D, it is much less susceptible to these effects; thus, if the computer must be separated by long distances from the sensor or instrument, it is better to place the A/D very close to the instrument and have a long digital signal connection to the computer.

3.3.2 Cabling

Several types of cabling are commonly used for interfacing tasks. In increasing order of cost and immunity from noise, the choices are single cables; flat cables; twisted pair conductors; coaxial cables; and triaxial cabling.

The simplest type of interfacing can be done using one or more single wires. The major problem with this is simply that your computer can end up being embedded in a bowl of spaghetti. Flat cable has roughly the same electrical characteristics but allows the cabling to be more organized. Both of these types of cable are suitable primarily when the signal-to-noise ratio is high; in practice, this means for signals in the 1 to 10 volt amplitude range. If you choose flat cable, you will find it convenient to use a single connector to it. One commonly used flat cable connection is 25-pin flat cable conductor with D-shell (RS-232C) connectors. Many commercial analog I/O plug-in cards for the IBM PC use 25-conductor flat cable for analog and digital I/O connections.

When the signal being transmitted is an analog signal with a relatively low signal-to-noise ratio or a high-frequency signal, other types of cable should be selected. Because most laboratory work involves signals under 1 MHz, a commonly-used alternative is shielded twisted pair cables, connected with banana jacks or similar devices. For higher frequencies either coaxial cable or triaxial cable should be used; these are often connected with BNC connectors. When any of these types of cables is used, proper grounding of the cables is essential. Grounding, and more details on these types of cables, are discussed in Chapter 4.

Whatever type of wiring you choose to use, be sure to label each connector at the time you install it; label the wire at both ends and every five feet or so to avoid confusion later.

3.3.3 External Amplifiers

As we mentioned earlier, it is often necessary to amplify a signal from an instrument or sensor in order to take full advantage of the resolution of the A/D. For example, a typical A/D converter for the IBM PC will use 12 bits of resolution in a fixed analog signal range of -10 V to +10 V. At best this A/D will be able to distinguish between analog voltages 5 mV apart (20 V / 2^{12}). A signal transducer such as a thermocouple will only produce small voltages, perhaps in the range of 0 to 30 mV. As a result the 12-bit A/D converter will provide only 6 out of a possible 4096 integers to describe the temperature being recorded by the thermocouple. If the thermocouple range covers 0 to 2000°C, then the A/D can only resolve 333°C (2000/6) differences in temperature! This is unsatisfactory. An external analog amplifier should be used to amplify the thermocouple signal to provide more resolution. An amplifier with a gain of 333 will amplify the thermocouple signal so that the full temperature range (0 to 2000°C) will correspond to 0 to 10 V at the A/D converter input. In this case, using an external amplifier the A/D can resolve 0.98°C (2000/2048) differences in temperature. This is clearly more satisfactory.

The most common type of external amplifier used is an **operational amplifier**. The **operational** part of the name comes from the fact that this type of amplifier was first developed to perform mathematical operations as a component of an analog computer. Two types of operational amplifiers are in common use: voltage amplifiers and transconductance amplifiers. A voltage amplifier produces an output voltage proportional to an input voltage. A transconductance amplifier produces an output voltage proportional to an input current.

Operational amplifiers are sold as integrated circuits in a DIP (dual inline package). They can be purchased for only a few dollars at any electronics hobby store. Single chips containing several amplifiers are available as well.

Most of the details of operational amplifier theory and utilization need not concern us here; the interested reader is referred to any of the "electronics for scientists" textbooks now available. In practice, we can view an operational amplifier as a device that *provides an output proportional to the difference between its two inputs*.

The characteristic of most interest for an operational amplifier is the voltage gain it produces. In the simple circuit shown in Figure 3.6, the gain (amplification factor) produced is approximately

$$Gain = (R_1 + R_2)/ R_2$$

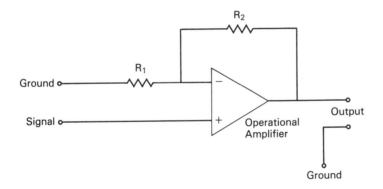

Figure   3.6.   Simple Operational Amplifier. In its simplest form the operational amplifier
circuit consists of an operational amplifier and two resistors. The gain can be
changed by changing the ratio of the resistances used. The - and + refer to the
inverting and non-inverting inputs of the operational amplifier, respectively.
The op amp also normally has connections for power and for a variable resistor
that is used to zero the op amp.

Two additional characteristics of operational amplifiers are high input
impedance (greater than 1 Mohm) and low output impedance (less than 100 ohms).
These are useful characteristics in the laboratory because many signal transducers
(thermocouples, for example) generate very small currents and cannot drive low
impedance loads. An amplifier connected to a thermocouple must provide a high
input impedance to appear invisible (like an open circuit) to the thermocouple. In
other applications, the output of the amplifier must be a relatively large current.
Hence, the output impedance of the amplifier is usually very small. Because of the
high input impedance and low output impedance, the amplifier provides current
amplification.

Another property of an operational amplifier that is useful to you in the laboratory is the rejection of common mode signals (DC biases). For example, suppose you
want to measure the 60 Hz sine wave ripple on a DC power supply. The voltage to be
measured might be a 200 mV peak-to-peak sine wave centered at a bias voltage of 20
V. Because an operational amplifier measures the difference between the two inputs,
the power supply output could be fed into one op amp input and a stable 20 V signal
(e.g., from a battery) fed into the other input. The result is that the two signals are
subtracted and only the difference is amplified by the op amp. In other words, the
common mode signal (the 20 V) is eliminated, and only the 200 mV peak-to-peak
signal is amplified. Operational amplifiers provide very large **common mode rejection
ratios** to provide for this situation.

We should also point out one very important practical consideration in using
amplifiers. An external amplifier is most useful if it is placed at or actually in the

casing of the instrument or sensor; this allows the distance traveled by a low-level signal to be minimized. Hence, in such an arrangement, the signal travels over the distance between the amplifier and the A/D as a high-level signal (e.g., 1 to 10 V) instead of a low-level signal, and this considerably reduces the effect of noise pickup during signal transmission.

3.3.3.1 Buffering

In addition to using external amplifiers strictly to amplify the signal, we sometimes use them for other useful properties. For example, we often use external amplifiers as **buffers**, that is to electrically isolate sensors and transducers from the PC data acquisition systems. This is done so that the noise or signals in the PC A/D converters do not affect the transducer responses. Other sensors that provide large signals such as piezo-electric transducers may damage the PC if they are not buffered or electrically isolated from the PC. Transducers that can't source much current such as thermocouples will not be able to drive A/D inputs directly. An external operational amplifier with unity gain (gain=1) will act as a buffer providing electrical isolation.

Finally, amplifiers are used as part of active filters, as described in Chapter 4.

3.3.4 Linearization

One other commonly-encountered problem in the laboratory is that some devices, notably thermocouples, thermistors and diodes, produce a non-linear response to a linear change of the property being measured. For example, a linear increase in the temperature measured by a thermocouple will yield a non-linear increase in the voltage produced by the thermocouple. There are two common methods to overcome non-linear transducer response.

First, you can measure with an A/D as if the transducer response were linear and then compensate for known non-linearities with appropriate software. This is a slow option because it means evaluating a high order polynomial between A/D conversions. Even using a software lookup table to perform the linearization can be rather slow, with the CPU of the IBM PC spending most of its time linearizing the results. In cases where the data rate is relatively slow and the computer is dedicated to one instrument, these may not be major problems, however.

Second, you can use an external amplifier with a non-linear gain designed to linearize the voltage generated by the non-linear transducer. This is the fastest and easiest way to measure utilizing non-linear transducers such as thermocouples because gain and linearization is provided at a small cost. For example, many thermocouple vendors also sell inexpensive boxes for linearizing and amplifying the output of each specific type of thermocouple. You should always test such devices to ensure that they are accurate across the entire temperature range, however.

3.3.5 Current and Resistance Measurements

So far, we have discussed using the A/D for applications involving measurement of voltages produced by sensors or instruments. Current, rather than voltage, measurements can be easily made by passing the source current through a resistor and measuring the voltage drop. Measurements of electrical resistance can be made in a related fashion, by applying Ohm's Law. Ohm's law can be written as

$$\text{Voltage (volts)} = \text{Current (amperes)} * \text{Resistance (ohms)}$$

As a result of this relationship, it can be seen that a voltage applied across a resistive load will establish a current. This current can be measured and used to determine the resistance or the load under test. The voltage applied across the resistive load must be able to source enough current to drive the resistive load. This may be a problem when measuring resistances below 1.0 ohm. For example, to measure a 0.01 ohm resistance, a 1.0 volt voltage source (e.g., a D/A) must source $1.0/0.01 = 100$ amperes to measure the correct resistance. In this case, and for the case of very large resistance measurements, it is appropriate to use a constant current source to drive the resistive load and measure voltage built up across the resistance with an A/D converter.

3.3.6 Noise Reduction

As previously mentioned, noise reduction is also a very important consideration when designing an interface to an analog instrument or sensor. It is so important, in fact, that we have devoted the next chapter to this topic. We highly recommend you read Chapter 4 before beginning to use your A/D.

3.4 PROGRAMMING THE A/D

As we noted in Chapter 2, the fastest way to learn about a laboratory board is to see some examples of its use, and to adapt them to your own laboratory. This is particularly important for the A/D, which is usually more difficult to program than the D/A. This is primarily because of the finite time required for the A/D conversion; we must first tell the A/D to initiate a conversion, and then wait until the A/D signals that the datum is actually ready to be read by the computer.

The examples shown here assume the use of the same data acquisition board used in Chapter 2, namely the Data Translation DT2801. The registers on the DT2801 used to program the A/D are shown in Table 3.1. The software in this chapter is again included in the Scientific Routines Diskette attached to this text (see Appendix B).

Table   3.1.   Registers on the DT2801 used by the A/D

Register[a]	Name	Address[b]	Meaning of Bits
1	Command	02ED	bits 7-4 should be 0000 bits 3-0 = 1111 for STOP bits 3-0 = 0001 for CLEAR ERROR bits 3-0 = 1100 for READ A/D IMMEDIATE
1	A/D Status	02ED	0=datum ready 2=previous command finished 3=previous byte was command 7=composite error
0	A/D Gain	02EC	1-0=gain code
0	A/D Channel	02EC	3-0=channel
0	A/D Datum	02EC	low or high byte of datum

[a] Only those registers and bit meanings are shown that are used in programming the A/D in the "READ A/D IMMEDIATE" mode.

[b] The addresses shown are for the factory configuration. If you have changed the hardware base address to another setting, the addresses shown should be adjusted accordingly. Addresses are in hexadecimal.

3.4.1 Collecting One Datum

Our first program is simply one to collect one datum from the A/D using polling. The program is listed in Program 3.1.

Clearly, this program requires some explanation. In lines 60 through 90 we calculate the address of each of the registers shown in Table 3.1 Notice that the base address is hardware switch selectable, so if you have modified the switch settings from the those set in the factory (e.g., because you have two or more of the same board in your PC), then you will need to change the base address from that given in the program.

In lines 120 through 150, we have the main program, which calls subroutines that perform the individual functions that we require. The first of these subroutines, in lines 190-220, inputs the necessary parameters. The board allows input from several different channels in sequence. There is actually only one A/D on the board, but the input lines (channels) are multiplexed into the single A/D. Hence, whenever the channel number is changed, we must allow a brief time for the "settling" of the multiplexer; this settling time is however normally of concern only at very high data collection rates.

```
10 REM PROGRAM ONE_DATUM
20 REM LABORATORY AUTOMATION USING THE IBM PC
30 REM Program to read one datum from the DT2801 A/D
40 REM ***********************************************************************
50 REM define all registers using their port addresses
60 BASEREG%=&H2EC                            'base address of board
70 COMMAND%=BASEREG%+1                       'command register
80 STATUS%=BASEREG%+1                        'status register
90 DATUM%=BASEREG%                           'data register
100 REM ***********************
110 REM                  Main program
120    GOSUB 190                             'get user parameters
130    GOSUB 250                             'set up board
140    GOSUB 330                             'read datum
150    PRINT "The datum is ";RESULT%
160 END
170 REM ***********************
180 REM Get parameters from user
190    CLS:INPUT "What is the A/D gain (1,2,4,8)? ",ADGAIN%
200    GAINCODE%=LOG(ADGAIN%)/LOG(2)         'convert gains 1,2,4,8 to 0,1,2,3
210    INPUT "What is the A/D channel? ",ADCHANNEL%
220 RETURN
230 REM ***********************
240 REM Set up board for correct operation
250    OUT COMMAND%,&HF                      'stop board
260    TEMP%=INP(DATUM%)                     'clear data register
270    'wait for command to finish, then clear errors
280    GOSUB 550
290    OUT COMMAND%,&H1
300 RETURN
310 REM ***********************
320 REM Read datum
330    GOSUB 550
340    OUT COMMAND%,&HC                      'give command READ A/D IMMEDIATE
350    GOSUB 520
360    OUT DATUM%,GAINCODE%                  'set gain
370    GOSUB 520
380    OUT DATUM%,ADCHANNEL%                 'set channel
390    'get datum, low byte then high byte
400    WAIT STATUS%,&H5                      'low byte ready?  check bits 0,2
410    LOW%=INP(DATUM%)                      'yes, so read in low byte
420    WAIT STATUS%,&H5                      'high byte ready?
430    HIGH%=INP(DATUM%)                     'yes, read it
440    REM convert datum to single word
450    RESULT%=HIGH%*256+LOW%
460    GOSUB 550
470    ERRORCHECK%=INP(STATUS%)              'check status register
480    IF (ERRORCHECK% AND &H80) THEN PRINT "Error while reading A/D"
490 RETURN
500 REM ***********************
510 REM Check to see if ready to set a register
520    WAIT STATUS%,&H2,&H2                  'wait for bit 1 to be reset
530 RETURN
540 REM Check to see if ready to send another command
550    WAIT STATUS%,&H2,&H2                  'wait for bit 1 to be reset
560    WAIT STATUS%,&H4                      'wait for bit 2 to be set
570 RETURN
```

Program   3.1.   Collect One Datum.

The DT2801 series boards also have software programmable gain, so we must specify a gain code. If you have a DT2805 board, you will need to change lines 190 and 200 to reflect the gains of 1, 10, 100, and 500 (and be sure to have the appropriate capacitor installed for operation at gains of 100 or 500, if necessary).

The DT2801 allows several modes of A/D use, the simplest of which we will illustrate here. The simplest mode is referred to as the READ A/D IMMEDIATE mode. In this mode, each A/D conversion is initiated by the software; in other modes, some sort of clock is used to trigger conversions, as illustrated in Chapters 9, 14, and 15.

The next subroutine (lines 25-300) is identical to the one we used in Chapter 2; it readies the DT2801 for commands. Note that the subroutines in lines 51-570 are also the same as those used in Chapter 2. In fact, these routines are used in all programs for the DT2801.

We are now ready to set up the board for the desired mode of operation. This is done by sending the READ A/D IMMEDIATE command (line 340) to the Command Register. Once this is done, the control logic of the board expects that the gain code and the channel be sent to the Data Out Register, in that order (lines 360, 380). Notice that each time, we must check to see that the board is ready for a new command or setting by waiting for bit 1 of the Status Register to go low (line 520). Commands sent to the Command Register must also wait for the previous command to finish (line 560), which is signalled by having bit 2 of the Status Register be set by the hardware. In any case, sending the READ A/D IMMEDIATE command and setting the gain code and channel automatically causes an A/D conversion to be initiated. Note that we may need to insert a pause to wait for the channel multiplexing to finish before initiating a conversion, depending upon the speed of our computer and the efficiency of the language we are using.

We now must "poll" the board to find out when the A/D conversion is complete. We do this on the DT2801 by waiting for bits 2 and 0 of the Status Register to be set (Table 3.1). We thus test for the data being ready, then read the low byte, test for data ready, and then input the high byte of the data (lines 400-430). The full 12-bit datum is then calculated (line 450). We then check bit 7 of the Status Register (lines 460-480) to see if any kind of error has occurred, and print out any error message, if appropriate, and the datum.

3.4.2 Collecting Multiple Analog Data

Of course, Program 3.1 is not all that useful. It allows us to collect only a single datum and does not store it for later use. Hence, Program 3.2 illustrates a more useful procedure, namely collecting multiple data.

Notice the similarities with Program 3.1. The differences include a dimensioned array (RESULT%) to hold the data. In addition, a FOR ... NEXT loop beginning in line 150 allows us to collect multiple data.

If you have already run this program, you may find it collects data fairly slowly. Removing all of the remark statements and combining several statements on a single line (separating the statements with colons) will make the program run somewhat faster. (Be prepared for nasty remarks from other programmers, however, if you do this!)

```
 10 REM PROGRAM MULTIPLE_DATA
 20 REM LABORATORY AUTOMATION USING THE IBM PC
 30 REM Program to read multiple data from the DT2801 A/D
 40 REM *********************************************************************
 50 DIM RESULT%(1000)
 60 REM define all registers using their port addresses
 70 BASEREG%=&H2EC                        'base address of board
 80 COMMAND%=BASEREG%+1                   'command register
 90 STATUS%=BASEREG%+1                    'status register
100 DATUM%=BASEREG%                       'data register
110 REM ************************
120 REM              Main program
130    GOSUB 230                          'get user parameters
140    GOSUB 320                          'set up board
150    FOR I%= 1 TO NDATA%                'data collection loop
160        GOSUB 400                      'read datum
170        PRINT "Datum ";I%; " is ";VALUE%
180        RESULT%(I%)=VALUE%
190    NEXT I%
200 END
210 REM ************************
220 REM Get parameters from user
230    CLS:INPUT "What is the A/D gain (1,2,4,8)? ",ADGAIN%
240    GAINCODE%=LOG(ADGAIN%)/LOG(2)      'convert gains 1,2,4,8 to 0,1,2,3
250    INPUT "What is the A/D channel? ",ADCHANNEL%
260    WHILE (NDATA% < 1 OR NDATA% > 1000)
270        INPUT "How many data do you wish to collect (1 to 1000)? ",NDATA%
280    WEND
290 RETURN
300 REM ************************
310 REM Set up board for correct operation
320    OUT COMMAND%,&HF                   'stop board
330    TEMP%=INP(DATUM%)                  'clear data register
340    'wait for command to finish, then clear errors
350    GOSUB 620
360    OUT COMMAND%,&H1
370 RETURN
380 REM ************************
390 REM Read datum
400    GOSUB 620
410    OUT COMMAND%,&HC                   'give command READ A/D IMMEDIATE
420    GOSUB 590
430    OUT DATUM%,GAINCODE%               'set gain
440    GOSUB 590
450    OUT DATUM%,ADCHANNEL%              'set channel
460    'get datum, low byte then high byte
470    WAIT STATUS%,&H5                   'low byte ready?  check bits 0,2
480    LOW%=INP(DATUM%)                   'yes, so read in low byte
490    WAIT STATUS%,&H5                   'high byte ready?
500    HIGH%=INP(DATUM%)                  'yes, read it
510    REM convert datum to single word
520    VALUE%=HIGH%*256+LOW%
530    GOSUB 620
540    ERRORCHECK%=INP(STATUS%)           'check status register
550    IF (ERRORCHECK% AND &H80) THEN PRINT "Error while reading A/D"
560 RETURN
```

Program   3.2 (Part 1 of 2).   Collect Multiple Data.

A much more satisfactory way to achieve a considerable increase in speed is by running a program in the compiled form. To do this you need a copy of a BASIC compiler. The BASIC compiler must usually be purchased separately, because it is not included with the computer as the BASIC interpreter typically is. A variety of such compilers are available, many with sophisticated editors and debugging systems included.

```
570 REM **************************
580 REM Check to see if ready to set a register
590   WAIT STATUS%,&H2,&H2              'wait for bit 1 to be reset
600 RETURN
610 REM Check to see if ready to send another command
620   WAIT STATUS%,&H2,&H2              'wait for bit 1 to be reset
630   WAIT STATUS%,&H4                  'wait for bit 2 to be set
640 RETURN
```

Program 3.2 (Part 2 of 2). Collect Multiple Data.

One last hint will help speed this program. BASIC is often very slow in printing information on the screen. If you eliminate line 170 from this program you will notice a considerable increase in speed (as much as 100-fold, depending upon the interpreter or compiler used).

EXERCISES

The following exercises will help you gain experience with the material presented in this chapter.

1.   Use Program 3.2 to collect data from an instrument that would normally output data to a strip-chart recorder (or from some other slowly-varying signal source).

2.   Modify Program 3.2 to simulate the data from an 8-bit A/D by "throwing away" the least-significant four bits of each datum. This can be done easily with the AND operation in BASIC. Have the program print both the 12-bit number and the "8-bit number" in decimal and compare the numbers taken from some instrument.

3.   Modify Program 3.2 to average data together. Have it only print the averaged data. Use the modified program to collect data from a signal generator. Compare what happens as you change the number of data points averaged and the frequency of the data collected (you may be surprised by some combinations -- if you are, read the following chapter).

4.   Find a small signal source (10 to 100 mV). Attach it to your A/D and collect data from it. Now, attach the signal source with a 50 foot wire and repeat the experiment. What happened? Attach an amplifier with a gain of 100 to the signal source and repeat the experiment. Are the results different if the amplifier is placed at the other end of the 50-foot wire?

4

NOISE DETECTION
AND REDUCTION
TECHNIQUES

As computer interface hardware gets cheaper, faster, and increases in function, the fundamental limits on performance and accuracy of data acquisition and control systems must be considered. One of the fundamental limits to the resolution and accuracy of data acquisition systems is *noise.* For example, it does not make sense to use a 16 bit A/D (0.01% accuracy) with your PC if the voltage to be measured contains 2 to 3% noise. In this situation a 12 bit A/D would ultimately provide the same effective resolution as the 16 bit A/D converter.

Noise can be described as any signal that interferes with the signal of interest. There are generally three sources of noise encountered in the laboratory. These are interfering signals, drift noise, and device noise.

Interfering signals include any analog waveforms that might otherwise be called **signals** but that happen in a particular application to interfere with the signal of interest. An example of this would be a laboratory located next to an elevator. Each time the elevator motor starts moving there is a large current surge in the elevator power line. The magnetic field generated by this current surge couples inductively into your electronic measurement apparatus. The result of this coupling is an erroneous measurement by your PC each time the elevator starts. A more common example of interfering noise is the 60 Hz coupling of power transmission lines for pumps, motors, and transformers with your lab experiment.

Drift noise is due to the finite stability of laboratory electronics over time. Amplifiers, voltage sources and other electronic components tend to be vary in stability when measured over a long period of time. Temperature and other environmental changes are often the cause of this drift instability. As a result, any experiment conducted in the laboratory for a long period of time may be subject to drifts and related instabilities. Drift therefore can be treated as a source of noise.

Device noise is common to any electronic circuit, component, detector, etc. Device noise is a caused by individual charge-carrying electrons in devices and conductors. The random collisions between electrons and other particles in conductors lead to random voltage variations sometimes called **shot noise**. Because device noise is due to the statistical assembly of electrons in conductors, it is easy to analyze with statistical methods.[1]

Because device noise is so easy to analyze, it gets most of the attention in the literature. However interfering signals and drift noise limit the ultimate sensitivity of most experiments. Interference, drift, and device noise will be discussed separately. First, however, we need to understand something about the detection and measurement of noise.

4.1 DETECTION OF NOISE

In order to maximize the resolution, frequency response, and performance of an automated data acquisition system, you must first classify the type of noise you wish to eliminate. You can then use different techniques to directly measure or estimate the noise level. Once you have classified the noise, you can choose among several methods in order to reduce this noise.

4.1.1 How to Classify Noise

There are essentially two variables that can be used to classify noise. These are the **amplitude**, or noise energy, and the **frequency distribution** of the noise. The noise amplitude is usually measured in units of volts. The frequency of the noise is measured in Hertz (cycles/second). Both noise and signal must of course be measured in the same units. By plotting the amplitude versus frequency characteristics of the composite signal and noise waveform, one can usually distinguish the noise from the signal of interest.

---

1   S.D. Senturia & B.D. Wedlock, **Electronic Circuits and Applications**, p. 550 (1975).

4.1.2 Frequency Domain Analysis

To investigate the noise in your data, it is often useful to plot signal amplitude (Y) versus the time (X). This is referred to as a **time domain representation** of the signal. For example, Figure 4.1 shows the time domain graph of a sine wave modulated with a high frequency noise component. In this case the frequency of the signal and noise are far enough apart that the noise amplitude may be read directly off the graph (1 volt). Figure 4.2 shows the display of an 8 Hertz sine wave modulated with a 10 Hertz noise component. In this case the noise component is too close in frequency to the signal to extract the noise amplitude from the graph.

Many signals in the laboratory contain a variety of frequencies of interest, and the noise is usually distributed over a wide range of frequencies. In order to determine the distribution of the noise amplitudes in frequency, the signal can be transformed to the **frequency domain**. There are laboratory instruments called **spectrum analyzers** that can be used to transform a time domain voltage into an amplitude versus frequency plot in real time. Spectrum analyzers are convenient in the case of the 8 Hertz sine wave modulated with the 10 Hertz noise component as described above.

A spectrum analyzer is typically very expensive ($5,000-$50,000). Because the IBM PC is abundant in the laboratory and can be used for a variety of functions other than noise analysis, it is the logical choice to replace the spectrum analyzer by emulating the function in software. The IBM PC can be used to digitize a time domain signal and transform the signal to the frequency domain using special software. This mathematical transformation implemented in software is called a **Fourier transform**. The Fourier transform is a mathematical operation performed on a sequence of numbers; it allows us to convert a time domain sequence of numbers into a frequency domain sequence, or vice versa. For example, Figure 4.3 illustrates a sequence of A/D integer numbers plotted in time. This sequence is a sine wave at a frequency of 50 Hz superimposed on a random noise signal. The noise component is clearly visible in Figure 4.3. Figure 4.4 illustrates the frequency distribution of this signal as a result of computing the Fourier transform. It is important to note that the difference between the two graphs in Figure 4.3 and Figure 4.4 is the X axis units. The X axis or abscissa of the first graph is time, and the X axis of the second graph is frequency. It is now easy to identify and separate the 50 Hz sine wave signal of interest from the background noise level. Software techniques for practical computation of the Fourier transform and its use as a digital filter are discussed later in this chapter (see page 80).

4.1.3 Signal-to-Noise Ratio

Of course, we would also like to measure not only the noise frequency but also the noise intensity. One of the most straightforward and frequently-used measures of noise intensity is the **signal-to-noise ratio** (**SNR**). The signal to noise ratio is defined as the ratio of signal power P_S to noise power P_N.

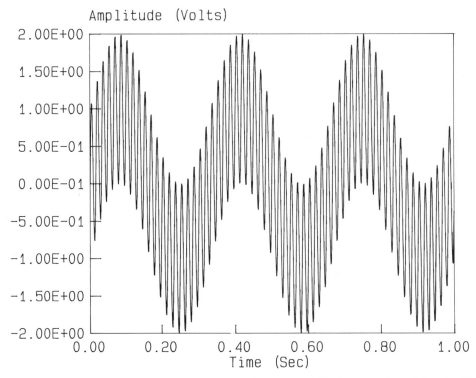

Figure 4.1. Time Domain Graph of Noisy Sine Wave. A 3 Hz sine wave is modulated with a high frequency noise source. The signal and noise are far enough apart in frequency to be able to distinguish between them visually.

$$\text{Signal to Noise Ratio (SNR)} = \frac{P_S}{P_N}$$

Low frequency signal or noise power can be defined in terms of voltage, current and resistance by Ohm's law (V=IR).

$$\text{Power (P)} = VI = \frac{V^2}{R} = I^2 R$$

The total noise contributed from several independent noise sources is equal to the sum of the values from each noise source. Because of this we can rewrite the signal to noise ratio.

$$\text{Signal to Noise Ratio (SNR)} = \frac{P_S}{P_{N1} + P_{N2} + \dots + P_{NX}}$$

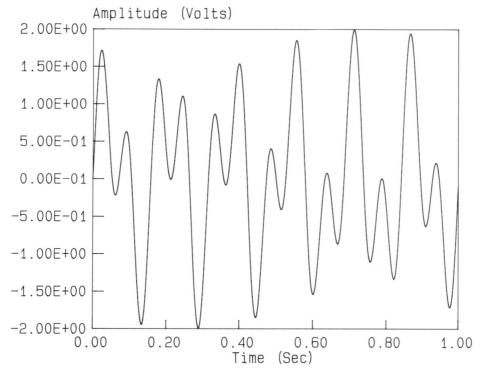

Figure   4.2.   Time Domain Graph of Interfering Noise.   An 8 Hz sine wave is modulated
with a 10 Hz noise source (sine wave). The frequencies of the two waveforms
are too close together to distinguish between them.

It is common notation to represent the signal-to-noise ratio in logarithmic or decibel
units notation since it is a power ratio.[2]

$$\text{Signal to Noise Ratio (SNR)} = 10 \log \frac{P_S}{P_N} \, (dB)$$

If the current is constant (same conductor) then voltages can be written since
they are easier to measure:

$$\text{Signal to Noise Ratio (SNR)} = \frac{V_S}{V_N}$$

One of the practical consequences of such a measurement might be to determine how our noise compares with the resolution of the A/D we are using. For
example, how much noise can we tolerate with a 12-bit A/D? Recall that for a 12
bit A/D with a ± 10 volt peak to peak amplitude signal, 5 mV of noise added to it
represents to minimum signal to noise ratio that can be tolerated to retain full 12 bit

2   S.D. Senturia & B.D. Wedlock, **Electronic Circuits and Applications** , p. 559 (1975).

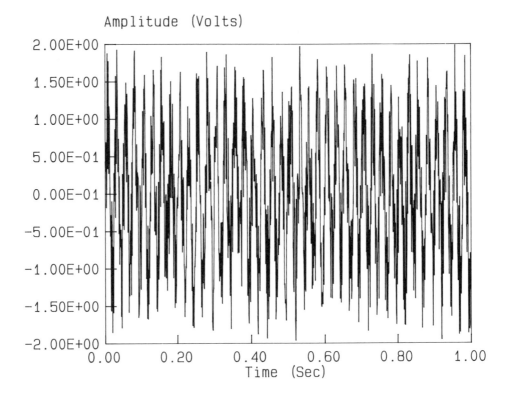

Figure   4.3.   Time Domain Graph of Noisy Sine Wave.   A 50 Hz sine wave is shown with
random noise added.

accuracy. Therefore in order to ensure 12 bit A/D accuracy the signal to noise ratio
must always be greater than or equal to :

$$SNR = 10 \log(\frac{V_S}{V_N}) = 10 \log (20/.005) = 36 \text{ dB (decibels)}$$

The larger the signal-to-noise ratio (SNR), the easier it is to detect the signal.
As a rule of thumb in the lab, the SNR must be larger than one to be able to detect
the signal.

4.1.3.1 Measurement of Signal-to-Noise Ratio (SNR)

There are two convenient ways to measure the signal-to-noise ratio directly. The first
is using an oscilloscope to measure the signal and noise power. The second is using
an A/D to measure these values. Each technique requires a subsequent calculation
of the signal to noise ratio.

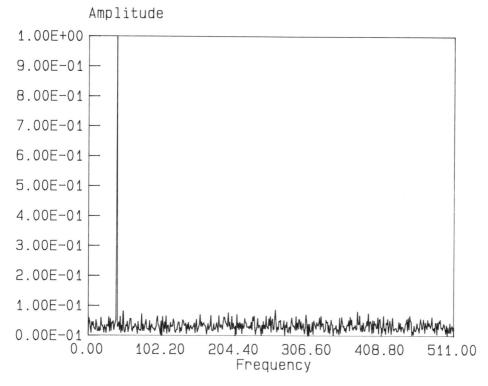

Figure 4.4. Frequency Domain Graph of Interfering Noise. The frequency distribution of the signal and noise in Figure 4.3 is plotted against amplitude. The spike represents the 50 Hz sine wave signal. This graph is produced using a Fourier transform.

Using an Oscilloscope to Record the SNR. By measuring a signal with an oscilloscope, you can directly read the signal amplitude or intensity and the noise amplitude directly off of the oscilloscope display. It is important that you adjust the time base setting (time step per display grid distance) of the oscilloscope to a sweep rate at which the signal peak voltage and noise peak voltage can be distinguished. This is not always easy, particularly when the frequency of the noise and frequency of the signal are close together. The noise frequency is relatively low in the case of D.C. drift (i.e. $< 10\,Hz$) and can be distinguished from most signals as a slow change in signal amplitude offset by the nearly constant noise voltage displayed on the oscilloscope. In the case of device or shot noise, the noise frequency is usually well confined to a narrow band of frequencies which enable the noise amplitude (**spikes**) to be read directly off the oscilloscope display in units of volts. Radio frequency noise (RF) is also easy to distinguish from most laboratory signals of interest since it is so high in frequency.

Using an A/D to Record the SNR. Although oscilloscopes are abundant in the lab, they are not always required to record the SNR. Your A/D may be used to record the SNR if the signal and noise frequencies are well below the maximum sampling frequency of your A/D. Measuring the SNR of a 500 Hz sine wave with 60 Hz noise interference from nearby fluorescent lights is a typical example of this. Recording 2000 data samples at a sampling rate of 30 KHz will yield 66 milliseconds of data. Plotting the sampled data should illustrate 4 cycles of 60 Hz noise superimposed on 33 cycles of 500 Hz signal. By measuring from the graph the peak-to-peak voltage amplitudes of the respective 60 Hz noise and 500 Hz signal, the SNR can be easily calculated.

4.2 INTERNAL SOURCES OF NOISE

As you might conclude from the discussion above, noise is not a major concern with low-precision measurements (for example, with 8-bit A/D converters in industrial settings). However sensitive laboratory measurements that demand 12 or 16 bit precision often require that you pay more attention to noise. Because noise sources are additive (i.e. the total noise in a data acquisition system is equal to the sum of all of the individual noise sources), it is important to analyze each electronic noise source independently to determine which if any is dominant and can be removed from the system. This in turn requires that you have some understanding of the various types of noise. In general, we shall look at the following types of noise:

■ Passive thermal noise.

■ Random shot noise.

■ Active thermal noise.

■ A/D quantization noise.

We shall then examine methods for reducing some of these types of noise.

4.2.1 Passive Thermal Noise Sources

One of the kinds of noise seen from electronic devices is **thermal noise**. Noise can arise from a variety of electrical devices, both **passive** devices such as resistors, capacitors and inductors, and **active** devices that require external power such as transistors and diodes.

Of the passive devices, electrolytic capacitors can introduce drift noise with changes in temperature, and inductors can act as antennae, generating noise currents because of their sensitivity to stray magnetic fields. Resistors however, are usually the dominant consideration when designing low noise interface circuits using passive circuit components.

In the resistor, currents of electrons collide randomly with the resistive material causing noise known as **thermal or Johnson**[3] noise. The average thermal noise voltage in a given frequency range can be defined as:

$$V_{TH} = \sqrt{4kTRB}$$

where

      k = Boltzmann's constant = $1.38x10^{-23}$ Joules/K
      T = temperature in Kelvins
      4kT = $1.6x10^{-20}$ Joules at room temperature
      R = value of the resistor in Ohms
      B = given frequency range in Hertz.

It should be noted that the noise voltage in a given bandwidth is independent of the frequency. In other words, plotting the noise voltage yields a straight line as illustrated in Figure 4.5. This flat noise frequency distribution is often called **white noise**. An example would be to calculate the rms thermal white noise noise voltage in a 10 KHz bandwidth across a 10 Mohm resistor:

$$V_{TH} = \sqrt{4kTRB} = \sqrt{(1.6x10^{-20})(1x10^{7})(10000)} = 40 \, uV.$$

An experiment using a 16 bit A/D converter in a 5 volt full-scale range provides a voltage resolution of $5 / 2^{16}$ = 76.3 uV. If the voltage is measured across a 10 MOhm resistor, it becomes evident that the resistor noise voltage can approach 1 bit of uncertainty in the A/D conversion.

4.2.2 Random Shot Noise

Because of the discrete charge of the electron and the collisions of these electrons inside the resistor, another type of noise known as **Shottky** (abbreviated shot) noise can be a factor in sensitive measurements. Since shot noise is the result of random electron collisions in a resistor subject to a dc current, the shot noise can be written as a current:

$$I_{Shot} = \sqrt{2qI_{dc}B} \quad \text{and} \quad V_{Shot} = I_{Shot}R$$

where $q = 1.6x10^{-19}$ Coulombs is the charge of an electron.

Like the thermal noise, shot noise is uniformly distributed in frequency (i.e. it is white noise). For example, assuming a voltage is measured over a 1 MOhm resistor in a 10 KHz bandwidth and $I_{dc} = 3nA$ then:

$$I_{Shot} = 3.1pA \text{ rms} \quad \text{and} \quad V_{Shot} = (3.1x10^{-12})(1x10^{7}) = 31 \text{ uV}.$$

---

[3]   M. Schwartz, **Information Transmission, Modulation, and Noise** (1970).

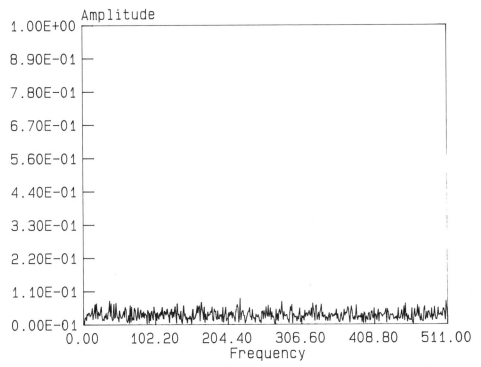

Figure 4.5. White Noise. White noise is noise arising at all frequencies. A common source of this is thermal noise arising from electron collisions in resistors.

Total Noise. The total noise in a resistor can be written as the vector addition of the shot and thermal noise components.

$$V_n = \sqrt{(V_{Th})^2 + (V_{Shot})^2}$$

By controlling the temperature and d.c. current flowing through a resistor, the shot noise voltage, and thermal noise voltage can be minimized. The rule of thumb to follow when designing low noise interface circuits is to choose resistors that minimize the d.c. currents to a shot noise level below the thermal noise level.

4.2.3 Active Thermal Noise Sources

It is evident from the previous discussion that temperature variations in passive circuit components like resistors can cause variations in the noise amplitude. It is also apparent from the thermal noise voltage equation that noise variations due to temperature are linear. Therefore as long as the temperature in the resistor does not change dramatically, the change in resistor noise will be small.

In active *semiconductor devices* (e.g. transistors and diodes), small changes in temperature can dramatically affect the noise characteristics of those devices. Since op-amps, A/D converters, and other measurement devices are composed largely of

transistors and diodes, this can be a dominant source of noise in modern instrumenta-
tion. The current flowing through a semiconductor diode can be written as :

$$I = I_S \left(e^{\left(\frac{qV}{kT} \right)} - 1 \right)$$

where

$I_S \cong 10^{-12}$ Amp. $= 10$ pA, the reverse saturation current
$q = 1.6x10^{-19}$ Coulombs, the electronic charge
V = the forward bias diode voltage
k=Boltzmann's constant= $1.38x10^{-23}$ Joules / K
T = the absolute temperature (Kelvins).

It is evident from the diode equation that a small linear change in temperature
will cause a large change (milliamps) in the diode current I. As an example, a diode
with a forward bias voltage of 0.65 volts subject to a 1 degree Centigrade change in
room temperature will undergo a 25 mA change in diode current I. This can represent a significant noise variation in circuits which employ diodes or transistors (a
transistor is composed of two back-to-back diodes). Because temperature changes
slowly with time, this type of semiconductor noise is often characterized as low frequency drift.

Many common op-amps composed of semiconductors are not designed for
ultra-low noise measurements. By incorporating temperature sensing and correction
circuitry, it is possible to compensate for the drift noise that results from these temperature variations. Precision op-amps and high precision A/D converters (14-16
bit) usually provide added circuitry to compensate for this drift noise internally.

It should be noted that a temperature-sensitive device can be exploited to act as
an accurate temperature sensor. Semiconductor temperature sensors based on this
principle are often used in the lab.

4.2.4 A/D Quantization Noise

One other type of commonly-encountered noise that is somewhat different from the
others we have examined is **quantization noise**. This type of noise is related to the
fact that we are making digital measurements of an analog signal.

Let's look at an example. A 12 bit A/D converter can digitize a $\pm$ 10 volt
signal with a resolution of roughly 5 mV. In other words, each sample of the $\pm$ 10
volt signal is **quantized** into 1 of 4096 (2^{12}) different integer levels that are each 5 mV
apart. Quantization noise in this case occurs when the A/D converter makes arbi-
trary decisions between two integer levels within a 5 mV step. For example, even if
the continuous analog signal is a constant d.c. signal then the digitized signal may
randomly jump between two integer values (around the d.c. value). This random
jumping represents quantization noise in the measurement. Since an A/D converter

will round the converted d.c. voltage up or down a maximum of 2.5 mV to match the 5 mV resolution, the quantization error equals 2.5 mV. In general, the quantization error of an A/D converter will always be:

$$\text{Quantization Error} = \frac{\text{A/D Converter Voltage Resolution}}{2}$$

Generally, we can not reduce quantization noise, unless we use a higher resolution A/D. The quantization noise can be filtered out later during data analysis. At that time, we treat the quantization noise as a high frequency noise source, which we can reduce with low pass digital filtering in software. This will be discussed in more detail later in this chapter.

4.3 EXTERNAL NOISE

So far, we have only discussed noise sources within the electronics of our own instruments. However, noise frequently is introduced from outside of our instrument -- from other instruments, nearby motors, fluorescent lighting, and so forth. Our goal then becomes to isolate our instrument from all of these external noise sources. One of the most potent weapons in this battle against external noise is an understanding of proper wiring practices. We shall look at the following methods of isolating our instruments from external noise:[4]

- Shielding
- Avoiding ground loops
- Guards and optical isolation
- Proper selection and connection of cables

Before examining these methods, however, let's discuss a few unifying principles that will help you understand these techniques.

Generally, you can divide external noise into two major categories: *electrical* and *magnetic*. The methods of dealing with these two types of noise are somewhat different. **Electrical noise** refers to currents induced by coupling external electric fields and the wiring in our instrument. In general, most of the methods for dealing with electrical noise are based on the principle of the **Faraday cup**, which is nothing more than a big conducting cup (or cage) that surrounds the object of interest and that is connected to ground; objects inside the Faraday cup are effectively isolated from external electrical fields.

Magnetic noise refers to the currents induced in the wiring of our instruments when they are placed near a *changing* external magnetic field (the same principle upon which an alternator produces electrical current). The usual technique for

---

4   S.D. Senturia & B.D. Wedlock, **Electronic Circuits and Applications** , p. 552 (1975).

dealing with magnetic noise is to isolate physically our instruments from large alternating magnetic fields found in electric motors.

For either type of noise *the smaller the signal of interest, the more important
the noise prevention.*

4.3.1 Shielding

Let's begin our examination of wiring practices with one of the most commonly-
encountered external noise sources, namely line frequency noise (50 or 60 Hz,
depending upon the country). This noise arises from the power lines and devices that
use that power. In other words, the line frequency alternating current gives rise to
currents in the wires of our laboratory instruments. Since power lines are ubiquitous,
so is the problem with line frequency noise.

Perhaps the best way to see this is with an example. Figure 4.6 illustrates the
laboratory line frequency interference noise problem. A signal of interest (e.g. a 10
mV chart recorder output from an instrument) is connected to an operational amplifier. The op-amp is powered by a $\pm$ 10V d.c. power supply.

The chart recorder $\pm$ 1mV signal cable couples capacitively with the electrical
fields from the power supply. Assuming this capacitance to be small (i.e. 1pF), a 40
uV 60 Hz interference signal could be superimposed on top of the chart recorder
signal driving the op-amp. Since the chart recorder signal may only have a 1000 uV
peak to peak amplitude, the 60 Hz noise represents 40 uV / 1000 uV = 4 percent of
the signal driving the op-amp!

A simple solution to this problem is the use of a **shield** as illustrated in
Figure 4.7. A shield is a piece of metal that surrounds the conductor driving the
op-amp. This shield is tied to the signal ground (e.g. chart recorder output). The
effect of the shield is to eliminate the capacitive coupling of the power line electric
field with the chart recorder signal. In the case of our example, Figure 4.6, 60 Hz
A.C. current will flow through the shield to signal ground rather than through the
op-amp input. A rule of thumb in the lab is to *ground the shield at the location of
greatest encountered interference*, which is usually at the transducer (e.g. chart
recorder output) end. The shield may not completely eliminate the capacitive coupling between the signal conductor and power line, but it can reduce this interference
signal by a factor between 100 and 1000.

4.3.2 Avoiding Ground Loops

The distance between an analog signal source or instrument and a remote computer
is an important consideration when designing a data acquisition system. All conducting cables have some finite resistance associated with them; the longer the cable,
the greater the resistance. Furthermore, "ground" is not an absolute value, but in
fact can vary dramatically from place to place. Hence, if we connect instrument
ground (at one potential) to the computer ground (at another potential), there will be
a voltage drop across the cable, equal to $V_{drop} = I_{cable}R_{cable}$. When the signal is

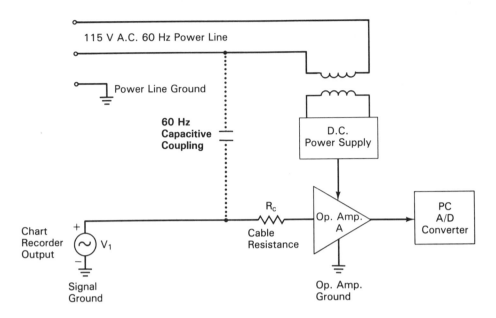

Figure 4.6. Laboratory Noise Interference. One type of external noise is that introduced from a 115V A.C. power line electric field capacitively coupling with a chart recorder output signal.

digitized by the A/D converter, the voltage drop across the cable will cause an erroneous offset in the A/D conversion. This condition is known as a **ground loop**. The current flow from a ground loop can often be detected with a sensitive current ammeter.

Ground loops can also arise in a variety of other ways, from either external electrical or external magnetic fields. For example, nearby motors generate stray alternating magnetic fields. These magnetic fields can induce substantial alternating currents in a long ground conductor cable. This type of ground loop is of particular concern since the measured noise (induced current) will vary with the magnetic field strength.

A ground loop can also occur when the PC A/D converter and the analog signal are connected to separate grounds. The two separate grounds will never be at exactly the same potential as illustrated in Figure 4.8A. That is, the voltage appearing at the input v_2 will consist of the signal at the A/D converter v_1 plus the difference in ground potentials, often called a **common mode voltage**, v_{cm}. If the common mode voltage v_{cm} is comparable to the magnitude of the signal, significant errors can be introduced in the measurement.

There are several ways to prevent ground loops. The best way to prevent ground loops is to minimize the distance between the analog signal transducer and

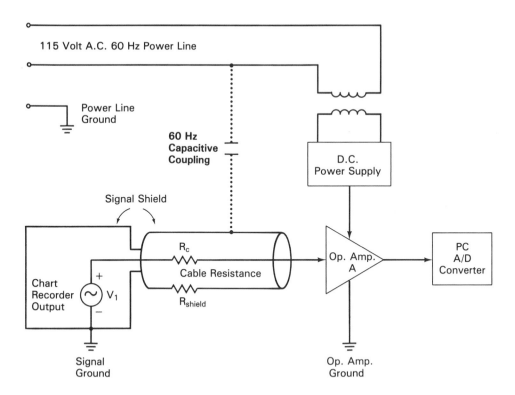

Figure   4.7.   Shield.   The use of a grounded shield eliminates the external noise introduced on the signal cable by the 50 or 60 Hz A.C. power line.

the PC. Another way of preventing a ground loop is to connect a heavy conductor (ground strap) between the two separate grounds, as illustrated in Figure 4.8B. Braided cable is often used because of its low inductance. Even though the ground loop still exists, the low impedance of the path will minimize the common mode voltage v_{cm}.

When setting up a PC with an A/D converter, ground loops can also be prevented by configuring the A/D using **differential input connections**, as opposed to the single ended input configurations commonly used. Also, choosing analog sensors that do not generate large currents can reduce ground loop noise. As an example, if a 10 meter cable is wired between a gas flow controller analog input (setpoint control) and a D/A analog output channel at the PC, if the resistance of the cable is = 0.1 ohm/meter, and if the gas flow controller input draws 15mA from the D/A analog output channel at the PC, then a voltage drop of

$$V_{drop} = (15mA)(0.1ohm/m)(10m) = 15mV$$

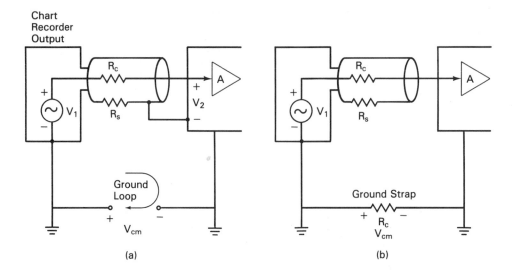

Figure 4.8. Ground loops. **A.** A ground loop is arises from a common mode voltage between the chart recorder output ground and the op-amp. **B.** A ground strap has been connected between the two ground locations to minimize the common mode voltage, V_{cm}.

will be seen across the cable as a ground loop. As a result, the gas flow controller might control at a value 1 to 3% less than that programmed by a D/A analog setpoint. Alternatively, using a gas flow controller that draws only 1 mA would reduce the voltage drop to 1 mV, which correspondingly reduces the error in the setpoint to less than 0.1%.

If the signal being measured is very small (i.e. microvolt range), or if the electrical interference noise is significant in the lab, more elaborate shielding schemes may need to be employed.

4.3.3 Guards and Optical Isolation

A **guard** is a shield that is connected directly to the common mode voltage V_{cm} between the signal source and A/D converter. This configuration, illustrated in Figure 4.9, is most effective in the isolation of noise. Any potential difference between the signal transducer ground and the A/D ground is effectively shielded by the guard connection. It is customary to define the quality of guarding in terms of a **common mode rejection ratio (CMRR)**. This is the ratio of V_{cm} to V_2. It is possible to achieve a CMRR of 10^6. This means that a 10 volt common mode voltage will produce only 10 microvolts of ground loop noise at the A/D input.

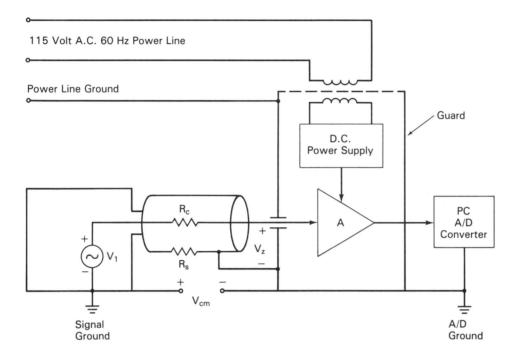

Figure   4.9.   Guard.   A guard provides the best isolation from external noise by improving the rejection of common mode ground loop voltages.

An alternate isolation device shown in Figure 4.10 called an **optoelectronic isolator**, is composed of a light emitting diode (LED) and photo-transistor mounted closely together in an integrated circuit package.   The light from the LED illuminates the base of the transistor which in turn controls the current conducting through the transistor.   Since there is no electrical conductor connecting the two circuits, the A/D converter attached to the PC is completely isolated from the analog transducer. Optical isolation circuits of this type are particularly useful for large signal (high voltage) applications when it is important not to have a large ground loop voltage damage the PC.

4.3.4 Proper Selection and Connection of Cables

4.3.4.1 Common Cabling Alternatives for Analog Signals

It may surprise you to learn that the type of cable you use may have an important effect on the amount of noise that enters your instrument.   Cables can provide various degrees of shielding from noise.   Since analog signals are particularly suscep-

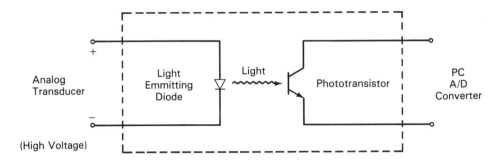

Figure 4.10. **Optoelectronic Isolator**. An optoelectronic isolator acts as a buffer between the analog transducer and the A/D convert of the PC, thereby blocking ground loops or high voltage noise.

tible to interfering noise, proper cable selection is especially significant when cabling to A/D's, D/A's and other analog devices.

There are two simple rules for cables: *keep them short and keep them shielded*. The reason for using short cables is that any long conductor, particularly an unshielded one, will act as an antenna picking up low frequency noise such as 60 Hertz line frequency noise from fluorescent lights and transformers. High frequency noise from switching power supplies can generate as much as 10 mV of noise voltage on an long unshielded conductor connecting an analog sensor to a PC A/D converter channel.

Single Cable. Of course, the simplest cable is just insulated single strands of cable. It has only one advantage: it is cheap. It is satisfactory only in low-noise situations where there is need for only a small number of connecting cables.

Flat Cable. Almost as simple is flat cable. It consists of many thin conductors running parallel to one another, and insulated with plastic. Flat cable is cheap and is usually sold in 100 ft rolls of 9, 25, or 40 conductor strips. It is easy to utilize since inexpensive tools are available to attach any type of connector you desire. Flat cable conductors are very thin so they should not be used to carry large currents (>100 mA). They are also not shielded and are therefore susceptible to noise. Flat cable is commonly used for large signal (e.g., 1 to 10 volts) analog connections between analog sensors or instruments and the PC. One commonly used flat cable connection is 25 pin flat cable conductor with D-Shell (RS-232-C type) connectors. Many commercial analog I/O plug in cards for the IBM PC use 25 conductor flat cable for analog and digital I/O connections.

Terminal Strip Connectors. Terminal strip connectors are quite common. Many commercial PC plug in peripherals for analog data acquisition have strips containing all the signal lines of interest. Screws mounted on these strips are provided so that bare wires may be interfaced to the A/D or D/A converter. Each screw is isolated with a plastic barrier to prevent short circuiting. The terminal strip connector is mounted on a printed circuit board which usually has a shielded flat cable connecting the printed circuit board (often called a signal distribution panel) with the PC data acquisition plug in card. Terminal strip connectors thus provide convenient connections, but may provide relatively little signal isolation, depending upon the cable attached to the terminal strip.

Twisted Pair Conductors. Twisted pair conductors can be used for differential, as opposed to single ended, analog (A/D, D/A) I/O connections. Twisted pair is more economical than coax and can be used where greater bandwidth and noise reduction than that provided by flat cables is necessary. A particularly useful form of twisted pair wiring is **shielded twisted pair**, which uses a shield (grounded at one end only!) to provide isolation from external electrical fields.

Coaxial Cable. Coaxial cable consists of two conductors: an inner conductor and an outer conductor that surrounds the inner conductor. Current flows in one direction on the inner conductor, and in the opposite direction on the outer conductor. As a result of this bidirectional current flow, the electromagnetic fields generated within the two conductors cancel each other. High frequency analog signals may be transmitted over coax cables without causing interference with other laboratory electronics, and with little loss of signal (attenuation). Coax cables are shielded so that laboratory electronic noise does not interfere with coax cable transmissions. A common type of coax cable used is the CATV 75 ohm cable type used for television and audio transmissions. BNC connectors are usually used with coax cable to provide convenient connections. Impedance matching connectors and terminators are available to improve the signal quality at high frequencies on coax cables.

Triaxial Cable (for current measurements). As the name suggests, triaxial conductor is similar to coaxial conductor, the difference being 3 instead of 2 conductors. Triaxial cables consist of an inner, outer, and guard conductor. The guard conductor may be grounded so as to provide better noise immunity for small signal transmissions. Triax cable is often used for small analog voltages and currents.

Practical Choices. Thus, for most common low-frequency situations, flat cable and terminal strip connecters are the cheapest and most commonly used connectors, although they provide little noise isolation. Twisted pair and twisted shielded pair provide good isolation from external noise. Coaxial and triaxial cabling provide the best isolation from external noise, although they tend to be more costly.

4.3.4.2 Cabling for Thermocouples, Thermistors, and Strain Gauges

Often the sensor must be located remotely from the PC A/D converter. This is especially true in factory environments where the measurements may be made hundreds of feet from the computer. The signals provided by temperature and pressure sensors (i.e. thermocouples and strain gauges) are small signals ($\pm$ 50mV). Referring back to the discussion on ground loops, it is obvious that these small signals could not survive transmission over hundreds of feet to a remote computer without any amplification.

Special signal transmitters are available for this particular situation. Low level signal transmitters are sold by manufacturers specifically for thermocouples and other small signal sensors. These signal transmitters convert the small sensor voltage into a standard current loop signal (4-20 mA) at the transmitter. These current signals can be transmitted over hundreds of feet to a receiver that converts the currents back into a small signal voltage for A/D conversion at the PC.

For safety reasons, it is sometimes not possible to run cables over long distances for interfacing. There are transmitter and receiver pairs sold for this type of application that operate using infrared light as a transmission medium. This allows the transmitter and receiver to be separated by hundreds of feet with no cables or wires between them. This is particularly useful when high voltages, magnetic fields, or high temperatures separate the sensor from the PC. These signal repeaters provide noise isolation and protection from harsh environments for analog signals.

4.3.4.3 Radio Frequency Signals

Low frequency signals (DC to 1 MHz) can often be transmitted over short distances (1 to 3 meters) using flat cables or shielded conductors. As the frequency of the signal extends above 1 MHz, however, careful consideration must be given to the cabling and interconnections used. High-frequency signals will radiate and capacitively couple with any nearby grounded conductors. To prevent this, the high frequency signal should always be transmitted using a coaxial, triaxial, or shielded twisted pair conductor to prevent attenuation or loss of signal due to capacitive coupling.

Balanced Transmission Lines. Any *high-frequency* conductor (digital or analog) should be balanced, or impedance matched. Whether the conductor is coaxial, triaxial, or twisted pair, a loss of signal can occur if the conductor is not impedance matched. The loss of signal strength (commonly called attenuation) occurs when the conductor or transmission line is terminated with an output impedance unlike that of the transmission line. A typical example of this is cable-television conductor. High frequency analog signals are transmitted over a coaxial cable conductor. The impedance of this conductor is a standard 300 Ohms. An impedance matching transformer is often used to couple the coax cable to the television. The purpose of the transformer is to terminate with cable with an impedance such that no signal attenuation occurs. The signal attenuation occurs by the reflection of the signal back towards the transmitter upon encountering an impedance different from that of the transmission line. This condition is known as a

Voltage Standing Wave, and can be measured in the laboratory using a Voltage Standing Wave Ratio meter (VSWR). The use of impedance matching transformers is common in all high frequency digital and analog signal transmission systems.

Impedance balancing is often required when interfacing with high frequency (high bandwidth) analog signals. Impedance balancing means that the impedance of the cable looks the same at all points along the cable. An unbalanced analog transmission line will attenuate (degrade) the amplitude of a high frequency (>50 Khz) analog signal. Coax cable (75 ohm CATV cable) is often used for these types of signals. Impedance matching transformers are available on the market so that coax cables may be terminated with the proper impedance (balanced) when connecting an analog instrument or sensor, and a PC analog I/O channel. Impedance matching transformers are also available for triaxial and twisted pair cables.

4.3.4.4 Digital I/O Connections

DI/DO Flat Cable. As we shall see in subsequent chapters, digital signals are much less susceptible to most forms of noise. This is because digital signals use two voltage levels (or frequencies) that are widely separated from one another to do their signaling. Hence, even rather large amounts of noise result in little degradation in the ability to distinguish between these two signal levels. As a result, almost all digital I/O systems use a cheap cable (i.e., one which provides little isolation from noise); the most common are flat cable for parallel digital I/O systems and twisted pair for serial digital I/O systems. At very high frequencies, however, noise may be more of a problem, and the methods of prevention used are those discussed in the preceding section.

4.4 ANALOG AND DIGITAL FILTERS

So far, we have looked at methods for preventing noise from reaching our PC-connected instrument. Most of these methods are quite effective. However, some noise may still be present in the signal transmitted from our instrument to the computer. How do we deal with the remaining noise?

The generic answer to this question is to use a **filter**, which is something that removes (filters out) unwanted data (i.e., noise) from a stream of data. In principle, then, we can use a filter to remove noise and leave just our signal of interest.

In practice, however, it is often difficult to clearly distinguish between noise and the data of interest; this choice is particularly difficult if the noise has similar properties (frequency, phase, and amplitude) to those of the signal. Hence, we must spend some time discussing how to design a variety of filters, and then you must choose those that are most applicable to your individual situation.

Filters can be implemented in either hardware or software; indeed, in most systems, it is best to have a combination of both types. Hardware filters that are external to the IBM PC usually have a faster response and can consequently filter signals in real time. Conversely, software filters implemented on the IBM PC are flexible because you can change the frequencies to be filtered in software.

Although you can filter signal amplitudes, phases, or other properties, by far
the most common filters are based upon *frequencies*. Hence, we usually describe
filters as falling into one of several categories based upon the frequencies affected by
the filter:

- **Low Pass Filters**. These filters pass low frequency components of the signal
  and filter out high frequency components beginning at a specific high frequency.
- **High Pass Filters**. These pass high frequencies and filter out low frequencies
  beginning at a specific low frequency.
- **Band Pass Filters**. These pass only those frequencies within a certain range
  specified by a low and high cutoff frequency.
- **Band Stop (Notch) Filters**. These filter out a certain range of frequencies specified by a start and stop frequency, passing all others.

4.4.1 Filter Characteristics

Ideally, a filter should eliminate all data outside of a specified frequency range. In
other words, there is a very sharp transition between the frequencies that are passed
and those that are filtered out. There are some types of software filters that provide
the ideal case. However, these filters tend to operate slowly and can only filter data
that has already been acquired by the PC. Most filters that work in real time must
respond quickly and do not usually eliminate all of the undesirable amplitude components above and below a specified frequency range. These are non-ideal filters.
Attributes common to both ideal and non-ideal filters include:

- **Cut-off Frequency**. This is the transition frequency at which the filter takes
  effect. It may be the high pass cut-off or low pass cut-off frequency.
- **Roll-off**. This is the slope of the amplitude versus frequency graph at the
  region of the cut-off frequency. The roll-off of a filter distinguishes an ideal
  filter from a non-ideal filter. An ideal low pass filter has a very sharp transition
  from passing all amplitudes below the cut-off frequency to passing no amplitudes above the cut-off frequency. A non-ideal low pass filter will pass some
  finite amplitude at frequencies just above the cut-off, and lower amplitudes at
  higher frequencies. For the more common non-ideal filters, the roll-off is
  usually measured on a logarithmic scale in units of decibels (dB). For example,
  Figure 4.11 illustrates a non-ideal low pass filter amplitude versus frequency
  plot. The amplitude (Y axis) is measured in units of decibels:

$$dB = 20 \log_{10}(\text{ amplitude in volts })$$

and the frequency (X axis) is measured in units of Hertz (cycles/second) on a logarithmic scale. The roll-off in this case is the slope of the curve in units of dB/decade.

The higher the roll-off, the better or more ideal the filter is. By adding multiple stages of the same filter (hardware or software) in sequence, you can increase the filter roll-off. This is called the **order** of the filter. For example a single stage filter is first order, a 2 stage filter is second order and so on.

4.4.2 Types of Filters

We can also distinguish filters by the way in which they are implemented. These filters can be implemented in either **hardware** or **software**. The two basic types of filters are **analog** or **digital**. Analog filters are generally non-ideal and can be implemented in software, although they are most commonly implemented in hardware (i.e., using capacitors, resistors, op amps and so forth). Digital filters are typically more ideal than analog filters and can be implemented in hardware, although they are usually implemented in software.

Digital hardware filters, are composed of programmable electronic components that allow you to change the filter characteristics (i.e. cut-off frequency, roll-off, etc.) using digital communications from an IBM PC or other computer. An example of this is a switched capacitor filter that digitally filters an analog signal and may be programmed from an IBM PC by using a D/A to generate a square wave whose frequency determines the cut-off frequency of the switched capacitor filter.

Digital software filters implemented on the IBM PC have several advantages. They can filter unwanted noise with great accuracy and precision. Digital software filters can also be modified easily to accommodate changing laboratory conditions where the noise characteristics change as time passes (i.e. temperature changes). However, the major disadvantage of digital software filters is that they are slow. Digital filtering can require significant computation time on a microcomputer like the IBM PC. Hence, a considerable amount of effort has been spent developing high-speed algorithms for software digital filters.

Analog hardware filters are composed of electronic circuits that are completely isolated from the IBM PC hardware. These circuits may be **passive** (i.e., include only resistors and capacitors) or they may be **active** (i.e., include operational amplifiers, resistors, and capacitors). These will filter the analog signal prior to A/D conversion. Active filters are generally preferred over passive filters since they provide a sharper roll-off and better stability. Passive filters are easier to construct and are considerably cheaper (i.e., no operational amplifiers that require an external power supply).

Analog software filters are essentially simulations of the effect of analog hardware filters, and are not commonly used in the laboratory.

The advantage of the analog filter is that unwanted noise can be removed from the signal with no computation required by the PC. The data acquisition cycle time is then simply the time required to perform an A/D conversion. Generally, you will

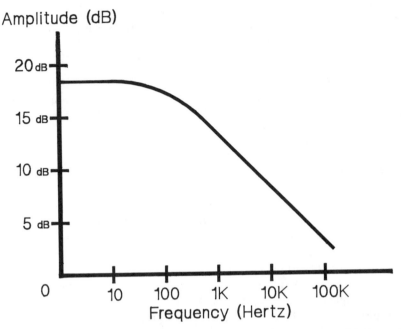

Figure 4.11. Non-Ideal Low-Pass Filter. A non-ideal low-pass filter with a cutoff frequency of 60 Hz and a rolloff of 5 dB/decade would have a response curve like the one shown.

probably find it useful to try building a few of the very simple first order filters, and then buy some of the more sophisticated filters for use in your system. Figure 4.12 illustrates the configuration of a first order active analog low-pass filter that includes a standard 741 operational amplifier. The numbers next to the 741 op. amp. represent the integrated circuit pin numbers. The cutoff frequency of this filter is approximately

$$f_{\text{Low Pass Cutoff}} = \frac{1}{2\pi R_2 C_2}$$

The low frequency gain (or amplification factor) of the low-pass filter is approximately R_2 / R_1.

Figure 4.13 illustrates the configuration of a first order active analog high pass filter that includes a standard 741 operational amplifier. The high-pass cutoff frequency of this filter is approximately

$$f_{\text{High Pass Cutoff}} = \frac{1}{2\pi R_1 C_1}$$

The high frequency gain of the high pass filter is $\dfrac{R_2}{R_1}$.

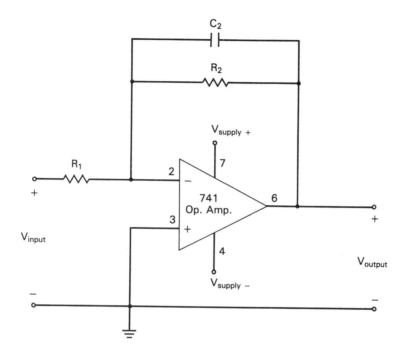

Figure   4.12.   Low Pass Analog Active Filter.   A first-order low-pass active analog filter can be constructed using two resistors, a capacitor, and an operation amplifier.

4.4.3 Anti-Aliasing Analog Filters

A/D converters are often connected to analog signals that contain frequencies higher than the sampling rate of the A/D converter. Many analog signal transducers like photodiodes and piezoelectric sensors generate high frequency noise often referred to as **harmonic distortion**. If these transducers are sampled using an A/D converter, the high frequency components in the signal will increase the amplitude of the low frequency components of the acquired data samples. This error is called **aliasing error** since the high frequency components of the signal appear erroneously in the sampled data as low frequency components.

An illustrative example of aliasing error is presented in Figure 4.14, in which a 10 Hertz sine wave is sampled at slightly less than 10 samples per second. The data samples acquired look like a low frequency d.c. signal due to the aliasing error in the measurement.

The **Nyquist** theorem states that any signal can be perfectly reconstructed as long as the signal is sampled at a rate of at least twice the highest frequency contained in the signal. A general rule of thumb in the laboratory is to use A/D conver-

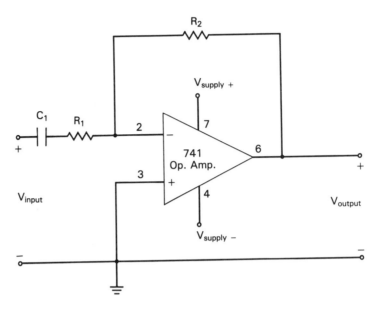

Figure 4.13. High Pass Analog Active Filter. A high-pass filter can be constructed with components very similar to those of the low-pass filter in Figure 4.12. Note, however, the difference in the placement of the capacitor in the two types of filters.

sion sampling frequencies that are at least 10 times higher than the highest frequency component in the signal. This is because the formula required to reconstruct the correct waveform from the sampled waveform that is sampled at only twice the highest frequency (called the Nyquist interpolation formula) is very computationally intensive and may not be feasible on a PC for large data sets.

To avoid aliasing errors, many PC A/D converters include anti-aliasing filters with the A/D converter circuitry. These are usually low-pass analog filters of the type illustrated in Figure 4.12. If your A/D does not have such a filter, then *you should always add a low-pass filter that has a cutoff frequency at least twice as high as the highest frequency in which you are interested*.

4.5 SOFTWARE FILTERS

The other general approach to filtering is to design software filters. Although, as we have mentioned, these filters may consume significant amounts of CPU time, they are often much more ideal than analog filters, because *they can be custom designed to match the filter to the data being analyzed*. In addition, as PC's become faster, this CPU usage becomes less of a burden. Generally, we recommend that you keep

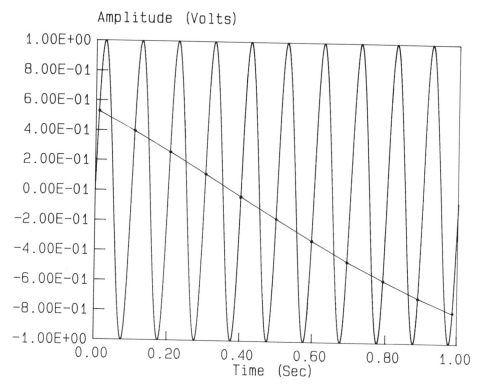

Figure 4.14. Aliasing. An aliasing error can be caused by sampling a 10 Hz sine wave at a frequency near 10 Hz, for example. The sampled waveform (small dots) thus appears as a very low frequency signal, rather than as 10 Hz.

the original (unfiltered) data stored on disk and do not modify it. In this way, the user can always return to the original (raw) data.

4.5.1 Bunching Software Filters

One of the simplest noise reduction techniques to implement in software is **bunching**, which is in essence a simple type of low-pass filter. In this technique, you reduce the number of data points in your data set by replacing each set of **n** consecutive points with a single point that is the sum or average of its neighbor points. For example, a data set of 1200 points might be bunched to produce 400 points by averaging points 1, 2, and 3 to produce a new point 1; points 4, 5, and 6 to produce new point 2, etc.

Bunching has several advantages. First, if the noise in the data is randomly distributed, then the noise in the data will be reduced (approximately by the square root of the number of points bunched). Second, bunching is very rapid, so that it can be done in real time (i.e., as the data are being collected) for all but the highest data collection rates. Third, the number of data bunched can be very easily varied with no change in your software.

The most obvious requirement for bunching is that you collect more points than are needed in the final data. Thus it might appear that you should always collect data at the highest rate allowed by your data collection board and bunch the data to produce the desired frequency of data points. However, you should be aware of several subtle problems.

First, it is certainly possible to do too much bunching. As suggested in the section on sampling theory (see page 73), you need to sample (after bunching) at least twice as frequently as the highest frequency component you wish to measure. In addition, you may need a large number of points to accurately define a curve being measured, to accurately integrate a peak, or other similar purpose.

Second, bunching should be done in addition to the hardware techniques we have described previously, not instead of them. For example, don't forget to sample at or above the Nyquist frequency (see page 73) and to use a low-pass filter to prevent aliasing (see page 73). Also, if you are sampling at well below the line frequency (50 or 60 Hz, depending upon the country), try to pick some multiple of the line frequency (e.g., 10 Hz rather than 9 Hz) so that sampling is always done at the same point in the noise waveform.

Third, use bunching sparingly. Its most frequent use has been because of the need to reduce data storage requirements or to speed data processing. Other techniques described below will often give much better noise reduction efficiencies while more accurately retaining the curve shape.

4.5.1.1 Moving Average (Convolution)

A software technique more sophisticated than the bunching technique described above is that of **moving averaging**, which is again inherently a low-pass filter. In a fashion similar to that of bunching, several points are averaged together. Unlike bunching, however, there is no reduction in the number of data points. So how it this done?

The answer is that in most data sets of scientific interest, there is a relationship between adjoining data points (i.e., you are measuring a continuous analog signal that is changing relatively slowly compared to the rate at which you are measuring the signal). Hence, one way to reduce the apparent noise in the data is to *average each point with several nearby points*. For example, to perform a 5-point moving average, we might average points 1, 2, 3, 4, and 5 to form a new point 3; points 2, 3, 4, 5, and 6 to form a new point 4; points 3, 4, 5, 6, and 7 to form a new point 5; etc.

Mathematically, we can express this process relatively simply.

$$X_n = \frac{\sum_{i=-q}^{+q} X_{n+i}}{2q + 1}$$

The number of points being smoothed together at each location is given by $(2q+1)$.

An example of software to perform a moving average smooth is given in the next section. However, it is primarily of interest as a special case of a more general set of smoothing functions.

$$X_n = \frac{\displaystyle\sum_{i=-q}^{+q} X_{n+i} F_i}{(2q + 1) \displaystyle\sum_{i=-q}^{+q} F_i}$$

Note that this is different from the moving average case in that a weighting factor, F, is included. We can make the filter even more general by placing fewer constraints on the denominator of the previous equation; for example, high-pass or edge enhancement filters can be produced (see Chapter 12).

By varying the weighting factors, F, we can produce a wide variety of smoothing functions. The interested reader is referred to several excellent articles on the subject.

4.5.1.2 Savitzky-Golay (Least-Squares)

One of the most useful of the techniques using this type of filter is named the **Savitzky-Golay** filter.[5] In this series of filters the weighting factors are chosen such that the effect is the same as fitting a least-squares line through the points. However, the process is much faster than a standard least-squares technique. An implementation of this filter in BASIC is given in Program 4.1.

One comment needs to be made about the Savitzky-Golay filter portion of Program 4.1. Specifically, the complexity of the filter implementation arises from the need to treat the end-points of the data set differently. For example, if you are performing a 25-point filter, the first 12 points and the last 12 points of the data set cannot be smoothed with this filter (recall that for a 25-point filter you need 12 points on each side of the current data point). Program 4.1 attempts to solve this problem by using a smaller filter on the end-points of the data set. If you do not wish to do this, you can replace lines 1150 to 1360 with the following much simpler routine:

```
1170 FOR I%= K% TO NEWNUM%-K%+1
1280    Y2!=Y%(I%)*FILTER!(PTSFILT%,1)
1290    FOR J%=1 TO K%-1
1300        Y2!=Y2!+(Y%(I%-J%)+Y%(I%+J%))*FILTER!(PTSFILT%,J%+1)
1310    NEXT J%
1320    Z%(I%)=Y2!/SUM!(PTSFILT%)
1330    IF Z%(I%)<YMIN% THEN YMIN%=Z%(I%)
1340    IF Z%(I%)>YMAX% THEN YMAX%=Z%(I%)
1350 NEXT I%
```

---

5   See A. Savitzky and M.J.E. Golay, **Analytical Chemistry**, 36:1627 (1964). Also, minor corrections were published in the same journal, 44:1906 (1972).

```
10 REM PROGRAM SMOOTH
20 REM LABORATORY AUTOMATION USING THE IBM PC
30 REM Program to illustrate bunching, moving-average, Savitzky-Golay smooths
40 REM Original data are in YORIG%, data after noise addition are in YNOISE%,
50 REM     data after bunching are in Y%, data ready for plotting are in Z%.
60 REM Requires graphics screen
70 REM ********************************************************************
80 DIM FILTER!(25,13),YORIG%(320),YNOISE%(320),Y%(320),Z%(320),SUM!(25)
90 REM                    Main program
100    KEY OFF
110    NUMDATA%=320                          'data sets contain 320 points
120    BUNCH%=1                              'initially, no bunching
130    NEWNUM% =320
140    GOSUB 310                             'read in smoothing coefficients
150    IGO%=1
160    WHILE IGO% < 7
170        ON IGO% GOSUB 540,680,770,930,1020,1220,300
180        IF IGO% > 3 THEN GOSUB 1510  'plot the data
190        IF IGO% > 3 THEN Y$=INKEY$ :IF Y$="" THEN 190    'wait for user
200        CLS :SCREEN 0:WIDTH 80
210        PRINT "Enter a 1 to read in a new data set"
220        PRINT "      a 2 to change the noise level"
230        PRINT "      a 3 to change the bunching factor"
240        PRINT "      a 4 to plot the bunched data"
250        PRINT "      a 5 to perform a moving average smooth"
260        PRINT "      a 6 to perform a least-squares (Savitzky-Golay) smooth"
270        PRINT "      a 7 to quit"
280        INPUT "Choice?",IGO%
290    WEND
300 END
310 REM ***************************
320 REM Read in all of smoothing coefficients
330    FOR I%=5 TO 25 STEP 2
340        FOR J%=1 TO (I%+1)/2
350            READ FILTER!(I%,J%)
360        NEXT J%
370        READ SUM!(I%)
380    NEXT I%
390 RETURN
400 REM ***************************
410 REM Savitzky-Golay coefficients
420 DATA 17.,12.,-3.,35.:            '5 pt filter coefficients
430 DATA 7.,6.,3.,-2.,21.                : '7 pt
440 DATA 59.,54.,39.,14.,-21.,231.      :    '9 pt
450 DATA 89.,84.,69.,44.,9.,-36.,429.   :     '11 pt
460 DATA 25.,24.,21.,16.,9.,0.,-11.,143. :     '13 pt
470 DATA 167.,162.,147.,122.,87.,42.,-13.,-78.,1105.:   '15 pt
480 DATA 43.,42.,39.,34.,27.,18.,7.,-6.,-21.,323.    :    '17 pt
490 DATA 269.,264.,249.,224.,189.,144.,89.,24.,-51.,-136.,2261. :'19 pt
500 DATA 329.,324.,309.,284.,249.,204.,149.,84.,9.,-76.,-171.,3059. :'21 pt
510 DATA 79.,78.,75.,70.,63.,54.,43.,30.,15.,-2.,-21.,-42.,805.      :'23 pt
520 DATA 467.,462.,447.,422.,387.,343.,287.,222.,147.,62.,-33.,-138.
530 DATA -253.,5175.                                            :'25 pt
540 REM ***************************
550 REM Create a sine wave to be used as the sample data set
560    INPUT "Enter the no. of sine waves to display across the screen ",NSINE%
570    Q%=320/6.38318/NSINE%              '=no. points / 2*pi /no. sine waves
580    YMIN%=32767:YMAX%=-32767
590    FOR I%=1 TO NUMDATA%                      'generate sine wave from -1000 to 1000
600        YORIG%(I%)=1000*SIN(I%/Q%)
610        YNOISE%(I%)=YORIG%(I%)
620        Y%(I%)=YORIG%(I%)
630        IF YORIG%(I%) < YMIN% THEN YMIN%=YORIG%(I%)
640        IF YORIG%(I%) > YMAX% THEN YMAX%=YORIG%(I%)
650    NEXT I%
660    NEWNUM%=NUMDATA%:BUNCH%=1:PTSFILT%=0    'reset parameters to defaults
670 RETURN
```

Program 4.1 (Part 1 of 3). Filters

```
680 REM ****************************
690 REM Add noise to the data set
700   INPUT "Enter the noise level (as a percent of the signal, 0-100) ",NOISEY%
710   N%=NOISEY%*10
720   FOR I%=1 TO NUMDATA%
730        YNOISE%(I%)=YORIG%(I%)+INT(RND*N%)
740        Y%(I%)=YNOISE%(I%)
750   NEXT I%
760 RETURN
770 REM ****************************
780 REM Bunch the data in YORIG% to create Y% array
790   PRINT "Enter the bunching factor.  It should be a power of 2."
800   INPUT "Enter a 1 for no bunching. ",BUNCH%
810   NEWNUM%=NUMDATA%/BUNCH%              'calculate new no. points to be plotted
820   K%=0
830   L%=BUNCH%-1
840   FOR I%=1 TO NUMDATA% STEP BUNCH%
850        Y2!=0
860        FOR J%=0 TO L%
870             Y2!=Y2!+YNOISE%(I%+J%)
880        NEXT J%
890        K%=K%+1
900        Y%(K%)=Y2!/BUNCH%
910   NEXT I%
920 RETURN
930 REM ****************************
940 REM Plot the bunched data without further smoothing
950   YMIN%=Y%(1):YMAX%=Y%(1)
960   FOR I%=1 TO NEWNUM%
970        Z%(I%)=Y%(I%)
980        IF Z%(I%)<YMIN% THEN YMIN%=Z%(I%)
990        IF Z%(I%)>YMAX% THEN YMAX%=Z%(I%)
1000   NEXT I%
1010 RETURN
1020 REM ****************************
1030 REM Perform moving average smooth
1040   YMIN%=Y%(1):YMAX%=Y%(1)
1050   INPUT "Enter number of points in moving average smooth (3-25) ",NAVG%
1060   PTSFILT% = NAVG%
1070   Z%(1)=Y%(1)
1080   FOR I%=2 TO NEWNUM%-1
1090        NAVG2%=NAVG%/2
1100        IF I% -1 < NAVG2% THEN NAVG2%=I%-1
1110        IF NEWNUM%-I% < NAVG2% THEN NAVG2%=NEWNUM%-I%
1120        Y2!=0
1130        FOR J%=-NAVG2% TO NAVG2%
1140             Y2!=Y2!+Y%(I%+J%)
1150        NEXT J%
1160        Z%(I%)=Y2!/(2*NAVG2%+1)
1170        IF Z%(I%)<YMIN% THEN YMIN%=Z%(I%)
1180        IF Z%(I%)>YMAX% THEN YMAX%=Z%(I%)
1190   NEXT I%
1200   Z%(NEWNUM%)=Y%(NEWNUM%)
1210 RETURN
1220 REM ****************************
1230 REM Perform second-order Savitzky-Golay smooth
1240   INPUT "Enter points in filter (5-25) ";PTSFILT%
1250   YMIN%=Y%(1):YMAX%=Y%(1)
1260   K%=(PTSFILT%+1)/2
1270   Z%(1)=Y%(1)                        'plot 1st 2 and last 2 points as is
1280   Z%(2)=Y%(2)
1290   FOR I%= 3 TO NEWNUM%-2
1300        PTSFILT2%=PTSFILT%:PTSFILT3%=PTSFILT%
1310        K2%=K%
1320        'treat ends of data differently
1330        IF I% < K% THEN PTSFILT2%=I%*2-1:K2%=I%
```

Program 4.1 (Part 2 of 3). Filters

```
1340        IF I% <= NUMDATA%/BUNCH%-K%+1 THEN GOTO 1370
1350            PTSFILT3%=(NUMDATA%/BUNCH%-I%)*2+1
1360            K3%=(PTSFILT3%+1)/2
1370        IF PTSFILT3% >= PTSFILT2% THEN GOTO 1400
1380            PTSFILT2%=PTSFILT3%
1390            K2%=K3%
1400        Y2!=Y%(I%)*FILTER!(PTSFILT2%,1)
1410        FOR J%=1 TO K2%-1
1420            Y2!=Y2!+(Y%(I%-J%)+Y%(I%+J%))*FILTER!(PTSFILT2%,J%+1)
1430        NEXT J%
1440        Z%(I%)=Y2!/SUM!(PTSFILT2%)
1450        IF Z%(I%)<YMIN% THEN YMIN%=Z%(I%)
1460        IF Z%(I%)>YMAX% THEN YMAX%=Z%(I%)
1470    NEXT I%
1480    Z%(NEWNUM%-1)=Y%(NEWNUM%-1)
1490    Z%(NEWNUM%)=Y%(NEWNUM%)
1500 RETURN
1510 REM ***************************
1520 REM Plot both unsmoothed and smoothed data on same screen
1530    SCREEN 1 :KEY OFF :CLS: COLOR 9,0
1540    LOCATE 1,1:PRINT "NOISE=";NOISEY%;"BUNCH%=";BUNCH%;"FILTER=";PTSFILT%
1550    VIEW (10,10)-(310,190),1,2
1560    WINDOW (1,YMIN%)-(NEWNUM%,YMAX%)
1570    FOR I%=1 TO NEWNUM%                  'plot original data as points
1580        PSET(I%,Y%(I%)),2
1590    NEXT I%
1600    PSET (1,Z%(1)),0                     'go to first point of next plot
1610    FOR I%=2 TO NEWNUM%                  'plot smoothed data as connected points
1620        LINE -(I%,Z%(I%)),3
1630    NEXT I%
1640 RETURN
```

Program   4.1 (Part 3 of 3).   Filters

It will prove very helpful to you if you try running Program 4.1 and varying the bunching factor, the number of points in the filter, and the noise level added to the data. You will notice that the moving average filter achieves more smoothing than the second-order Savitzky-Golay filter, but it also is less faithful to the shape of the original curve. Hence, you will need to design your filter carefully to match your data rate and noise levels. Cram, Chesler and Brown have measured and described these effects in an excellent paper that will provide some guidelines for you in designing an appropriate filter.[6]

4.5.2 Fourier Transform-Based Digital Filtering

The Fourier transform differs from other digital filters (software or hardware) in that it provides *ideal* frequency filtering of acquired data. In other words, the roll-off or slope of the filter at a particular cut-off frequency is infinite. Another difference between the Fourier transform and other digital filters described earlier is that the Fourier transform requires that all data samples be acquired and stored prior

---

6   See S. P. Cram, S. N. Chesler and A.C. Brown III, **J. Chromatography**, 126:279 (1976).

to filtering. For this reason the Fourier transform is generally not used for real time filtering of noisy data.

Recall from our earlier discussion (see page 51) that the Fourier transform is a mathematical operation performed on a sequence of numbers. In our case, it may be a sequence of A/D integer numbers that have been converted into floating point values. The Fourier transform involves converting the sequence of real numbers from the time domain into the frequency domain. The output of the Fourier transform is an array of complex numbers. The array has two components associated with each element in the array, that is the **real part**, and the **imaginary** part. The magnitude (or amplitude) and phase of the signal can be derived from the real and imaginary part respectively. The magnitude or frequency domain plot can be calculated for each corresponding point in the time domain as:

$$\text{Magnitude (i)} = \sqrt{(\text{Real (i)})^2 + (\text{Imaginary (i)})^2}$$

The process of ideal filtering involves multiplying the resulting magnitude array with another array with 1s in the bandpass frequency array locations, and 0s in the frequency array locations to be filtered out. The filtered magnitude array is then converted back in real and imaginary parts and an **inverse Fourier transform** is performed to obtain the final filtered time domain sequence of numbers. (The forward and inverse FFT can be performed using code in Program 4.2, below; the technical details of this process are outside the scope of this text.)

An important consideration is the frequency resolution of the Fourier transform. In other words, to how many cycles per second, or Hertz does a magnitude array element correspond? The answer depends on the spacing in time of the acquired data samples. The frequency resolution is

$$\text{Frequency Spacing Between Points} = \frac{1}{N\Delta t} \quad \text{Hertz per step}$$

where N is the number of points in the time domain A/D array and Δt is the spacing in time between A/D acquired samples or 1 / A/D sampling rate. For example, if the A/D sampling rate is 1 KHz, and 1000 A/D points are acquired then the frequency resolution of the Fourier transform filter will be 1 Hertz per step.

The theory behind the practical computation of the Fourier transform is beyond the scope of this discussion and the reader is referred to other references on digital signal processing for a more detailed discussion of this. However, using the Fourier transform for ideal removal of noise (e.g. removing all noise components at 60 Hertz) can be practically accomplished with a BASICA program provided for the IBM PC, such as Program 4.2.

This program consists of a main test program that will read in a disk data file that consists of N floating point or integer ASCII data points (i.e. A/D values). The Fourier transform test program will first compute the real and imaginary parts of the Fourier transform, and then compute the magnitude and phase components and display them on the screen. It should be noted that the number of points N must be an even integer power of 2 (i.e. $64 = 2^6$).

```
10 REM PROGRAM FFT
20 REM LABORATORY AUTOMATION USING THE IBM PC
30 REM Fourier transform for N data points where N is even power of 2
40 REM    The data are stored in two arrays FREAL and FIMAG with real and
50 REM    imaginary parts.  Computation is done in place with the transform
60 REM    overwriting the original arrays FREAL, FIMAG.  The direction of the
70 REM    transform is determined by the value of ISIGN (+1 or -1).
80 REM    If ISIGN = +1, the results are divided by N.  The magnitude
90 REM    and phase are stored in arrays FMAG and FPHASE, respectively
100 REM *******************************************************************
110 DEFINT I-N
120 DIM FREAL(1000), FIMAG(1000), FMAG(1000), FPHASE(1000), W(1000)
130 CLS : KEY OFF
140 REM              Main program
150    GOSUB 230                         'get user input
160    GOSUB 300                         'read data file
170    GOSUB 390                         'display current data
180    GOSUB 470                         'compute and display FFT real, imag.
190    GOSUB 750                         'compute magnitude, phase
200 END
210 REM *******************************
220 REM Get user input
230    INPUT "Enter number of points (must be a power of 2) "; N
240    INPUT "Enter the name of the input data file      "; FILESPEC$
250    N1 = N - 1: N2 = N \ 2            'adjust array indices
260    DX = 6.283185 / N                 'frequency spacing between points
270 RETURN
280 REM *******************************
290 REM Read data file
300    OPEN FILESPEC$ FOR INPUT AS #1    'initialize the function
310    FOR I = 0 TO N1                   'get data
320       INPUT #1, FREAL(I)
330       FIMAG(I) = 0!                  'set imaginary part = 0 before FFT
340    NEXT
350    CLOSE                             'close all files
360 RETURN
370 REM *******************************
380 REM Display current data
390    CLS : PRINT "Real part of input data"
400    MIN = FREAL(1): MAX = FREAL(1)
410    FOR I = 0 TO N1
420       PRINT USING "##.##^^^^ "; FREAL(I);
430    NEXT
440 RETURN
450 REM *******************************
460 REM Perform FFT (compute real and imaginary parts)
470    PRINT "Performing Fourier transform....Please be patient"
480    TIME = TIMER                      'Check the timer before FFT
490    ISIGN = -1                        'ISIGN=-1 convert time domain -> frequency domain
500    '                                 'ISIGN=+1 convert frequency domain -> time domain
510    GOSUB 1000                        'Call the Fourier Transform subroutine
520    'Display the real and imaginary parts
530    TIME = TIMER - TIME 'Check the timer
540    MIN = FREAL(1): MAX = FREAL(1)
550    FOR I = 0 TO N1
560       IF FREAL(I) < MIN THEN MIN = FREAL(I)
570       IF FREAL(I) > MAX THEN MAX = FREAL(I)
580       IF FIMAG(I) < MIN THEN MIN = FIMAG(I)
590       IF FIMAG(I) > MAX THEN MAX = FIMAG(I)
600    NEXT I
610    CLS : LOCATE 1, 1: PRINT "Real "
620    FOR I = 0 TO N1                   'display the transform (real)
630       PRINT USING "##.##^^^^ "; FREAL(I);
640    NEXT
650    INPUT "", Y$
660    CLS : LOCATE 1, 1: PRINT "Imaginary "
670    FOR I = 0 TO N1                   'display the transform (imag)
```

Program 4.2 (Part 1 of 3). Fourier Transform

```
680      PRINT USING "##.##^^^^ "; FIMAG(I);
690   NEXT
700   PRINT : PRINT "Execution time for "; N; " points = "; TIME; " seconds"
710   INPUT "", Y$
720 RETURN
730 REM ******************************
740 REM Calculate the magnitude and phase
750   FOR I = 0 TO N1
760      FMAG(I) = SQR(FREAL(I) ^ 2 + FIMAG(I) ^ 2)
770   NEXT I
780   FOR I = 0 TO N1
790      FPHASE(I) = ATN(FIMAG(I) / FREAL(I))
800   NEXT I
810   CLS : LOCATE 1, 1: PRINT "Magnitude"
820   FOR I = 0 TO N1                          'display the frequency magnitude
830      PRINT USING "##.##^^^^ "; FMAG(I);
840   NEXT
850   INPUT "", Y$
860   MIN = FPHASE(1): MAX = FPHASE(1)
870   FOR I = 0 TO N1
880      IF MIN < FPHASE(I) THEN MIN = FPHASE(I)
890      IF MAX > FPHASE(I) THEN MAX = FPHASE(I)
900   NEXT I
910   CLS : LOCATE 1, 1: PRINT "Phase "
920   FOR I = 0 TO N1                          'display the phase
930      PRINT USING "##.##^^^^ "; FPHASE(I);
940   NEXT
950   INPUT "", Y$
960 RETURN
970 REM ******************************
980 REM Fourier Transform Subroutine
990 REM routine to calculate the N'th roots of 1
1000   FOR I = 0 TO N2 - 1
1010      THETA = 3.141593 * (I) / N2
1020      W(2 * I) = COS(THETA)
1030      W(2 * I + 1) = SIN(THETA)
1040   NEXT
1050   IREV = 1
1060   FOR I = 1 TO N
1070      IF I >= IREV GOTO 1100
1080         SWAP FREAL(I - 1), FREAL(IREV - 1)
1090         SWAP FIMAG(I - 1), FIMAG(IREV - 1)
1100      ML = N
1110      WHILE IREV > 0
1120         ML = (ML + 1) \ 2
1130         IREV = IREV - ML
1140      WEND
1150      IREV = IREV + ML + ML
1160   NEXT I
1170   IP1 = 1
1180   WHILE IP1 < N
1190      FOR I1 = 0 TO IP1 - 1
1200         NW = N * I1 / IP1
1210         WR = W(NW)
1220         WI = ISIGN * W(NW + 1)
1230         ISTEP = IP1 + IP1
1240         FOR IA = I1 TO N1 STEP ISTEP
1250            IB = IA + IP1
1260            TMPR = WR * FREAL(IB) - WI * FIMAG(IB)
1270            TMPI = WR * FIMAG(IB) + WI * FREAL(IB)
1280            FREAL(IB) = FREAL(IA) - TMPR
1290            FIMAG(IB) = FIMAG(IA) - TMPI
1300            FREAL(IA) = FREAL(IA) + TMPR
1310            FIMAG(IA) = FIMAG(IA) + TMPI
1320         NEXT IA
1330      NEXT I1
```

Program   4.2 (Part 2 of 3).   Fourier Transform

```
 1340          IP1 = IP1 + IP1
 1350     WEND
 1360     IF ISIGN = -1 THEN RETURN
 1370     FOR I = 0 TO N1
 1380          FREAL(I) = FREAL(I) / N
 1390          FIMAG(I) = FIMAG(I) / N
 1400     NEXT
 1410 RETURN
```

Program 4.2 (Part 3 of 3).   Fourier Transform

The execution of this program is not very fast because it is designed to run in the Basic interpreter. There are numerous commercial subroutine libraries on the market that provide or make use of Fourier transform subroutines. These mathematical subroutine packages can be used to analyze signals, noise, and filters on an IBM PC. Some of these packages make use of extra hardware (sold by the manufacturer) to acquire the data to be analyzed.

Program 4.3 illustrates a simple Basic program to calculate sine waves. The program will compute sine waves of different frequencies with noise imbedded and store the resulting waveforms on disk in ASCII format. This program can be used to generate interesting test cases for the Fourier transform program.

4.5.3 Chapter Summary

We have discussed noise detection, analysis, and reduction in some detail. How can you best use all of this information in the laboratory? Although the answers may vary in different situations, there are some general steps to follow when attacking the noise problem.

The first step is to isolate your instruments from external noise induced by nearby motors, and high voltage sources. Shielding noise sensitive instruments using a Faraday cup and shielded cables is a good idea. Minimizing the distance of your instruments to a common ground will help to avoid ground loop noise.

The use of filters is usually a second step to eliminate lingering noise. Hardware analog filters (i.e. low pass active) are the cheapest and most effective filter type to implement. In particular, usually you should have a low-pass filter to avoid aliasing. Real-time software filters are effective for lower frequency drift noise but do not adequately compensate for noise sources spread across a wide range of frequencies.

Fourier transform based filtering is the last step in the noise reduction process and the most exact. Although it requires some programming sophistication and can not be accomplished in real time, Fourier transform filtering is the ultimate solution to the removal of noise after all else fails.

Understanding and applying each of these techniques in the proper sequence will ensure that you get the most out of your laboratory data acquisition system.

```
10 REM PROGRAM TEST_FFT
20 REM LABORATORY AUTOMATION USING THE IBM PC
30 REM This program creates a multiharmonic sine wave buried in white noise
40 REM     SIGMA=noise power,  GAIN=sinusoid gain,  FREQ=sine wave freq. (Hz)
50 REM ***************************************************************
60 DEF FNCLOCK=VAL(RIGHT$(TIME$,2))
70 DEF FNSINE=SIN((FREQ(J)*T*2*PI))
80 PI=3.141593
90 CLS:LOCATE 8,20
100 REM               Main program
110    GOSUB 160                         'get user input
120    GOSUB 320                         'create data, store in file
130 END
140 REM ******************************
150 REM Get user input
160    PRINT "This program creates a multiharmonic noisy sine wave ";
170    LOCATE 9,20
180    PRINT "for test purposes in the time frame 0 to 1 second ";
190    LOCATE 11,20:INPUT "Enter the number of sine harmonics ";HARM
200    FOR J=1 TO HARM                   'fill in the frequency array
210       LOCATE 13,25
220       INPUT "Enter a sine wave harmonic frequency (Hz) ";FREQ(J)
230       LOCATE 13,1:PRINT STRING$(79,32) 'Clear the line
240    NEXT J
250    LOCATE 15,25:INPUT "Enter the sine Amplitude gain       ";GAIN
260    LOCATE 17,25:INPUT "Enter the white noise gain          ";SIGMA
270    LOCATE 19,25:INPUT "Enter the number of points          ";N
280    LOCATE 21,10:INPUT "Enter drive:filename.filetype       ";D$
290 RETURN
300 REM ******************************
310 REM Create data and store in a sequential file
320    OPEN D$ FOR OUTPUT AS #1          'open output file
330    CLS
340    FOR I=0 TO N
350       T=I/N                          'time in seconds
360       RANDOMIZE (FNCLOCK)            'reseed random number generator
370       RAND=(RND-.5)*SIGMA            'random variable (-.5 to .5 with gain)
380       VALUE=0                        'Initialize value
390       FOR J=1 TO HARM
400          VALUE=VALUE+(FNSINE)*GAIN 'sine with gain buried in white noise
410       NEXT J
420       PRINT "time = ";T;"   value = ";VALUE
430       PRINT#1, USING "####.######";VALUE  'write time,sine in noise to disk
440    NEXT I
450    CLOSE #1
460 RETURN
```

Program   4.3.   Sine Wave Generation

EXERCISES

1. Use an A/D to collect data from a sine wave generator tuned to 500 Hz. Collect 100 data samples at 10 Hz, 100 Hz, 500 Hz, 1000 Hz, 5000 Hz. Plot all of the data on a single graph. For which frequencies do you expect to see evidence of aliasing? Can you suggest a method for detecting aliasing?

2. Try the same experiment as above, except with a low-pass filter attached between the sine wave generator and the A/D. Does this make any difference? What can you say about the filter characteristics (i.e. roll-off, cut-off frequency) from this experiment?

3.   An A/D is used to digitize a signal in a $\pm$ 10 volt range. Ignoring internal noise sources (i.e. quantization noise), how much noise voltage may be tolerated to maintain a minimum signal to noise ratio of 30 dB?

4.   Monitor the voltage output of an instrument with an oscilloscope. How can you tell what frequencies are present in your data from the oscilloscope? When is it more difficult to separate the noise frequencies from the signal frequency?

5.   Collect data from an experiment with the A/D in the single-ended mode. Now collect data using the A/D in differential input mode. What happens to the signal? Under what conditions would you expect to get different data from each mode?

6.   Use a sensitive ammeter to test for ground loops in your system. Try using this as a method of determining how much effect each of the grounding techniques discussed in this chapter improves the signal-to-noise ratio. At what current levels do you expect ground loops to become a problem?

7.   Try collecting a 10 mV signal from an experiment using first unshielded conductor and then coaxial cable. Measure the noise in each case. What works best in eliminating the noise observed in your laboratory? Now try the same thing with a 10 V signal. Describe any differences.

8.   Plot the magnitude of the Fourier transform of a signal collected from your instrument. How high of a signal frequency can you detect? Try adding a low-pass filter to the circuit below the highest detectable frequency and repeat the measurement. What can you say about the quality of the filter by comparing the two measurements?

9.   Modify the moving average software filter program to operate on real-time incoming data from a sine wave generator and plot the resulting waveform. Increase the frequency of the sine wave generator at intervals of 5-10 Hz. Can you accurately measure the cut-off frequency and roll-off of the software filter? How might you increase the cut-off frequency of the filter?

5

DIGITAL INPUT AND OUTPUT

So far, we have only discussed methods for communicating with analog devices. Beginning with this chapter, we turn our attention to various types of digital techniques for communicating with laboratory instruments. In general, these digital techniques are used whenever the instrument we are collecting data from already has its own circuitry to digitize analog signals, or when we are sending data to a digital portion of an instrument. Digital techniques are usually much simpler than analog techniques to implement.

Digital techniques are usually divided into two classes: **parallel** and **serial** input/output (I/O). In the parallel I/O techniques, several bits of information are transferred simultaneously; usually, 8 or 16 bits are sent or received. In order to transfer this data, there must be as least as many wires as there are bits being sent. In the serial techniques, typically only one bit is sent per unit time; thus, at a given moment in time there is only one wire actually carrying the signal in a typical serial device. In both techniques, however, there are often a number of other wires to provide grounding, "handshaking," and other functions, as we shall see.

In Chapter 6, we shall discuss a sophisticated parallel technique, the IEEE-488 protocol. In Chapter 7, we discuss a frequently-used serial technique, the RS-232C protocol. However, we suspect you will be less confused if we start by discussing the simplest of the parallel techniques, usually called **digital input/output** or **binary I/O**. It is implemented on almost all manufacturers' general-purpose laboratory data acquisition boards.

5.1 SIMPLE DIGITAL I/O

In simple digital I/O, each signal wire is used to communicate with a separate device or section of a device. For example, the DT2801 card has 16 lines to perform digital I/O; the lines are configurable in groups of 8 for either input or output. Each of these 16 lines can control or sense signals from a separate device.

Usually, simple parallel I/O is designed to drive or accept signals from TTL type devices. In practice, this means two things: first, "on" means signals in the range of 2.2 to 5 V, and "off" means signals near ground (0 to 1.2 V). Second, only very small currents can be used, typically 1 mA or so. The limitation to small currents usually means that simple parallel I/O circuits must include provisions for "buffering" the signals so that the currents generated or accepted by the digital I/O chip are not too large.

In our laboratories, the most common applications of simple parallel I/O are to sense switch closures and to actuate lights and relays. Let's examine each of these.

5.1.1 Sensing Switches

We often wish to have the computer monitor switch closures. For example, the experimentalist may wish to start the experiment by depressing a button or toggling a switch on the instrument. To synchronize the start of data collection with the triggering of the instrument, we often will connect the switch to the digital I/O section of the laboratory board, as shown in Figure 5.1.

Notice in Figure 5.1 that the source voltage (5 V) is not fed directly to the digital input; this is to avoid the problem of excess current we have mentioned. Instead, a resistor is used to limit the current flow.

One of the major problems encountered when using switches is that mechanical switches exhibit considerable "bounce." Thus, when you close a switch, you will observe not a single transition from open to closed, but actually many such transitions. Bounce is particularly pronounced on older switches whose contacts have corroded.

There are three ways of dealing with bounce. First, if you are simply testing for a single switch closure, you may be able to ignore it. For example, if you are using a switch closure to indicate the start of an experiment, bounce can be ignored. However, if your experiment includes numerous switch closures (e.g., human subjects pushing buttons to indicate their responses to psychological tests), then you will need to "debounce" the data so that you record each switch closure as occurring only once.

The preferred way to debounce switch closures is using a commercially-available "debouncing" chip. Figure 5.2 shows the theory of operation of such a chip. This approach is quite satisfactory for a wide variety of switching speeds. A second method is to use software to accomplish debouncing. This approach has the advantage of being independent of the switch hardware chosen. In practice, this is accomplished by requiring that the switch test as "closed" for a defined period of

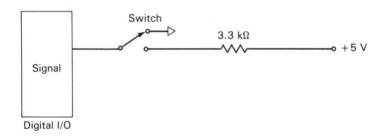

Figure   5.1.   Sensing a Switch Closure.   A simple circuit is shown for sensing switch closures using digital I/O.

time before it is considered closed. The software must then test for the switch being
reopened for a certain period of time before it can be considered closed again.

5.1.2 Actuating Lights and Relays

The other common application for simple digital I/O is actuating lights and other
"on/off" type devices. Again, the major limitation is the small current that can be
sourced by digital I/O devices. Hence, two approaches are commonly used in constructing circuits of this type: either using devices designed for TTL-level inputs, or
interposing buffering devices capable of sourcing more current.

First, let's take a look at the use of digital I/O with TTL-level devices. Two
common devices of this type are light-emitting diodes (LEDs) and digital relays. A
simple circuit for driving an LED is shown in Figure 5.3 as an example.

The critical feature of interest for both LEDs and digital (solid-state) relays is
that they draw very little current and hence can be driven directly by the digital I/O
chip. The digital relay can then be used to switch on or off devices that draw much
larger currents. Digital relays are available in a number of sizes to match the current
loads of various external devices. For example, we might use the digital relay to
switch a valve open or closed, or to initiate a spectrometer scan, or some other useful
task.

The second approach is to use **buffered I/O**. A common application of this
approach is to use digital I/O to actuate standard (non-solid-state) relays. These
relays may draw currents far greater than those that can be directly supplied by the
digital I/O board. Hence, any of a number of current-buffering devices can be interposed between the digital I/O board and the relay. In Figure 5.3 we have illustrated
a common buffer, based upon a power transistor. Note that this chip also inverts the
signal, so that the relay is turned on when the digital I/O bit corresponding to it is
turned off. You should be careful to note the limits of the buffer you use, however.
For example, you may need to use a buffered small relay, which in turn drives a

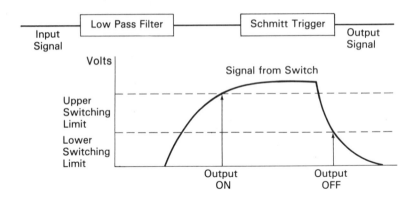

Figure 5.2. Debouncing a Switch Closure. Shown is a simple circuit for debouncing switch closures. The smoothed output from the filter triggers the Schmitt device when the output is above the upper switching limit voltage.

larger relay, if your application requires substantial current. Note also that relays may be sources of substantial electrical noise as they open and close, so that they should not be placed close to low-voltage analog signal-carrying wires. Also, if you use many relays, you may wish to investigate boards containing 16 or 32 relays specifically designed for computer control.

5.1.3 Programming Simple Digital I/O

As we have mentioned, programming digital I/O is particularly simple. We find it simpler to use hexadecimal numbers in programming digital I/O than it is to use decimal numbers, because we are trying to set or read each of the binary bits on the digital I/O board. Hence, you may find it useful to review Appendix A if you have not already done so. As in previous chapters, we will use the DT2801 board for our examples. Table 5.1 shows the registers of the DT2801 that are used in programming digital I/O.

Our first program, Program 5.1, is a useful program for testing whether the digital I/O section of the DT2801 board is functioning properly, and it also serves to illustrate how digital I/O can be performed. It is designed to allow you to individually test each of the 16 I/O lines. The DT2801 has 2 sets of 8 lines that can be configured as either input or output, but not both at the same time. Thus, it is often convenient to configure one set of 8 lines for input and the other set of 8 lines for output. It is also possible to configure all 16 lines for either input or output, if desired. The manufacturer refers to each set of 8 lines as a "port."

In this example, we shall configure 8 lines as input and 8 lines as output. Hence, you should connect wires between the 8 lines of port 0 and the same-

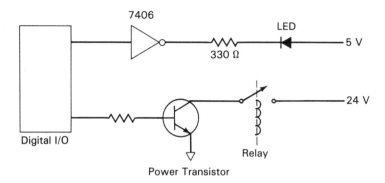

Figure   5.3.   LED and Digital Relay.   Digital I/O can be used to directly actuate both
LEDs and digital relays. The 7406 inverts the output so that a 1 will turn on
the LED.

numbered lines in port 1; i.e., line 0 in port 0 to line 0 in port 1, etc. Then run the
program, which alternates (toggles) the output of the selected bit between 0 and 1
and then reads the value from the digital input of the same bit.

In BASICA, this is accomplished quite easily. As usual, we begin in lines
330-380 by initializing the board, and then we use the SET DIGITAL PORT FOR
INPUT and SET DIGITAL PORT FOR OUTPUT commands to configure the two
8-bit ports for input and output, respectively (lines 410-550). This configuration
needs to be done only once per program.

Since Program 5.1 is designed to look at only one bit at a time, we must first
design a **mask** that allows us to look at just that one bit. In effect, this mask is simply
a 16-bit word with the bit of interest set to 1 and all of the rest of the bits zeros. We
will use only the low 8 bits of the mask in this example. Notice that we use the XOR,
or exclusive or, operation in BASIC to change the value of the bit in order to create a
test value.

We then use the WRITE DIGITAL OUTPUT IMMEDIATE command (lines
580-590) and write this test value to the digital I/O port we have configured to be an
output port (lines 600-640). Of course, we must write the entire byte, not just the bit
of interest.

We then read data from the input port using the READ DIGITAL INPUT
IMMEDIATE command (lines 710-830). Finally, we change the desired output bit
back to its original value and read it one more time. This should allow us to see the
bit change between its two values.

Table 5.1.   Registers on the DT2801 Used for Digital I/O

Register[a]	Name	Address[b]	Meaning of Bits
1	Command	02ED	bits 7-4 should be 0000
			bits 3-0 = 1111 for STOP
			bits 3-0 = 0001 for CLEAR ERROR
			bits 3-0 = 0100 for SET PORT FOR INPUT
			bits 3-0 = 0101 for SET PORT FOR OUTPUT
			bits 3-0 = 0110 for READ INPUT IMMEDIATE
			bits 3-0 = 0111 for WRITE OUTPUT IMMEDIATE
1	DIO Status	02ED	1=datum full
			2=previous command finished
			3=previous byte was command
			7=composite error
0	DIO Datum	02EC	1 byte of datum
0	Port Select	02EC	bits 1,0 = 00 for port 0 only
			bits 1,0 = 01 for port 1 only
			bits 1,0 = 10 for both ports

[a] Only those registers and bit meanings are shown that are used in programming the D/A in the WRITE D/A IMMEDIATE mode.

[b] The addresses shown are for the board at its factory-default location. You may select other base addresses using jumpers on the DT2801. All addresses are in hexadecimal.

Before leaving this program, we should also point out that the DT2801 is used in much the same fashion as we have seen in previous programs, including checking of the board between instructions and commands. Notice that we have include a new routine at line 930 to check to be sure the data are ready to be read from the board.

Although Program 5.1 is useful for testing digital I/O on the DT2801 board, it is not useful for much else. So let's see a much more useful program that is typical of those we frequently use in our labs. In Program 5.2, we use digital I/O as we might if, for example, we wished to turn on an instrument and start the computer-based data collection at the same instant. We thus use it to sense a switch closure; upon switch closure the program actuates a relay (hooked to an instrument in our example) and lights an LED to indicate that the experiment has commenced. Program 5.2 might thus conveniently serve as a subroutine in a larger program that would begin taking data when the switch is closed.

In Program 5.2 the switch is attached to the digital input bit 0, the relay to output bit 0, and the LED to output bit 1, as shown in Figure 5.4. It is then very

```
10 REM PROGRAM TEST_I/O
20 REM LABORATORY AUTOMATION USING THE IBM PC
30 REM Program to test digital I/O using the DT2801
40 REM You should connect bit N of Port 0 to pin N of Port 1 for N=0 to 7
50 REM    Port 0 is configured as output, Port 1 as input
60 REM CAUTION:  make sure no signal-producing devices are connected to port 0
70 REM ***************************************************************
80 REM Define all registers, etc.  Change register addresses if not adapter 0.
90 BASEADD%=&H2EC                     'base address of board
100 COMMAND%=BASEADD%+1               'command register
110 STATUS%=BASEADD%+1                'status register
120 DATUM%=BASEADD%                   'data register
130 REM *******************************
140 REM         Main program
150    GOSUB 330                      'set up board
160    GOSUB 410                      'set up for DIO
170    OUTBYTE%=0:BIT%=0              'start with all bits set to 0
180    CLS:PRINT "If the bit is working, you will see the bit set to "
190    PRINT "    both 1 and 0.  If the bit value stays the same "
200    PRINT "    then it is not working."
210    PRINT:INPUT "What bit do you wish to test (0-7, 8 to stop)? ",BIT%
220    WHILE BIT% > -1 AND BIT% < 8
230        LOWBIT%=2^BIT%             'calculate bit mask for that bit
240        GOSUB 580                  'toggle the output on that bit
250        GOSUB 710                  'input on that bit
260        GOSUB 580                  'toggle bit back to original value
270        GOSUB 710                  'read it again
280        INPUT "What bit do you wish to test (0-7, 8 to stop)? ",BIT%
290    WEND
300 END
310 REM *******************************
320 REM Set up board
330    OUT COMMAND%,&HF               'stop board
340    TEMP%=INP(DATUM%)              'clear data register
350    REM wait for command to finish, then clear errors
360    GOSUB 890
370    OUT COMMAND%,&H1               'clear errors
380 RETURN
390 REM *******************************
400 REM Set up for digital I/O
410    GOSUB 890
420    OUT COMMAND%,&H4               'write SET DIGITAL PORT FOR INPUT command
430    GOSUB 860
440    OUT DATUM%,1                   'input on port 1
450    GOSUB 890
460    ERRORCHECK%=INP(STATUS%)       'check for error
470    IF (ERRORCHECK% AND &H80) THEN PRINT "Error writing to Port 1":STOP
480    GOSUB 890
490    OUT COMMAND%,&H5               'write SET DIGITAL PORT FOR OUTPUT command
500    GOSUB 860
510    OUT DATUM%,0                   'output on port 0
520    GOSUB 890
530    ERRORCHECK%=INP(STATUS%)       'check for error
540    IF (ERRORCHECK% AND &H80) THEN PRINT "Error writing to Port 0":STOP
550 RETURN
560 REM *******************************
570 REM Toggle output on specified bit
580    GOSUB 890
590    OUT COMMAND%,&H7               'send WRITE DIGITAL OUTPUT IMMEDIATE command
600    GOSUB 860
610    OUT DATUM%,0                   'output on Port 0
620    GOSUB 860
630    OUTBYTE%=OUTBYTE% XOR LOWBIT%  'XOR toggles the value
640    OUT DATUM%,OUTBYTE%            'output byte
650    GOSUB 890
660    ERRORCHECK%=INP(STATUS%)       'check for error
670    IF (ERRORCHECK% AND &H80) THEN PRINT "Error writing to Port 0":STOP
```

Program   5.1 (Part 1 of 2).   Testing Digital I/O.

```
 680 RETURN
 690 REM ********************************
 700 REM Input on specified bit
 710    GOSUB 890                        'wait for board to be ready
 720    OUT COMMAND%,&H6                  'send READ DIGITAL INPUT IMMEDIATE command
 730    GOSUB 860
 740    OUT DATUM%,1                      'input from port 1
 750    GOSUB 930                         'wait for datum to be ready
 760    J%=INP(DATUM%)                    'get low and high bytes
 770    LOW%=J% AND LOWBIT%               'mask off all but desired bit
 780    IF LOW% >0 THEN RESULT%=1 ELSE RESULT%=0
 790    PRINT "Bit ";BIT%;" set to ";RESULT%
 800    GOSUB 890
 810    ERRORCHECK%=INP(STATUS%)          'check for error
 820    IF (ERRORCHECK% AND &H80) THEN PRINT "Error reading from Port 1":STOP
 830 RETURN
 840 REM ********************************
 850 REM check to see if ready to set a register
 860    WAIT STATUS%,&H2,&H2              'wait for bit 1 to be reset
 870 RETURN
 880 REM check to see if ready to send another command
 890    WAIT STATUS%,&H2,&H2              'wait for bit 1 to be reset
 900    WAIT STATUS%,&H4                  'wait for bit 2 to be set
 910 RETURN
 920 REM check to see if datum is ready
 930    WAIT STATUS%,&H5
 940 RETURN
```

Program   5.1 (Part 2 of 2).   Testing Digital I/O.

simple to wait for the switch closure using the WHILE loop beginning in line 650, and then actuate both the relay and the LED simultaneously by outputting a value (hex 3) that corresponds to sending a 1 to both bits 0 and 1.

Note that this program is very similar to Program 5.1, except for lines 640-770, where we wait for a particular bit to be toggled by the user.

5.2 I/O WITH HANDSHAKING

You may have noticed that all of our examples of digital I/O so far have involved simple on/off devices. What happens if we try to use such techniques for more complex devices? For example, what happens if we are trying to read the output of a digital voltmeter? Well, the major difficulty we encounter involves synchronization between the device and our computer, or more generally between the two devices that are communicating.

With a digital voltmeter, for example, we may wish to use the computer to program the various functions of the voltmeter and to send data from the voltmeter to the computer. How do we keep the voltmeter from sending data at the same time the computer is sending it instructions? Or, how do we keep the computer from reading the voltmeter just as it is changing from one reading to another?

Several solutions have been developed to this problem of synchronization. The complexity of these solutions is related to the speed of information transferal. It is

```
10 REM PROGRAM SIMPLE_PARALLEL_I/O
20 REM LABORATORY AUTOMATION USING THE IBM PC
30 REM Program to sense a switch closure using the DT2801
40 REM When the switch is closed, an LED is lit and a relay actuated.
50 REM Note:  The switch should be connected to Port 1, bit 0, the
60 REM        relay to Port 0, bit 0 and the LED to Port 0, bit 1
70 REM CAUTION:  make sure no output devices are connected to port 0
80 REM ***********************************************************
90 REM Define all registers, etc.  Change register addresses if not adapter 0.
100 BASEADD%=&H2EC                'base address of board
110 COMMAND%=BASEADD%+1           'command register
120 STATUS%=BASEADD%+1            'status register
130 DATUM%=BASEADD%               'data register
140 REM *******************************
150 REM          Main program
160   GOSUB 270                   'set up board
170   GOSUB 350                   'set up DIO
180   DEVICE%=0                   'turn off all outputs
190   GOSUB 520
200   PRINT "Please turn the switch on."
210   GOSUB 640                   'wait for the switch
220   DEVICE%=2^1+2^0             'turn on LED at bit 1, relay at bit 0
230   GOSUB 520
240 END
250 REM *******************************
260 REM Set up board
270   OUT COMMAND%,&HF            'stop board
280   TEMP%=INP(DATUM%)           'clear data register
290   REM wait for command to finish, then clear errors
300   GOSUB 830
310   OUT COMMAND%,&H1            'clear the command register
320 RETURN
330 REM *******************************
340 REM Set up digital I/O
350   GOSUB 830
360   OUT COMMAND%,&H4            'write SET DIGITAL PORT FOR INPUT command
370   GOSUB 800
380   OUT DATUM%,1                'input on port 1
390   GOSUB 830
400   ERRORCHECK%=INP(STATUS%)    'check for error
410   IF (ERRORCHECK% AND &H80) THEN PRINT "Error writing to Port 1":STOP
420   GOSUB 830
430   OUT COMMAND%,&H5            'write SET DIGITAL PORT FOR OUTPUT command
440   GOSUB 800
450   OUT DATUM%,0                'output on port 0
460   GOSUB 830
470   ERRORCHECK%=INP(STATUS%)    'check for error
480   IF (ERRORCHECK% AND &H80) THEN PRINT "Error writing to Port o":STOP
490 RETURN
500 REM *******************************
510 REM Output on specified bit(s)
520   GOSUB 830
530   OUT COMMAND%,&H7            'send WRITE DIGITAL OUTPUT IMMEDIATE command
540   GOSUB 800
550   OUT DATUM%,0                'output on Port 0
560   GOSUB 800
570   OUT DATUM%,DEVICE%          'output byte
580   GOSUB 830
590   ERRORCHECK%=INP(STATUS%)    'check for error
600   IF (ERRORCHECK% AND &H80) THEN PRINT "Error writing to Port o":STOP
610 RETURN
620 REM *******************************
630 REM Input from switch
640   LOW%=0
650   WHILE LOW% = 0              'wait for switch to be toggled
660       GOSUB 830               'wait for board to be ready
670       OUT COMMAND%,&H6        'send READ DIGITAL INPUT IMMEDIATE command
```

Program   5.2 (Part 1 of 2).   Digital I/O for LED, Switch and Relay.

```
680        GOSUB 800
690        OUT DATUM%,1              'input from port 1
700        GOSUB 880                 'wait for datum to be ready
710        J%=INP(DATUM%)            'get low and high bytes
720        LOW%=J% AND 1             'mask off all but bit 0
730        GOSUB 830
740        ERRORCHECK%=INP(STATUS%)  'check for error
750        IF (ERRORCHECK% AND &H80) THEN PRINT "Error reading from Port 1":STOP
760   WEND
770 RETURN
780 REM ********************************
790 REM check to see if ready to send next command or to set a register
800    WAIT STATUS%,&H2,&H2          'wait for bit 1 to be reset
810 RETURN
820 REM check to see if ready to send another command
830    WAIT STATUS%,&H2,&H2          'wait for bit 1 to be reset
840    WAIT STATUS%,&H4              'wait for bit 2 to be set
850 RETURN
860 REM check to see if datum is ready
870    WAIT STATUS%,&H5
880 RETURN
```

Program 5.2 (Part 2 of 2). Digital I/O for LED, Switch and Relay.

possible to provide some synchronization during serial transfers, but the most common methods of synchronization involve **handshaking** protocols on parallel I/O.

In essence, handshaking involves each device sending an electrical signal, via a special line other than those used to transmit data, saying it is ready for data transfer to occur. This is easy to envision if we imagine two polite individuals talking on the telephone; we will signify these individuals as P1 and P2. If P1 wants to talk, he says, "I'd like to talk." P2 then says, "OK, go ahead," and P1 begins talking. When P1 is finished talking, then P2 can similarly request the chance to talk, and P1 can tell P2 to go ahead (complete the handshake).

However, difficulties can arise with a simple handshake. For example, what happens if both P1 and P2 decide to request the opportunity to talk at exactly the same time? Or what happens if P1 desperately needs to talk while P2 is talking?

Indeed, the situation for electrical signals is even more complex than that for the phone conversation described above. Not only is there a problem synchronizing data transmissions, but even such things as the voltage level of the signals, the sex of the connectors (male or female), the number of data lines, the maximum distance of signal transmission, and other characteristics must all be defined. Thus, while it is possible to implement handshaking protocols using simple digital I/O, it is rarely done.

Instead, laboratory scientists almost always use one of a small number of data transmission protocols: RS232 or RS422 for serial transmissions, IEEE-488 for short-distance high-speed parallel transmissions, and one of several proprietary techniques for connecting together many instruments over large distances into what are often termed **local area networks**. Hence, we shall examine the most useful of these in the succeeding chapters. It is useful to remember, however, that these techniques

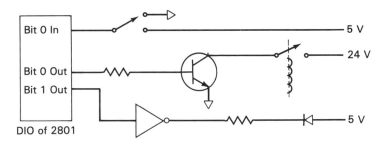

Figure 5.4. Typical Digital I/O Experiment.   Upon switch closure, the relay is actuated
and the LED lit.

all are inherently similar in that they transfer information in digital, rather than
analog form.

EXERCISES

1.   Write and implement a program that will sense a series of 16 switches, and
     depending upon whether the switches are on or off, light one of 16 LEDs.

2.   Write and implement a program that will count from 0 to 32767, using 16
     LEDs to display the resulting numbers in binary.

3.   Set up a system to use the digital I/O section of your data acquisition board to
     turn on and off the lights in your office. What do you need to do to buffer the
     digital I/O board from the lights?

6

IEEE-488 (GPIB)

One of the most useful data transmission techniques in the laboratory is a high-speed parallel interface originally designed by Hewlett-Packard. This technique was accepted as a standard by the Institute of Electrical and Electronic Engineers in 1975, 1978, and 1987, and hence is generally referred to by the designation IEEE-488. It is also often referred to by other names: Hewlett-Packard Interface Bus (HP-IB) and General Purpose Interface Bus (GPIB) are two of the more common. The international standard is the International Electrotechnical Commission (IEC) Publication 625, which is essentially the same as the IEEE-488 standard except for the type of connector used. To further complicate matters, the IEEE 1987 standard actually creates two standards, IEEE-488.1 (the IEEE 1978 standard) and IEEE-488.2. To simplify matters, we will generally refer to the entire set of standards as GPIB. By whatever name, it has been widely adopted by manufacturers of scientific instruments, particularly for many of the more expensive instruments.

The major advantages of the GPIB standard are that it is very rigidly defined so that various implementations of it are compatible (unlike RS-232C, where lack of compatibility is often a serious problem). It allows multiple instruments to be connected via a single adapter in the computer, which lowers the cost. It has a single type of connector, so that there is never any problem with having the wrong connector. And finally, it is a high speed interface, with data transfer rates of up to 1 Mbyte/sec, depending upon the exact hardware and cable configuration used.

The major disadvantage of the GPIB standard is that it adds substantially to the cost of an instrument, so it is only infrequently used with low-cost instruments. In addition, the high speed capability is only infrequently used; most GPIB instruments send data at rates much less than the standard permits; rates of 1 Hz to 10 kHz are common.

The IEEE-488 standard has a carefully designed set of hardware specifications. Among the more important of these are the following:

- A maximum of 15 devices can be attached to a single network.

- The total cable length connecting the devices cannot be more than 20 meters without additional buffering. The length of cable to each individual device is also limited to 2 meters.

- The number of lines in the cable is 24, with the function of each line clearly defined. Eight of the lines are for data transmission, while 8 are for hand-shaking and 8 are for grounding and shielding. The 16 data and handshaking lines are collectively the **bus** of the GPIB.

- The devices can be connected in almost any convenient arrangement (Figure 6.1).

- The maximum data rate is 1 Mbyte/sec.

Because the IEEE-488 standard allows multiple instruments on the same bus or network, the question of control of the bus becomes important. The standard requires that there be only one active **controller** at any time; usually this is the laboratory computer. The controller literally controls all interactions on the bus because it alone is allowed to determine the sequence of communications on the bus.

The standard further requires that only one device send data at a time; this avoids the problems having the messages sent by two instruments interfering with one another. The device sending information is referred to as a **talker**. One of the functions of the controller is thus to designate who is the talker at any given time. The talker is sometimes the controller, but it can also be any of the instruments connected to the bus.

In addition, the standard allows any or all of the devices on the bus (including the controller) to be a **listener**. Thus, it is possible for multiple devices to receive commands or information at the same time. This is again determined by the controller, which can designate any combination of the devices to be listeners.

The controller addresses each device, and sometimes even different sections of the same device, by a specific address, which must be in the range of 0 to 31. This address is switch selectable on most instruments, although some have a factory-set GPIB address. In addition, the standard allows the use of **secondary addresses**, which address separate functions at a given primary address. Typically, only the more sophisticated devices use the secondary addresses.

This may all be made clearer by a simple example. Suppose we have a computer that we wish to use to communicate with a digital volt meters (DVM) at GPIB address 4. We would then install an adapter card in the computer that would make it a GPIB device, and connect the adapter card to the instrument using special GPIB cables.

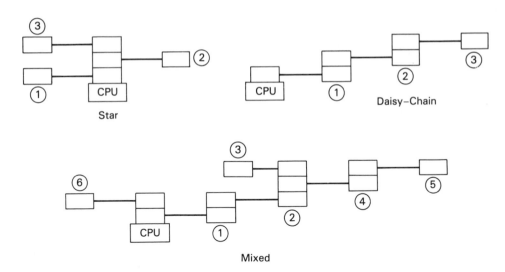

Figure 6.1. GPIB Connections. GPIB devices can be connected in almost any arrangement of serial (daisy-chained) or star-shaped connections. The devices being connected are represented by numbers; the connecting plugs by disks.

We might then use software that would do the following:

1. Designate the computer to be the controller.
2. Have the controller send commands to the DVM to tell it to listen.
3. Have the controller, acting as a talker, send commands to the DVM to initialize it and set it to prepare to read voltages in kilovolts.
4. Have the controller to send a command to the DVM to initiate a reading
5. Have the controller designate the DVM as a talker and itself as a listener.
6. Have the controller wait until the DVM sends a datum.
7. Store the datum.

It would also be possible, in this example, to tell two DVMs the same thing at the same time; for example, to tell both to send a datum to the computer. However, since we can only make one device at a time the talker, how do we find out that one or both of the DVMs have a datum ready? To do this, we often use another feature of the IEEE-488 standard, namely the **service request**. One of the 16 lines in the GPIB cable is a service request line. This can be used to signal the controller that the device wants to be a talker. The controller can issue a special command, **enable service request**, which allows individual instruments to signal the controller, via the service request line, that they wish to become a talker. When the controller receives such a request, it then performs **serial request poll**, during which each device is polled to discover whether it was the one that issued the serial request. That device can then be made the talker and send its data to the computer.

One other question that might occur to you is how does the computer know when the talker has finished talking. One common way for the talker to provide this is to assert another of the 16 lines, the **end or identify (EOI)** line. This signifies the end of a multi-byte transmission.

In addition to the above features, the IEEE-488 standard allows a number of other activities. However, most of these are only infrequently used.

6.1 IEEE-488 STANDARDS

We are now ready to discuss the IEEE standards for GPIB in detail. For some applications, and particularly in finding hardware faults, you may find this level of detail useful. If this is your first contact with GPIB, however, you may wish to skip this section and return to it after you have had some experience with GPIB programming.

6.1.1 GPIB Bus

As we have mentioned, the IEEE-488 standard specifies that the signal bus consists of 16 lines and that an additional 8 lines are used for logical ground returns and shielding. Three types of signal lines are used, as illustrated in Figure 6.2.

1. **Data bus**. Data and messages are sent along this 8-wire portion of the bus. The data bus is thus essentially an 8-bit parallel bus used to send information one byte at a time. The bus is bidirectional.

2. **Handshake**. The IEEE-488 standard requires a 3-wire handshake. The three wires are:

    ■ **DAV (Data Valid)**. This line is asserted (turned on) to indicate that data on the 8 data lines is available and valid.

    ■ **NRFD (Not Ready for Data)**. This line is asserted by a device until it is ready to accept data.

    ■ **NDAC (Not Data Accepted)**. This line is asserted by a device until it has accepted a set of data.

3. **Bus management**. Five lines are used to manage the information flow on the data bus. The five wires are:

    ■ **ATN (Attention)**. The Attention line is asserted by the controller. When asserted, data on the data bus are treated as commands. When this line is not asserted, data on the data bus are device-dependent messages such as readings from the device.

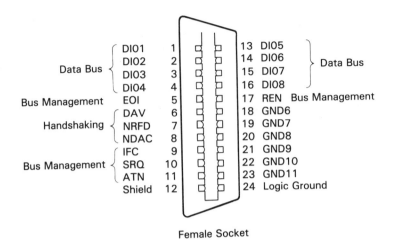

Female Socket

Figure 6.2. GPIB Bus. The IEEE-488 standard uses a 16-wire bus. Eight of the wires are the data bus, three are used for handshaking, and five are used for bus management. The arrangement shown is the IEEE-488 rather than the IEC-625 standard.

- **IFC (Interface Clear)**. This line can be used by the controller to "clear" devices. This puts them in some known state, usually the same one that the device has when it is first turned on.

- **SRQ (Service Request)**. This is used by a device to request service from the controller. Notice that this is a request, not a demand, that the controller can be programmed to honor or ignore.

- **REN (Remote Enable)**. This line is used by the controller to place a device in the "on-line", or remote mode. In the remote mode, the device is usually not programmable from its front panel. This line is often used to prevent accidental front-panel reprogramming of the device by a user.

- **EOI (End or Identify)**. This line is used by a talker to indicate the end of a transmission. It is used by a controller (with the ATN line asserted) to initiate a parallel or serial poll.

How are all of these lines used? For simplicity, we'll look first at the case where there is one controller (the IBM PC) and one device (such as the DVM) that will, for the moment at least, be used only as a listener.

A typical series of interchanges along the bus might then goes as shown in Figure 6.3.

In this series of interchanges along the bus, most of the commands are sent by asserting a single wire (ATN, DAV, etc.). Some of these one-wire commands are

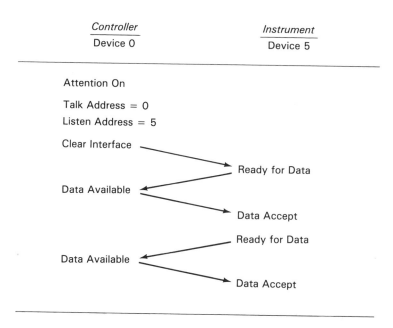

Figure   6.3.   Typical GPIB Interchange between a Controller and a Talker.

sent by unasserting a particular wire. For example, the DATA READY command shown above is actually sent by unasserting the NRFD line.

However, for some types of controller commands and for all transmissions between talkers and listeners, more than a single wire is required.

6.1.1.1 Multiwire Transmissions

All of the multiwire transmissions use the 8-wire data bus. In general, they can be divided into controller commands and data transmissions by a talker.

Controller Commands. Commands sent by the controller are distinguished by having the ATN line asserted while they are being sent. In general, the controller commands are 7-bit ASCII codes. The format of these codes is illustrated in Table 6.1.

Universal commands are those controller commands that apply to all devices on the bus, whereas **addressed commands** are those that apply to only those devices addressed as listeners when the command is issued. **Unaddressed commands** are used to clear the bus of talkers or listeners. These three types of commands are shown in Table 6.2.

For example, suppose we wish to clear a device (this usually returns the device to the status it was when it was first turned on). We have two ways in which we can do this. First, we can issue the universal command **DCL**, which clears all the devices on the bus. More often, however, we wish to clear just a single device. To do this, we first unaddress all listeners (i.e., make certain that no one is designated a listener) by issuing the **UNL** command. We then select a specific device or set of devices to listen; for example, to address device 6 to listen, we issue the bit pattern 00100110 (38 decimal). We then issue the addressed command **SDC**, which clears only the device selected to listen, in this case device 6.

In addition to the commands we have already discussed, there are several other **secondary commands** used primarily with parallel polling or with secondary addresses. These are beyond the scope of this text, but the interested reader is referred to the full IEEE standard.

Table 6.1.   GPIB Controller Command Format

		Bits					
6	5	4	3	2	1	0	Command
0	0	D_4	D_3	D_2	D_1	D_0	Universal Commands
0	1	D_4	D_3	D_2	D_1	D_0	Listen Addresses
		except					
0	1	1	1	1	1	1	Unlisten Command
1	0	D_4	D_3	D_2	D_1	D_0	Talk Addresses
		except					
1	0	1	1	1	1	1	Untalk Command
1	1	D_4	D_3	D_2	D_1	D_0	Secondary Commands
		except					
1	1	1	1	1	1	1	Not defined

6.1.1.2 Transmissions from Talker to Listener

In a manner similar to that used for commands, individual devices acting as talkers send data along the 8-wire data bus, with the three handshaking lines being used to control the sending of each byte of data. Notice that the ATN is not asserted; only controller commands are accompanied by the ATN signal.

6.1.1.3 Messages to Multiple Devices

As we have implied, it is also possible to send data from one talker to many listeners simultaneously. To do this, the controller addresses the various devices as listeners. All of the listeners must then unassert the NRFD line prior to the transmission of each byte, and all must unassert the NDAC line to indicate that the datum has been received before the talker can proceed to the next byte of the transmission.

Table 6.2. Universal, Addressed, and Unaddressed Commands

Command	Code[a]	Purpose
LLO (local lockout)[1]	17	Disable front panel of device
DCL (device clear)[1]	20	Clear all devices to predefined state
PPU (parallel poll unconfigure)[1]	21	Sets all devices to ignore parallel poll
SPE (serial poll enable)[1]	24	Enable serial poll mode
SPD (serial poll disable)[1]	25	Disable serial poll mode
SDC (selective device clear)[2]	4	Clear addressed devices
GTL (go to local)[1]	2	Return device to local control
GET (group execute trigger)[2]	8	Initiate a preprogrammed action
PPC (parallel poll configure)[2]	5	Assign lines for a parallel poll
TCT (take control)[2]	9	Addressed device becomes active
UNL (unlisten)[3]	63	Unaddress all current listeners
UNT (untalk)[3]	95	Unaddress the current talker

[a]Decimal value
[1]Universal command
[2]Addressed command
[3]Unaddressed command

6.1.2 Software Standards

The IEEE 488.1 standard has no requirement for the data format, while the newer IEEE 488.2 does. The IEEE 488.2 standard is an ASCII 7-bit code. It also allows 8-bit binary and 8-bit binary floating point formats (the latter, the IEEE 754-1985 standard). Within the 7-bit ASCII code format, several data types are allowed, including integer, hexadecimal, characters, and binary data, among others.

However, unlike the situation for controller transmissions, there is no universal standard for device message *content*. The IEC-625 and IEEE-488.2 specifications do suggest the format of the messages sent by talkers, but do not specify the meaning. For example, the message "D3" may mean "Set to volts" to one device, "Send stored data" to another, and nothing at all to a third device. Hence, you must consult the users' manual for the GPIB device you are interfacing to obtain the meaning of the device-specific messages.

6.1.2.1 Examples

Let's look at some typical examples. For example, we might send the following message to our HP3478A digital volt meter:

F1CRLF

where CRLF is the ASCII carriage return and line feed characters in sequence. In this case, F is a header that signifies that a "measurement function" follows, and 2 is the data. The CRLF combination is a separator that tells the volt meter that the message has ended and should be acted upon. In the case of the HP3478A volt meter, a command of F1 sets the volt meter to the AC volts function.

We can send a series of such commands in a single message; for example:

F1R2N5CRLF

sends three commands in a single message; in this case, the 3478A is set to DC volts (F1), 300 volts range (R2) and 5 1/2 digit display (N5). Again, the CRLF is used to terminate the message. Notice that in this particular application, there are no separators between message units. However, the HP3478A understands the message with separators such as commas or spaces:

F1 R2 N5 CRLF

or

F1,R2,N5 CRLF

The separators often provide increased legibility to the user, and so you may wish to use them if your instrument permits.

The DVM may also be requested to send back data; if it does so, it may send back a message such as

-000.000E+0CRLF

to indicate that it has read zero volts. In this case, there is no header; the data is the -000.000E0, and the CRLF is the separator.

Of course, much more sophisticated messages are possible that still conform to the IEEE standard. Few applications require such sophistication.

6.2 PROGRAMMING GPIB

Generally, you can program GPIB devices at one of three levels of operation:

- Low-level. In this approach, the programmer has direct control over all of the lines of the GPIB bus. This allows the most flexibility, but requires a detailed understanding of the GPIB bus and the protocol for running it. This is also sometimes referred to as adapter-level programming.

■   Middle-level. The programmer uses only a small number of more general purpose commands. These allow almost the full GPIB functionality, but without your having to understand the often intricate details of the bus protocol. It is often referred to as device-level programming.

■   High-level. The programmer uses a library of routines specifically designed for the instrument being controlled and the computer controlling it. These are sometimes available from the manufacturer of the laboratory device, or they may be constructed by the programmer using middle-level commands.

The most common of these is the middle-level approach, because it allows flexibility and yet is relatively easy to learn. It is the approach we shall use in the following discussion. We illustrate this approach using the IBM GPIB adapter and software developed by National Instruments. Naturally, GPIB boards from other manufacturers may be programmed quite differently, so you should consult the appropriate programming guide before proceeding.

There is a general pattern for efficiently designing software for the GPIB interface. The steps are:

1.   Configure the interface software.
2.   Program the interface using the interactive software supplied by the vendor.
3.   Write your own software, utilizing the commands tested with the interactive programming.

6.2.1 Configuring the GPIB Interface Software

The National Instruments (and many other vendors') software uses a technique that is rather different from the software we have designed in the remainder of this book. In order to provide an easy technique for supporting several different programming languages, the GPIB-PC software uses a combination of a **device driver** and a **language-specific interface**. The device driver is in essence a program that is loaded into memory when the computer is first turned on, and that remains resident in memory until the computer is turned off. In this case, the device driver is called GPIB216.COM. It is loaded into memory by having a statement in the CONFIG.SYS file in the root directory of the disk from which your system boots (normally A or C). This statement is of the form:

$$DEVICE=\backslash GPIB\backslash GPIB216.COM$$

where we assume that you have placed all of the GPIB software in a subdirectory called GPIB.

However, we highly recommend that you rename the file GPIB216.SYS and then use:

DEVICE=\GPIB\GPIB216.SYS

in CONFIG.SYS. The reason for this is that if you inadvertently type

GPIB216

as a command in DOS, nothing will happen. If you leave the file with the extension .COM, then DOS will try to run the program and your computer will "hang" and require rebooting.

The device driver GPIB216.SYS performs two functions: it contains all of the software necessary to make the GPIB adapter and the GPIB devices connected to it perform their various GPIB operations. In addition, it contains a user-defined table of all of the device-specific parameters.

The IEEE-488 standard leaves a number of options to instrument manufacturers; in particular, the exact content of messages transmitted from talker to listener is not specified. Hence, there are several device-specific parameters that must be specified in order for two GPIB devices to communicate with one another properly. A separate program (called IBCONF) is provided to allow you to set these parameters.

These parameters include the following:

- The address of each device. As mentioned above, each device on the bus has its own unique address, in the range 0 to 31.

- Any secondary addresses. If an instrument uses secondary addresses for specific functions, then these must be specified.

- Time-out limits. To prevent the software from "hanging" when it tries to access a device that is turned off, malfunctioning, or otherwise unavailable, it is useful to give a time-out limit. If the device does not respond during this time period, an error is generated, which your program can then process.

- Message termination. Depending upon the manufacturer, one or more techniques may be used to terminate transmission of data from an instrument. These techniques include:

    1. Transmission of a specific character at the end of each string of data. The ASCII line feed character (10) is often used for this purpose.
    2. Use of the EOI line of the GPIB bus to signal the end of a transmission.
    3. Terminate upon end count. Some devices always send a predefined number of bytes of information in each transmission.

In addition, at the time of installation of your GPIB interface card, you usually need to specify parameters relating to the operation of the interface, such as its address, interrupt level, and DMA channel use. These parameters must also be set on the board using hardware switches. Normally, with only one GPIB board, you should use the factory-default setting for the address, and then pick interrupt and DMA settings that are not used by any other board in your computer. Disk drives are the most common users of DMA, and interrupts are also commonly used, so you will need to check the user manuals for each of the boards to avoid possible conflicts. Interrupt level 7 and DMA channel 1 are good choices if you have no conflicts with these.

Figure 6.4 and Figure 6.5 illustrate some typical parameter settings for the adapter and a Hewlett-Packard 3478A digital volt meter.

6.2.2 Interactive Programming of GPIB

Before writing a program in BASIC or another high-level language for a GPIB device, you normally should first use the interactive programming facility often supplied by the vendor. In the case of the National Instruments (and IBM) software, this is a program called IBIC. This allows you to test the various commands you will issue to the device before you put them in a program.

The first thing to test is whether you have all of the device and adapter parameters set correctly. In particular, do you have the address set properly for the device? Many devices will respond to a "device clear" instruction by clearing their displays and resetting to some default parameters. Hence, it is often a good command to send first. If you are using the National Instruments GPIB software to address the Hewlett Packard HP3478A, for example, then you can type the following:

```
IBIC
IBFIND HP3478A
IBCLR
```

Note that this assumes you have already configured the interface software, as described above, by assigning the name HP3478A to that device. If the device does not appear to respond, then check the hardware setting of the address of the device against that assigned during the configuration process.

Next, if the device can function as a listener, try writing a command to it. With the National Instruments software, this is done with an IBWRT command. For example, we can make the HP3478A display a message on its LCD display by typing:

```
IBWRT "D2HELLO"
```

assuming that we have already issued the commands described above. In this case, the **D2** is a command to write a string on the display panel. Of course the IEEE-488

```
Primary GPIB Address .......... 30
Secondary GPIB Address ........ NONE
Timeout setting ............... T30s
EOS byte ...................... 00H
Terminate Read on EOS ......... no
Set EOI with EOS on Write ..... no
Type of compare on EOS ........ 8-bit
Set EOI w/last byte of Write .. yes
GPIB-PC Model ................. PC2A
Board is System Controller .... yes
Local Lockout on all devices .. no
Disable Auto Serial Polling ... yes
High-speed timing ............. no
Interrupt jumper setting ...... 7
Base I/O Address ......... .... 02E1H
DMA channel ................... NONE
Internal Clock Freq (in MHz) .. 8
```

Figure   6.4.   GPIB Adapter Parameters.   These are the set of parameters we use in our own laboratory to configure the GPIB-PC software for the **IBM GPIB** adapter (or equivalently, the National Instruments **PC-IIA** GPIB adapter). You may need to change the settings, depending upon the other adapters used in your PC.

standard does not specify the syntax of commands, so the commands vary tremendously from manufacturer to manufacturer and even among instruments from the same manufacturer.

If a simple command does not work, you may wish to try the following:

■ Check the case of the command being sent. Note that upper and lower case are not necessarily going to produce the same results. For example, the HP3478A does not respond to command to, IBWRT "d2HELLO".

■ Check that an end-of-command message is not needed. For example, the HP3478A expects an ASCII 0D (hex) to terminate a text command.

■ Check that the desired device is addressed correctly.

Check that you can read information from the device, assuming that it can function as a talker. For example, we might get the current reading from the HP3478A by typing:

IBRD 1000

```
Primary GPIB Address .......... 23
Secondary GPIB Address ........ NONE
Timeout setting .............. T10s
EOS byte ..................... 00H
Terminate Read on EOS ......... no
Set EOI with EOS on Write ..... no
Type of compare on EOS ........ 8-bit
Set EOI w/last byte of Write .. yes
```

Figure   6.5.   GPIB Device Parameters.   These are the parameters we typically use with a Hewlett-Packard HP3478A digital volt meter. Note in particular that the address used may vary depending upon the switch settings on the back of the HP3478A.

This command causes 1000 bytes of information to be read, or as many bytes as are received prior to an end-of-data command (depending upon how the device is configured).   In the case described here, it causes only 13 bytes to be read from the HP3478A, where the first 11 bytes represent the current reading displayed on the LCD display of the meter, and the last two bytes are the ASCII codes for a carriage return/line feed.

Next, if the device supports serial polling, issue a serial poll command to the device by typing:

IBRSP

This will return a message message that is device-specific, except that if bit 6 is set in the byte returned by this device, then it has requested service. For example, issuing a serial poll to the HP3478A causes it to return a byte indicating its status; a typical response would be to return a value of 65 (bits 0 and 6 set), indicating that it is ready to send another datum.

Once you have determined that all of the general functions of the device can be used, you then should examine the users' manual of the instrument you are attempting to interface and try issuing the same sequence of commands that you would issue during normal operation of the instrument.

6.2.3 Writing a GPIB Program

Writing a program for the GPIB adapter is relatively easy, once you have used the interactive programming facility to determine the sequence of commands to send. We have found that almost all programming of the GPIB card can be accomplished with the following commands or their equivalents(where the command given here is the one used by the National Instruments software):

- **IBFIND**. This allows you to specify the devices with which you are going to communicate.
- **IBCLR**. This clears the GPIB device you are currently addressing. Clearing the device usually resets it to some specific set of conditions.
- **IBRD**. This reads a set of data from the current device.
- **IBWRT**. This writes a set of data to the current device.
- **IBRSP**. This initiates a serial poll of the current device.

Three less-commonly used commands are:

- **IBTRG**. This triggers the current device. The effect of the trigger is device-dependent, but often triggering causes the device to initiate data collection or some other preprogrammed operation.
- **IBLOC**. This places the current device in local mode so that its panel settings can be manually changed.
- **IBWAIT**. This causes the program to wait for one or more events. The events are usually requests for service from a device, but can be other events such as time-outs or GPIB errors.

In all of the above commands, the "current device" is the device specified in the command. The device is specified not by its name, but by a number determined by the IBFIND command.

6.2.3.1 Simple GPIB Program

These commands are of course easier to understand if we can see an example program. Program 6.1 illustrates many of these commands.

This program is designed to read voltages from two Hewlett-Packard 3478A digital volt meters. Each of the two meters must be connected to the GPIB board and the GPIB software configured for the two meters. In this case, we have chosen to call one of the meters HP3478A and one HP3478B; otherwise the configuration parameters for each of the meters are those shown in Figure 6.5.

As we have mentioned, the GPIB216.SYS program remains in memory and contains code for all of the allowed GPIB commands. However, in order to access that code from BASIC, we must use an additional file, namely BIB.M for the BASICA interpreter and CBIB.OBJ for IBM Compiled BASIC.

In Program 6.1, lines 60 to 110 are required in order to use the GPIB software with the BASICA interpreter. Lines 60 and 70 contain a number, 59741, that is system dependent; you should consult the GPIB software manual for instructions on how to determine this number for your system. Note that the CLEAR statement

```
10 REM PROGRAM GPIB_DVM
20 REM LABORATORY AUTOMATION USING THE IBM PC
30 REM Program to read data from two HP 3478A digital volt meters using GPIB
40 REM     Adapter with National Instruments' GPIB-PC software
50 REM ********************************************************************
60 CLEAR    ,59741!          ' BASIC Interpreter -- See GPIB software manual for
70 IBINIT1 = 59741!         '     method to determine value for lines 60, 70
80 IBINIT2 = IBINIT1 + 3
90 BLOAD "\GPIB\bib.m",IBINIT1
100 CALL IBINIT1(IBFIND,IBTRG,IBCLR,IBPCT,IBSIC,IBLOC,IBPPC,IBBNA,IBONL,IBRSC,
    IBSRE,IBRSV,IBPAD,IBSAD,IBIST,IBDMA,IBEOS,IBTMO,IBEOT,IBRDF,IBWRTF)
110 CALL IBINIT2(IBGTS,IBCAC,IBWAIT,IBPOKE,IBWRT,IBWRTA,IBCMD,IBCMDA,IBRD,IBRDA,
    IBSTOP,IBRPP,IBRSP,IBDIAG,IBXTRC,IBRDI,IBWRTI,IBRDIA,IBWRTIA,IBSTA%,
    IBERR%,IBCNT%)
120 REM ***** If you are using the BASIC compiler, comment out lines 60-110 ***
130 COMMON IBSTA%, IBERR%, IBCNT%
140 DIM DEV%(2)
150 rem              Main program loop
160   GOSUB 460                             'initialize GPIB devices
170   GOSUB 380                             'get user input
180   STARTTIME!=TIMER                      'get starting time
190   LOCATE 1,1:PRINT "POINT","DVM 1","DVM 2"," "
200   FOR I%=1 TO MAXPTS%                   'start of data input loop
210     PRINT I%,                           'print point number
220     WHILE ELAPSTIM! < DURAT!*I%         'check if desired time elapsed yet
230        CURTIM!=TIMER
240        ELAPSTIM!=CURTIM!-STARTTIME!
250     WEND
260     DEVICE%=DEV%(1)
270     GOSUB 630                           'read first meter
280     PRINT DATUM!,                       'print datum from first DVM
290     DEVICE%=DEV%(2)
300     GOSUB 630                           'read second meter
310     PRINT DATUM!                        'print datum from second DVM
320     Y$=INKEY$:IF Y$<>"" THEN GOSUB 790  'check for user abort using Alt-Q
330   NEXT I%
340   PRINT "Run finished!"
350 END
360 REM ****************************************
370 REM User input
380   CLS:INPUT "Enter the number of pairs of points to collect ",MAXPTS%
390   INPUT "Enter the time between data points in seconds ",DURAT!
400   KEY OFF:CLS:LOCATE 25,1:PRINT "Type an <Alt-Q> to abort data collection.";
410   LOCATE 1,1:PRINT "Type any key to begin data collection."
420   Y$=INKEY$: IF Y$="" THEN 420
430 RETURN
440 REM ****************************************
450 REM Initialize GPIB devices
460   CLS:PRINT "Initializing GPIB devices..."
470   RD$=SPACE$(80):DEV%(1)=0:DEV%(2)=0          'initialize variables in CALL
480   DEVNAME1$="HP3478A":DEVNAME2$="HP3478B"     'name these devices
490   'open first GPIB device
500   CALL IBFIND (DEVNAME1$,DEV%(1))
510   IF DEV%(1) < 0 THEN GOSUB 690
520   CALL IBCLR (DEV%(1))
530   IF IBSTA% < 0 THEN GOSUB 740
540   'open second GPIB device
550   CALL IBFIND (DEVNAME2$,DEV%(2))
560   IF DEV%(2) < 0 THEN GOSUB 690
570   CALL IBCLR (DEV%(2))
580   IF IBSTA% < 0 THEN GOSUB 740
590   PRINT "Successfully initialized GPIB devices..."
600 RETURN
610 REM ****************************************
620 REM Read from one GPIB meter
630   CALL IBRD(DEVICE%,RD$)
640   IF IBSTA% < 0 THEN GOSUB 760
```

Program   6.1 (Part 1 of 2).   GPIB for Two Digital Volt Meters.

```
 650   DATUM!=VAL(RD$)                       'VAL function removes non-numerics
 660 RETURN
 670 REM ****************************************
 680 REM GPIB error routines
 690   PRINT "Unable to open device "
 700   PRINT "Please check CONFIG.SYS to make sure that you have"
 710   PRINT "  included GPIB.SYS as a device.  Also, use IBCONF
 720   PRINT "  to check that you have configured the device properly."
 730 STOP
 740   PRINT "Unable to clear a GPIB device."
 750 STOP
 760   PRINT "Unable to read data from a DVM."
 770   PRINT "Please turn on all devices and try again"
 780 STOP
 790 REM ****************************************
 800 REM Check for Alt-Q
 810   IF MID$(Y$,1,1)<>CHR$(0) THEN RETURN  'Alt-Q generates ascii 0, 16
 820   IF MID$(Y$,2,1)<>CHR$(16) THEN RETURN
 830   PRINT "Run terminated by user."
 840 END
```

Program   6.1 (Part 2 of 2).   GPIB for Two Digital Volt Meters.

must precede all other executable statements in the program. Line 90 loads a machine-language routine that provides the interface between BASICA and GPIB216.SYS; notice that the BIB.M routine is assumed to be in the GPIB subdirectory.

If wish to run this program with compiled BASIC instead of the BASICA Interpreter, then you should comment out lines 60 to 110 and type the following:

BASCOM PROG61;
LINK PROG61+CBIB,,,BASRUN20;

where CBIB is the machine-language routine that provides the interface between compiled BASIC and the GPIB software. Notice, however, that line 130 must be included for either interpreted or compiled BASIC.

This is the first time we have used the CALL statement. It is similar to the BASIC GOSUB statement, in that it transfers control to a subroutine. However, there are two differences between GOSUB and CALL. First, CALL allows us to communicate with an assembly language routine. In this case, the routines are the BIB.M or CBIB.OBJ provided by the software vendor. GOSUB, on the other hand, can only be used with other BASIC subroutines. (In Compiled BASIC, the CALL statement can also be used with separately compiled BASIC subroutines, but that need not concern us here.)

The second difference is that, unlike GOSUB, CALL uses **formal parameters**. These are variable names that describe the variables to be used in the assembly language routine. This allows the assembly routines to use variables without knowing what the user named them in the main program. For example, we can use

$$\text{CALL IBFIND (B\$,C\%)}$$

or

$$\text{CALL IBFIND (DEVICE\$,DEV\%)}$$

to CALL the IBFIND routine. The only criteria, as defined by the software manufacturer, are that the parameters need to be of the same type, number and order as those in the subroutine.

IBM BASIC and Microsoft BASIC require that variables used in CALL statements be initialized before the CALL is performed. Thus, in lines 470-480 of Program 6.1, several variables are initialized; these variables are used in subsequent CALL statements.

The next activity to be programmed for the two volt meters is to ascertain that the current GPIB parameter configuration has an entry for each meter. This is done with the **IBFIND** command, which has two parameters: a string variable that names the device and an integer variable that after IBFIND is called contains a "unit descriptor." If the unit descriptor is less than zero, then an error has occurred; usually, the error is that you have forgotten to configure for that device, or have not put a DEVICE=\GPIB\GPIB216.SYS statement into CONFIG.SYS.

If the unit descriptor is larger than zero, it can be used during the remainder of the program to refer to that device. For example, if the unit descriptor for the first voltmeter is 7, then all subsequent GPIB commands referring to that meter use 7, rather than HP3478A, as the "name" of that device.

Thus, lines 500 to 590 of Program 6.1 perform an IBFIND for each of the two volt meters, and do an IBCLR to initialize the device. After each GPIB operation, the program checks to see if an error has occurred.

Once the two DVMs have been correctly identified by IBFIND and cleared, then the program begins its data collection loop. After a suitable period of time has elapsed, the program reads a datum first from one DVM, then the other. This is done with a CALL to the IBRD routine (lines 630-660). The IBRD routine returns a string of characters; in the case of the HP3478A meter, this is a string of 11 bytes containing the current reading on the HP3478A LCD display. (Note: the IEEE-488 standard does not require that the data be in ASCII. Some instrument manufacturers return the data in binary, which must then be decoded. However, the IBRD routine returns the data as a string, whether it is an ASCII string or some other type of string.) The program proceeds until the requested number of data are collected or until the user strikes the Alt-Q key combination.

6.2.3.2 GPIB using Polling

As we mentioned, the IEEE-488 standard supports a concept termed **polling**. Particularly in an environment where several GPIB devices require asynchronous service, serial polling is often used. Parallel polling is also allowed by the IEEE-488 standard, but is much less often used. In either case, the GPIB adapter conducts a poll by asking which device needs service. In a serial poll, each device is asked individually; in a parallel poll, a limited number of devices is asked at one time.

```
10 REM PROGRAM GPIB_DVM_SERIAL
20 REM LABORATORY AUTOMATION USING THE IBM PC
30 REM Program to read data from two HP 3478A digital volt meters using GPIB
40 REM   Adapter with National Instruments' GPIB-PC software and serial polling
50 REM ********************************************************************
60 CLEAR   ,59741!               ' BASIC Interpreter -- See GPIB software manual for
70 IBINIT1 = 59741!              '    method to determine value for lines 6 0, 70
80 IBINIT2 = IBINIT1 + 3
90 BLOAD "\GPIB\bib.m",IBINIT1
100 CALL IBINIT1(IBFIND,IBTRG,IBCLR,IBPCT,IBSIC,IBLOC,IBPPC,IBBNA,IBONL,IBRSC,
    IBSRE,IBRSV,IBPAD,IBSAD,IBIST,IBDMA,IBEOS,IBTMO,IBEOT,IBRDF,IBWRTF)
110 CALL IBINIT2(IBGTS,IBCAC,IBWAIT,IBPOKE,IBWRT,IBWRTA,IBCMD,IBCMDA,IBRD,
    IBRDA,IBSTOP,IBRPP,IBRSP,IBDIAG,IBXTRC,IBRDI,IBWRTI,IBRDIA,IBWRTIA,
    IBSTA%,IBERR%,IBCNT%)
120 REM ****** If you are using the BASIC compiler, comment out lines 60-110 ***
130 COMMON IBSTA%, IBERR%, IBCNT%
140 DIM DEV%(2)
150 SRQ$="M01"                              'code for service request on data ready
160 REM *************************************
170 REM              Main program
180   GOSUB 380                             'get user input
190   GOSUB 450                             'initialize GPIB devices
200   LOCATE 1,1:PRINT "POINT","DVM 1","DVM 2"," "
210   CALL IBWRT(DEV%(1),SRQ$)              'turn on service request on data ready
220   CALL IBWRT(DEV%(2),SRQ$)
230   'Data input loop
240   FOR I%=1 TO MAXPTS%
250     PRINT I%,                           'print point number
260     DEVICE%=DEV%(1)
270     GOSUB 620                           'get datum from first DVM
280     PRINT DATUM!,                        'print datum from first DVM
290     DEVICE%=DEV%(2)
300     GOSUB 620                           'get datum from second DVM
310     PRINT DATUM!
320     Y$=INKEY$:IF Y$="" THEN GOSUB 810 'check for user abort using Alt-Q
330   NEXT I%
340   PRINT "RUN FINISHED!"
350 END
360 REM *************************************
370 REM Get user input
380   CLS:INPUT "Enter the number of pairs of points to collect ",MAXPTS%
390   KEY OFF:CLS:LOCATE 25,1:PRINT "Type an <Alt-Q> to abort data collection.";
400   LOCATE 1,1:PRINT "Type any key to begin data collection."
410   Y$=INKEY$: IF Y$="" THEN 410
420 RETURN
430 REM *************************************
440 REM Initialize GPIB devices
450   CLS:PRINT "Initializing GPIB devices..."
460   RD$=SPACE$(80):DEV%(1)=0:DEV%(2)=0       'initialize variables in CALL
470   DEVNAME1$="HP3478A":DEVNAME2$="HP3478B"  'name these devices
480   'open first GPIB device
490   CALL IBFIND (DEVNAME1$,DEV%(1))
500   IF DEV%(1) < 0 THEN GOSUB 710
510   CALL IBCLR (DEV%(1))
520   IF IBSTA% < 0 THEN GOSUB 760
530   'open second GPIB device
540   CALL IBFIND (DEVNAME2$,DEV%(2))
550   IF DEV%(2) < 0 THEN GOSUB 710
560   CALL IBCLR (DEV%(2))
570   IF IBSTA% < 0 THEN GOSUB 760
580   PRINT "Successfully initialized GPIB devices..."
590 RETURN
600 REM *************************************
610 REM Get datum from DVM using serial polling
620   WHILE STATUS% AND 65 <> 65            'loop until dvm is ready
630       CALL IBRSP(DEVICE%,STATUS%)        'poll first dvm
640   WEND
```

Program   6.2 (Part 1 of 2).   GPIB Using Serial Polling.

```
650    CALL IBRD(DEVICE%,RD$)
660    IF IBSTA% < 0 THEN GOSUB 780
670    DATUM!=VAL(RD$)                       'VAL function removes non-numerics
680 RETURN
690 REM **********************************
700 REM GPIB error routines
710    PRINT "Unable to open device "
720    PRINT "Please check CONFIG.SYS to make sure that you have"
730    PRINT "  included GPIB.SYS as a device.  Also, use IBCONF
740    PRINT "  to check that you have configured the device properly."
750 STOP
760    PRINT "Unable to clear a GPIB device."
770 STOP
780    PRINT "Unable to read data from a DVM."
790    PRINT "Please turn on all devices and try again"
800 STOP
810 REM **********************************
820 REM Check for Alt-Q
830    IF MID$(Y$,1,1)<>CHR$(0) THEN RETURN   'Alt-Q generates ascii 0, 16
840    IF MID$(Y$,2,1)<>CHR$(16) THEN RETURN
850    PRINT "Run terminated by user "
860 END
```

Program   6.2 (Part 2 of 2).   GPIB Using Serial Polling.

Program 6.2 illustrates the use of serial polling. In order for serial polling to occur, you must do three things:

■   Enable automatic serial polling by the adapter. This is done as part of the configuration process described above.

■   Enable service requests by the device. This is a device-specific process. For example, in the case of the HP3478A DVM, you can enable generation of service requests whenever the DVM has a datum ready by sending the message *M01* to the meter with the IBWRT command

■   Conduct a serial poll of the device of interest. This is done using the GPIB function IBRSP. The second parameter of the CALL to IBRSP is a status variable. By testing the value of this parameter, you can tell whether the device has requested service, and, if so, for what reason. The status variable is actually a byte of data returned from the device; if bit 6 of that byte is set (on) then the device has requested service. The meaning of the remaining bits is device-specific. For example, the HP3478A sets bit 6 and bit 0 to to indicate that it has a datum ready to read.

Hence, Program 6.2 is very similar to Program 6.1 except that it sends the command to enable serial polling in lines 210 to 220. Then in lines 620 to 640, it performs a serial poll, and tests to see if both bits 6 and 0 are set by ANDing the status byte with the value 65. When it finds that the first DVM has requested service, then it reads the datum with an IBRD statement. This process is then repeated for the second DVM.

A different method for accomplishing the same thing is to use the following lines instead of lines 620 to 640 in Program 6.2:

```
620 MASK%=&H1800        'wait for SRQ and device requiring service
630 CALL IBWAIT(DEVICE%,MASK%)   'wait for first dvm
640 CALL IBRSP(DEVICE%,STATUS%)  'clear SRQ
```

IBWAIT is used to wait until the DVM has requested service; here, IBWAIT waits until the bits specified in MASK% are set. In this case, it waits for bits 11 and 12, which are set whenever the device has issued a service request (bit 12) and the controller has detected it by a serial poll (bit 11).

EXERCISES

1. Write a program using GPIB to control a programmable power supply or similar device. Be sure to add tests for values that are outside of the allowable range for the device.

2. Write a program to collect data from the GPIB device of your choice. Be sure to include all of the necessary commands for setting the device to acquire data at the desired rate and with the desired user-selected operating parameters.

3. If you are an advanced programmer, try writing a program using the low-level (adapter-level) commands that accomplishes the same task as Program 6.1. Then modify it to use parallel polling of the meters.

4. Obtain a bus analyzer, or use a bus analysis program. Watch the commands being sent to a device by the PC acting as a controller. In particular, watch what happens when a command such as IBWRT is sent. What low-level commands are used to implement the command? Make a chart diagramming the interchange that occurs, similar to that used in Figure 6.2.

5. Develop a library of high-level commands (described on page 107) for a device of your choice. Then write a program which will give the user a menu of the various functions performed by this library (see Chapter 10 for hints on menu writing).

6. If you are an advanced user of GPIB, write a program to implement parallel polling for three devices of your choosing.

7. What would be the advantages of having a separate "box" for data acquisition that passes data upon command to the PC? Such boxes often contain signal conditioning and other special purpose hardware controllable using GPIB. Can you think of any disadvantages to such an arrangement? Investigate one of the commercially available systems of this type (especially in the area of very high-speed devices).

7

SERIAL COMMUNICATIONS

As we discussed in Chapter 5, communications between laboratory devices can be by either serial or parallel techniques. In Chapter 6 we discussed a popular parallel technique; in this chapter, we describe commonly-used serial techniques, particularly RS-232.

In serial techniques, we transmit data one bit at a time (rather than one byte at a time as used in most parallel techniques). In principle, then, we need only one signal wire and one return wire; the electrical connections used in serial transmission techniques can thus be very simple and inexpensive. In practice, it turns out that serial techniques are indeed inexpensive. However, the lack of rigorously defined standards for serial interfaces makes them more complex to install than parallel interfaces in many cases.

7.1 CHOICES FOR SERIAL TECHNIQUES

Recall that in the case of the GPIB standard, there are rigid specifications for all of the electrical and mechanical connections. However, in the case of serial systems, there are choices for cabling, signal levels, timing, physical connections, electrical connections, and transmission rates, as well as the signal protocols and so forth. Let's examine each of these briefly.

7.1.1 Cabling

As we have mentioned, only one signal wire is absolutely required for serial transmissions. However, usually two signal wires are used in order to provide bidirectional transmissions. The terminology often used here is that serial transmissions can be **simplex** (one unit transmits and one listens), **half duplex** (either unit can transmit while the other listens, but both cannot transmit simultaneously), or **full duplex** (both can transmit simultaneously). Simplex is only rarely used; hence serial standards usually require as a minimum two signal wires and a return wire. In addition, as we shall see, there are usually several wires used for sensing interface conditions or other types of signaling.

The type of cable used can also vary considerably. Depending upon the transmission speed and distance, serial cables can be simple twisted cable, shielded twisted pairs, coaxial cable, or fiber optic cable, as discussed in Chapter 4. In addition, optical or other coupling techniques may be used to further reduce noise problems. The more sophisticated techniques are of course more expensive.

Before leaving the topic of cables, we should point out that you should carefully distinguish between serial techniques used within a single laboratory and techniques used to communicate between laboratories. Generally speaking, interlaboratory communications require much higher speeds and longer cables, and therefore are much more susceptible to noise problems. Hence, we shall discuss interlaboratory techniques in this chapter, and high-speed intralaboratory techniques separately, in Chapter 13, where we describe **local area networks**. The remainder of our discussion is thus oriented primarily toward relatively low-speed intralaboratory communications, i.e., between a microcomputer and a laboratory instrument over distances of less than 100 feet at rates under 20 kbits/sec. Usually twisted cable or shielded twisted cable is used in these circumstances.

7.1.2 Signal Modulation and Timing

For the inexpensive communications protocols usually used for direct low-speed serial communications in the laboratory, the information content is sent as a series of transitions between two voltage levels. TTL level transitions are usually too sensitive to noise for transmission of signals over moderate distances, so common serial techniques use two voltage levels differing by 10 or more volts. Thus, transitions between -12 and +12 volts are used in typical implementations of the RS-232 standard, for example. However, it may be less obvious that many signal modulation schemes are possible. For example, one possibility is simply to consider one voltage level a logical zero and another voltage level a logical 1. A second possibility is to consider a negative transition between levels to be a 0 and a positive transition to be a 1.

In most such schemes, however, it is important that the sending and transmitting units operate at the same frequency; i.e., that some sort of timing information is transmitted between the two units. Two basically different techniques are used for transmitting this information during asynchronous transmissions and synchronous transmissions, respectively.

In **asynchronous transmissions**, the information is sent at random intervals. Hence, in this technique, timing information is usually sent immediately prior to each set of information (e.g., prior to each byte of data). A commonly-used version of this technique is to send a **start bit** before each byte of data, and one or more **stop bits** after each byte of data. The start bit essentially signals the beginning of a message, and allows synchronization of the clocks in the sending and receiving units. The stop bit informs the receiving unit that 'he transmission has been completed.

In **synchronous transmissions**, on the other hand, timing data is constantly sent along the wires connecting the two devices. The advantage to this technique is that it is more efficient, because no start and stop bits need to be sent. However, it is rarely used in the laboratory except to communicate with mainframe computers.

7.1.3 Additional Information Sent with Data

In addition to the data and the start and stop bits, there may be other information sent as part of serial transmissions. For example, many serial systems, particularly those sending data in ASCII format (American Standard Code for Information Interchange), use a **parity bit** to permit limited error checking to occur. In this case, a single bit is added to the transmitted character so that the sum of the bits transmitted is even (**even parity**) or odd (**odd parity**). The receiving unit then can check each character transmitted to see if it has the desired parity; if not, then an error has occurred during the transmission process.

A more sophisticated alternative to using parity bits, and one that is often used with long transmissions, is to divide the transmitted stream of characters by a fixed value and transmit the result as an additional datum appended to the data stream. Again, however, this technique is very infrequently used in the laboratory.

Finally, in some cases where there are multiple units connected on the same serial network, additional information may be sent, such as the destination of the message, character count, and other control information.

7.2 RS-232-C PROTOCOL

Although there are many other choices available, there are basically only two commonly-used serial protocols in the laboratory with the IBM PC: RS-232-C and RS-449. A third choice, **current loop** is only very infrequently used and hence we shall not consider it here.

RS-232-C is by far the most common, in part because it is the older. The RS stands for Recommended Standard of the Electronic Industries Association. RS-232-C can be a very confusing standard to use. This is in part because the standard is not very rigid, and in part because RS-232 has been utilized for purposes never intended by those who set the standard.

The only way to understand RS-232 is to realize that the standard was designed to allow computers to be interfaced to terminals via commercial telephone lines. The device used to translate signals between the telephone lines and the serial line coming from the computer is a **modem** (contraction of *mo*dulator- *dem*odulator). Almost all of the terminology associated with RS-232 reflects this relationship between the computer and the modem.

Since modems are used with the IBM PC and other modern microcomputers to permit using a PC as a remote terminal to a mainframe computer, let's look at the electrical connections used. Figure 7.1 shows a typical arrangement, with a terminal and modem at one site connected by telephone wire to a modem and computer at another site. The RS-232 standard is thus built around the idea of communications between a piece of data communications equipment (**DCE**), namely the modem, and a piece of data terminal equipment (**DTE**), namely the computer or terminal. The serial connection at each site is essentially the same; in Figure 7.2 we show the details of this connection. Other wires than those shown in Figure 7.2 are defined in the RS-232-C standard, but are only rarely used. Notice that the naming conventions are from the point of view of the computer or terminal rather than the modem.

Two signal wires, **transmit** and **receive**, are used to allow full duplex transmissions. The **signal ground** wire carries the signal return. All of the other wires are used to assess the state of the telephone lines.

If the telephone lines are half-duplex, then signals can be sent in one direction only. In this case, the **request to send** line is used by the computer (or terminal) to indicate that it has a character ready to send. After the modem has changed from receive to send mode, it uses the **clear to send** line to indicate that it is ready to send the character. If the full duplex mode is used, then the clear to send and request to send lines are kept in a constant state that allows either transmission or receipt of characters from the modem.

Two signals indicate the fact that the terminal and the modem are powered on and ready to use; these are the **data terminal ready** and the **data set ready** signals, respectively. The **carrier detect** signal indicates that the remote connection is currently active (across the telephone wire); the **ring detector** indicates that a ringing tone is present on the telephone line (i.e., that an attempt is being made to establish a telephone connection between the terminal and the computer).

7.2.1 RS-232-C without a Modem

Unfortunately for us, the RS-232-C interface has been adapted by a number of manufacturers to a variety of non-modem applications. For example, it is common to use RS-232 connections between a microcomputer and a plotter or printer, or between two computers, or between an instrument and a computer. In these **null-modem** cases, which are of more interest to us in the context of this textbook than the uses of a modem, many complications arise.

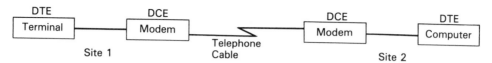

Figure   7.1.   Typical Remote Terminal System. In its original use, RS-232 involves a remote terminal that communicated across a telephone wire to a computer system. Modems are used to perform the conversion between the digital computer signals and the signals carried on the telephone lines. In this system, the terminal and computer were both DTE type equipment, while the modem was DCE.

The simplest variation of RS-232 is shown in Figure 7.3. In this case, we simply use the minimum set of lines (transmit, receive and signal ground) to connect two devices (e.g., a computer and a printer). Remembering that all of the other lines are related to the status of the modem and telephone lines anyway, this arrangement is simple and cheap.

However, as shown in Figure 7.4, another equally simple arrangement is sometimes required. In this arrangement, the transmit and receive wires are crossed between the two devices being connected. In a situation of the type pictured in Figure 7.4, it is possible that we may be trying to communicate between two devices, both of which have been defined by their manufacturers as DTE type, or both DCE type, or one may be DTE and one DCE. Hence, we may need to exchange two of the wires at one end if both devices are of the same type. If other wires than the three basic ones (2, 3 and 7) are used, then these are often wired as shown in Figure 7.5. This permits some or all of the modem signal wires (often incorrectly called "handshaking" wires) to be configured to allow continuous signaling; i.e., so that there is no waiting for the signals normally generated by a modem.

To further complicate matters, there is no such thing as a standard RS-232 connector. The most commonly used connector is a 25-pin D-shell connector, but this may be either male or female. Other types of connectors (e.g., a 9-pin connector on the PC/AT) may be used.

What may really make your job difficult, however, is that many manufacturers have chosen to use lines such as data set ready and clear to send with devices not containing a modem. Hence, these lines serve other purposes, which may be entirely unique to that device (e.g., one manufacturer uses some of the lines to determine the transmission rate). Hence, the process of connecting together two RS-232 devices often means spending considerable time reading the instruction manuals of the two devices to determine exactly which lines are tied together, which sex connectors are required, etc. If you plan to do this more than once or twice in your lifetime, you may wish to buy a simple device, often called a **break-out box** to allow easy reconfiguration of the lines during installation. Most break-out boxes have LED's to indicate

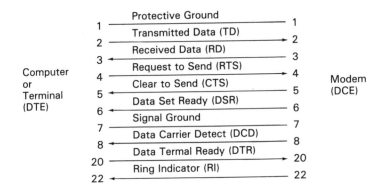

Figure   7.2.   RS-232-C Connections.   The wires used in most common RS-232 connections involve the wires shown.   Additional wires are sometimes used; however, for most laboratory applications, no additional wires are required.

which lines are asserted, which may help you considerably. (For example, you can determine whether the device is DTE or DCE; devices asserting the request to send line, for example, are likely to be DTE, whereas devices asserting the data set ready line are likely to be DCE.)

Several other features of the RS-232-C standard are of practical interest. For example, RS-232 uses start and stop bits to send timing information, as shown in Figure 7.6. It also allows, but does not require, parity bits. Hence, the normal transmission of information includes one start bit followed by either 7 or 8 data bits and none or one parity bit (allowing no parity or even or odd parity, respectively). One or two stop bits are sent after the data and parity bits. Common data transmission rates are 110 Hz (or 110 **baud**, as it is often called), 300 Hz, 600 Hz, 1.2 kHz, 2.4 kHz, 4.8 kHz, 9.6 kHz and 19.2 kHz. Because each byte of data requires start and stop bits, and there are usually 8 bits for data plus parity, the rate of data transfer can be divided by 10 to calculate the approximate rate in bytes per second.

Another major practical consideration is the allowable distance between connection points. Generally, this is a function of the electrical and magnetic noise in the environment, the type of cabling used, proper grounding, the transmission rate, and the error rate that is acceptable in your environment. Hence, it is difficult to give exact rules. However, RS-232 is certainly satisfactory for within-lab transmissions, less satisfactory for transmissions over substantial distances within the same building, and rarely satisfactory between buildings. The easiest test is simply to send data at or above the transmission rate you desire and look for errors during transmission. If you get several errors, then either decrease the rate or shorten the cables or use cable with better shielding.

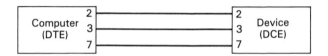

Figure   7.3.   Minimum RS-232 Connections.   A simple three-wire connection is all that is required for some laboratory applications. If one of the devices is DTE and one is DCE, then the wiring is as shown here.

7.3 RS-449 STANDARD

The major disadvantage of the RS-232 standard is its low speed (9.6 or 19.2 kHz maximum in most implementations). A second serial-interface standard has been defined that allows rates up to 2 MHz. This is the RS-449 standard, which is in essence the RS-232 interface expanded to provide better isolation of the signal from external noise. The RS-449 standard requires the use of more wires, and uses a 37-pin connector (plus nine-pin connectors for each additional "secondary" channel).

Two methods of electrically connecting devices are permitted under the RS-449 standard. The first, referred to as RS-423, is effectively the same as RS-232. Hence, in this mode, transmission rates are limited to a maximum of 20 kHz, but it becomes possible to connect two devices, one of which is RS-232-C and one of which is RS-423, on the same cable.

In the second mode, additional connections are used to define a standard referred to as RS-422. In this "balanced" mode, the devices can send data at much higher rates than 20 kHz, but both devices must be RS-422.

With its additional wires and more complex circuits, the RS-449 is more expensive than RS-232-C. Hence, it has not been widely used. In any case, those with needs for high-speed transmissions are increasingly moving toward local area networks (Chapter 13) that have yet higher transmission rates and that can be used to connect many devices on a single cable. Therefore, we will not discuss RS-449 further.

7.4 PROGRAMMING SERIAL INTERFACES

If you have entirely figured out the electrical connections to the RS-232 interface, then usually the programming of the interface is relatively simple. Because a serial interface is included on almost all PCs, most programming languages include functions for accessing the serial interface easily. BASIC is no exception to this rule. The IBM PC RS-232 interface can be extensively programmed from BASIC. Before we do so, however, we need to gather the following information about the device that we are connecting to the IBM PC:

Figure   7.4.   Serial Communications between Two DTE Devices.   If both devices are DTE, or both are DCE, then wires 2 and 3 must be reversed from the normal configuration.

- Data transmission (baud) rate.
- Parity (even, odd or none).
- Number of data bits (7 or 8).
- Number of stop bits (1 or 2).
- Whether the device should be sent a carriage return character at the end of each transmission.
- Whether the device transmits a signal indicating its data buffer, if any, is full.
- Whether it requires the PC to ignore some interface lines.

Most of these can be determined from the manufacturer's literature on the device being interfaced and are set during the OPEN command described below. However, some devices, particularly mechanical devices such as plotters and printers, have buffers (memory) that allow them to store data if it is being sent faster than the device can process it. Many systems with buffers use some sort of protocol to signal when the buffer is in danger of overflowing, as discussed below. This requires more extensive programming.

7.4.1 Simple Serial Program

A simple program for the RS-232-C interface of the IBM PC is shown in Program 7.1. This program uses BASIC's OPEN COM command, which has the format:

$$\text{OPEN COMn:b,p,d,s,o AS \#n LEN=e}$$

In this expression, **n** is an integer (usually 1 or 2) that indicates the number of the serial port (RS-232 interface) being utilized; this number is typically determined by a jumper on the interface card. The transmission rate is **b**, which can range from 110 to 9600 bits/second. The parity is given at position **p**, and can be N (none), E (even) or O (odd). Two other much less frequently used parity options are S (space) and M (mark), in which the parity bit is used but always set to a 0 or 1, respectively. The next position, **d**, gives the number of data bits, which is normally either 7 or 8 but which can also be 5 or 6; **s** gives the number of stop bits, which is usually 1 but

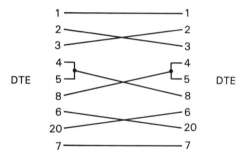

Figure   7.5.   Serial Devices with No Modem.   Two devices connected together without a modem, and where both are DTE or both are DCE, may require some or all of the modem signal lines to be connected as shown.

may be 2. One or more options may be listed at position **o**, separated by comma; these include: **RS, CS[n], DS[n], CD[n], LF**, and **PE**.

The first four options control use of the Request to Send (RTS), Clear to Send (CTS), Data Set Ready (DSR), and Carrier Detect (CD) lines, respectively. The default is for each of these lines to be used (as they would with a modem, for example). Specifying any one of these four options with **n** not given, or with **n** set to zero turns off checking of that line (or use of that line, for **RS**). If **n** is specified, it gives the time in milliseconds that the line is checked before a timeout error is generated. The defaults are CS1000, DS1000, and CD0; if RS is specified, then CS0 is the default. This means that normally, the CTS and DSR lines are checked and an error generated in 1 second if the lines are not sensed; the carrier detect line is normally ignored.

The **LF** option allows you to automatically insert the linefeed character (ASCII code 10, 0A hex) after any output line containing a carriage return character (ASCII code 12, 0C hex). Similarly, on input, it terminates the read operation whenever a carriage return character is encountered. This option is useful primarily for use with printers, but some other types of devices also require linefeed characters to be output to them (e.g., some plotters).

The PE option enables parity checking on input; the default is no parity checking. You can use this option if you wish to check the parity on data being received by the PC.

The value of **n** gives the file number (e.g., 1) by which the serial port will be known, in a fashion very similar to that of disk files (Chapter 10). If more than one serial port is open at a given time, then each port must be referenced by a different file number.

Finally, **e** is the buffer length, i.e., the maximum number of characters that the PC can store at one time while awaiting the processing of the characters by your program (see below).

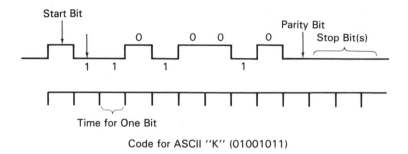

Code for ASCII "K" (01001011)

Figure   7.6.   Timing Diagram for RS-232-C.   Transmissions sent using RS-232 require stop and start bits to signal the beginning and end of the transmission, respectively. In addition, there may be a parity bit to allow parity checking.

To send information to the serial port, we use the PRINT command, again following the syntax used with the PRINT command for disk files:

PRINT #n,"string"

where **n** is the file number used in the OPEN COM statement and **"string"** is the string of ASCII characters to be sent. Note that we can use the STR$ function in BASICA to build strings containing non-printing characters. For example, if we wish to send word HELLO followed by the carriage-return and line-feed characters to a device using the serial port, we could use the command:

PRINT #1,"HELLO"+STR$(13)+STR$(10)

where 13 and 10 are the ASCII codes for carriage-return and line-feed, respectively, and the plus sign is used to concatenate the two strings.

In a similar fashion, we can input data from a serial device using the INPUT command:

INPUT #n,a$

where **n** is the file number using in the OPEN COM command and **a$** is a string variable.

Because RS-232-C is an asynchronous protocol, we need to be able to find out when there is data to be read from the serial interface. To do this, we can use the **LOC** function of BASICA. This function tests for data that has been sent to the PC. To avoid the possibility of the device sending data to the PC before we have read the previous data, BASICA uses a **buffer**. All data are automatically moved from the serial interface to a region in memory (the buffer) as soon as they are received. The

```
10 REM Program Serial
20 REM LABORATORY AUTOMATION USING THE IBM PC
30 REM Program to send data to a simple serial interface
40 REM ***************************************************************
50    SCREEN 0:CLS:WIDTH 80
60    INPUT "Enter serial port number (1 or 2) ",PORT$
70    INPUT "Enter baud rate ",BAUD$
80    INPUT "Enter parity (E, O or N) ",PARITY$
90    INPUT "Enter number of data bits (7 or 8) ",DATABIT$
100   INPUT "Enter number of stop bits (1 or 2) ",STOPBIT$
110   INPUT "Enter options (CS, RS, etc.) ",OPTIONS$
120   ARGUMENT$="com" + PORT$ + ":"+BAUD$+","+PARITY$+","+DATABIT$+","+STOPBIT$
130   IF OPTIONS$<>"" THEN ARGUMENT$=ARGUMENT$+","+OPTIONS$
140   PRINT "opening port using the argument ";ARGUMENT$
150   OPEN ARGUMENT$ AS #1
160   INPUT "Do you wish to use the device without carriage returns? ",Y$
170   IF Y$="y" OR Y$="Y" THEN WIDTH #1,255
180   PRINT "Type the <*> key to quit "
190   Y$=""
200   WHILE Y$ <> "*"
210      WHILE LOC(1) > 1                   'check if input data
220            A$=INPUT$(LOC(1),#1)         'get data
230            PRINT A$
240      WEND
250      IF Y$<> "" THEN PRINT #1,Y$
260      Y$=INKEY$
270   WEND
280 END
```

Program   7.1.   Simple Serial Interface

INPUT statement is then used to move data from the buffer into our program's data space. The size of the buffer is determined by the /C option used when starting BASICA; in other languages, it may be determined within the equivalent of the OPEN statement. Particularly for high data rates (e.g., 9600 baud), you may need to increase the buffer size to prevent overflow.

Hence, the LOC function simply tests the number of characters in the buffer. When there is one or more characters in the buffer, we read them into our program using the INPUT$ statement. Hence, a typical program segment might be:

$$\text{WHILE (LOC(1) > 1)}$$
$$\text{B\$ = INPUT\$(LOC(1),\#1)}$$
$$\text{PRINT B\$}$$
$$\text{WEND}$$

which would store the data in the string B$ as it arrives and print it on the screen. In this example, we have simply printed the characters. In other applications, you may need to test for a certain character (e.g., a carriage-return character) or combination of characters to determine when sufficient characters have been received.

As a very simple example, in Program 7.1, we simply ask the user to provide the parameters for the OPEN statement (lines 60-140), and open the serial port (line 150). We then check if the device requires that carriage returns not be sent between data (e.g., many plotters), in which case we use the WIDTH statement to indicate this fact (lines 160-170). We then use a program segment (lines 190-270) which simultaneously prints incoming data on the screen and sends whatever data the user types on the keyboard. It continues to do this until the user types an asterisk on the keyboard.

If the data you are receiving are numbers and you wish to store them or use them in mathematical functions, you may need to convert the ASCII string to an integer or decimal value using the VAL function:

$$VALUE! = VAL(B\$)$$

For example, if you receive the string "32.66", the VAL function would store 32.66 as VALUE!. Alternatively, if you are receiving binary data, rather than ASCII characters, (normally this requires requesting 8 data bits in the OPEN COM command) you may need to convert them from string to binary data. This might be done with the MID$ and CVI commands:

```
J%=0
FOR I%=1 TO 200 STEP 2
    J%=J%+1
    A%(J%)=CVI(MID$(B$,I%,2))
NEXT I%
```

This sequence looks at each pair of bytes in a string and converts them to an integer value, much as we do with direct access disk file input (Chapter 10). Similar functions, CVS and CVD, can be used to read binary strings representing single or double precision floating point numbers, respectively.

7.4.2 Buffers

As we have mentioned, many devices such as plotters have buffers (small amounts of memory) into which they immediately load all data, and then subsequently remove the data for processing. This is because these devices are often mechanical devices that cannot process (e.g., print or plot) as fast as they receive the data are received. Because the buffers are of limited size, the devices need to be able to stop data from being sent whenever the buffer is almost full. Hence, such devices usually use either one of the modem status lines or, preferably, a software handshaking protocol such as the XON/XOFF protocol. In the XON/XOFF protocol, for example, the device sends an XOFF signal (ASCII code 19) when the buffer is almost full, and an XON signal (ASCII code 17) when it is again ready to receive data. A short program to do this is illustrated in Program 7.2.

This program is essentially similar to Program 7.1. However, in line 100, we define a new variable, XFLAG. This variable is set to 1 whenever we first detect that the buffer is getting too full. Thus, in line 120, we test to find out the number of

```
10 REM PROGRAM XON/XOFF
20 REM LABORATORY AUTOMATION USING THE IBM PC
30 REM Program to send data on a serial interface with XON/XOFF protocol
40 REM ***********************************************************************
50 SCREEN 0:CLS:WIDTH 80
60 REM                   Main program
70    GOSUB 310                          'get user input
80    GOSUB 410                          'open COM port
90    PRINT "Type the <*> key to quit "
100   Y$="":XFLAG=0
110   WHILE Y$<>"*"
120       WHILE LOC(1) > 0               'check if input data
130           'Input buffer overflow imminent?  If so, send XOFF
140           IF (XFLAG=0) AND (LOC(1)>128) THEN XFLAG=1:PRINT #1,XOFF$;
150           'If have sent XOFF, see if time to send XON
160           IF (XFLAG=1) AND (LOC(1)<=128) THEN XFLAG=0:PRINT #1,XON$;
170           A$=INPUT$(1,#1)            'get one character input
180           IF A$=XOFF$ THEN OFFFLAG=1    'test if XOFF
190           IF A$=XON$ THEN OFFFLAG=0     'test if XON
200           PRINT A$;                  'print on screen
210       WEND
220       WHILE Y$<>"" AND OFFFLAG <> 1 'check if user wants to send data
230           PRINT #1,Y$;              'send it
240           Y$=""
250       WEND
260       Y$=INKEY$                     'get new keyboard input, if any
270   WEND
280 END
290 REM ****************************
300 REM Get user input
310   INPUT "Enter serial port number (1 or 2) ",PORT$
320   INPUT "Enter baud rate ",BAUD$
330   INPUT "Enter parity (E, O or N) ",PARITY$
340   INPUT "Enter number of data bits (7 or 8) ",DATABIT$
350   INPUT "Enter number of stop bits (1 or 2) ",STOPBIT$
360   INPUT "Enter options (CS, RS, etc.) ",OPTIONS$
370   INPUT "Do you wish to use the device without carriage returns?",CR$
380 RETURN
390 REM ****************************
400 REM Open COM port
410   ARGUMENT$="COM"+PORT$+":"+BAUD$+","+PARITY$+","+DATABIT$+","+STOPBIT$
420   IF OPTIONS$<>"" THEN ARGUMENT$=ARGUMENT$+","+OPTIONS
430   PRINT "Opening port using the argument ";ARGUMENT$
440   OPEN ARGUMENT$ AS #1
450   IF CR$="y" OR CR$="Y" THEN WIDTH #1,255
460 RETURN
```

Program   7.2.   Serial Interface using XON/XOFF.

characters in the input buffer (i.e., characters received but not processed by our program).   When this number exceeds 128 (line 140), we look at XFLAG.   If XFLAG is zero (meaning that no XOFF has already been sent), then we send XOFF and change XFLAG to a 1; hence, no more data will be sent to us until we reduce the number of characters in the buffer to less than 128.   Similarly, in line 160, we look to see if XON needs to be sent; it does if XOFF has been sent (XFLAG=1) and if there are less than 128 characters in the buffer.

We also must process XON and XOFF messages received from the sending device.   Again, we use a flag; if OFFFLAG is set to one (indicating receipt of an

XOFF message), then we do not send any characters until an XON message is received (lines 220-250).

7.4.3 Plotter Program

We are now ready to look at a typical program for a serial device. This is a program that controls a plotter with a serial interface that uses a RS-232-C interface. Notice that the program uses the XON/XOFF protocol when sending data only, because the plotter, sends almost nothing back to the computer. Also, the program converts all numerical data to ASCII before sending them to the plotter. We have extracted this program from a larger program, similar to those in Chapter 10, that performs plotting. The program is shown in Program 7.3. In essence, this program reads in a data file (lines 180-280), and then initializes the plotter by setting the starting location (origin) at a point 1.5 inches in each direction from the lower left-hand corner of the piece of paper (lines 310-390). It then scales the data to fit the graph, draws a box around the region that will contain the graph, and labels the axes. However, it is important to note that we have tried to put the plotter-specific code in separate subroutines, which can then be modified for the plotter of your choice.

7.4.4 Hints on Using Serial Interfaces

Before we move to our next topic (timers and counters), we should give you several hints about serial interfaces. First, there are many programs available commercially that may help you if you are planning to use serial interfaces. For example, some serial programs allow you to send the data to a buffer in the PC rather than directly to a device. The data are then sent form the buffer to the device only as the device is ready for them. This practice, referred to as **spooling** the data, is especially useful for output devices such as printers and plotters, because you can perform other tasks, or even other programs, while the data are being sent to the device from the buffer. These programs often also can be configured to handle XON/XOFF or other protocols.

Second, you will need to be particularly careful about using serial interfaces over long distances at high data rates. Most users of serial interfaces eventually try this, only to find that they receive or send bad data randomly interspersed among their good data. Although using the noise-reduction techniques discussed in Chapter 4 can help extend the distance data can be sent, you may find it much better to consider a local area network, as described in Chapter 13, which is more expensive but which has been designed with longer distances in mind.

Third, as you may have gathered, we basically don't prefer serial interfaces. We think that where possible, you will find it much more satisfactory to use GPIB than serial interfaces. In general, we find GPIB to be easier to install, program, and maintain than serial, plus it supports much higher data rates and multiple devices on a single GPIB adapter. However, there are still many commercially-available scien-

```
10  REM PROGRAM SERIAL_PLOTTER_DRIVER
20  REM LABORATORY AUTOMATION USING THE IBM PC
30  REM Program to plot data using a Houston Instruments DMP-7 plotter
40  REM    as a serial plotter, with XON/XOFF
50  REM ********************************************************************
60  DIM COMP$(10),A%(1200),MODE$(2),PEN2$(2)
70  SCREEN 0:CLS:WIDTH 80
80  REM               Main program
90     GOSUB 180                        'open data file
100    GOSUB 310                        'set up for plotter use
110    GOSUB 420                        'scale data to fit graph
120    GOSUB 1430                       'plot the graph
130    INPUT "Do you want to label the axes?";Y$
140    IF Y$="Y" OR Y$ = "y" THEN GOSUB 570
150 END
160 REM ***************************
170 REM Open data file and read data
180    PRINT "FILES CURRENTLY ON DISK:"
190    FILES "*.*"
200    PRINT:INPUT "WHAT IS THE NAME OF THE FILE TO BE ANALYZED";FILN$
210    OPEN FILN$ FOR INPUT AS #1
220    INPUT #1,N%                      'get number of points in data file
230    FOR I=1 TO N%                    'input data points
240        INPUT #1,A%(I)
250        Y(I)=A%(I)
260    NEXT I
270    CLOSE #1
280 RETURN
290 REM ***************************
300 REM Set up for plotter use
310    GOSUB 900                        'start plotter
320    X0%=PTPERINCH*1.5                'define origin at 1.5 inches from margin
330    Y0%=PTPERINCH*1.5
340    GOSUB 1670
350    INPUT "Enter the height and width of plot.",H,W
360    H%=PTPERINCH*H                   'plotter is ptperinch points per inch
370    W%=PTPERINCH*W
380    XFACT=W%/(N%-1)                  'compute scaling factor in X direction
390 RETURN
400 REM ***************************
410 REM Find min, max data values so can scale data
420    YMIN=A%(1)                       'find min, max Y values
430    YMAX=A%(1)
440    FOR I=1 TO N%
450        IF A%(I)< YMIN THEN YMIN=A%(I) ELSE IF A%(I) > YMAX THEN YMAX=A%(I)
460    NEXT I
470    YFACT=H%/(YMAX-YMIN)             'factor for scaling plot in Y direction
480 RETURN
490 REM ***************************
500 REM Draw box around graph
510    X0%=W%
520    Y0%=H%
530    GOSUB 1140
540 RETURN
550 REM ***************************
560 REM Label axes
570    INPUT "Enter x-axis label ";MESSG$   'label the x axis
580    X0%=W%/2                         'center label
590    Y0%=-.75 * PTPERINCH
600    SIZE$=MEDIUM$                    'use lettering size 2
610    ORIEN$=UPRIGHT$                  'no rotation of lettering
620    GOSUB 1580
630    INPUT "Enter y-axis label ";MESSG$   'label the y axis
640    X0%=-.75 * PTPERINCH             'center label
650    Y0%=H%/2
660    SIZE$=MEDIUM$                    'use lettering size 2
670    ORIEN$=TILT270$                  'rotate lettering by 270 degrees
```

Program 7.3 (Part 1 of 3). Serial Plotter Driver

```
680    GOSUB 1580
690    CLOSE #3
700 RETURN
710 '
720 REM SUBROUTINES FOR USE WITH HOUSTON DMP-7 PLOTTER.
730 '
740 REM ****************************
750 REM Write string of characters to plotter
760    E$="_ "                          'end of string code
770    PEN2%=0                          'lift pen to move to starting location
780    MODE%=0
790    GOSUB 850                        'position pen
800    S$="S"+ORIEN$+SIZE$+" "          'set size, orientation
810    PRINT #3,S$;MESSG$;E$;           'plot string
820 RETURN
830 REM ****************************
840 REM Position pen at x0%,y0% using absolute position mode
850    PRINT #3, MODE$(MODE%) + PEN2$(PEN2%);   'lift pen
860    GOSUB 1300                       'go to desired location
870 RETURN
880 REM ****************************
890 REM Start plotter
900    MODE$(0)="A "                    'absolute address mode
910    MODE$(1)="R "                    'relative address mode
920    PEN2$(0)="U "                    'pen up
930    PEN2$(1)="D "                    'pen down
940    PTPERINCH=200                    'number of plotter points / inch of paper
950    MEDIUM$ ="2"                     'medium size print
960    UPRIGHT$ = "1"                   'upright text orientation
970    TILT270$ = "4"                   'text rotated by 270 degrees
980    OPEN "com1:9600,E,7,1,RS" AS #3  'plotter is connected to com1
990    WIDTH #3,255                     'set so not send carriage returns
1000   PRINT #3, "H;: ;: H A O U ";     'wake up plotter and init pen position
1010 RETURN
1020 REM ****************************
1030 REM Draw line between x0%,y0% and x1%,y1%; note x0%,y0%=x1%,y1% upon return
1040    MODE%=0                         'lift pen and move to initial location
1050    PEN2%=0
1060    GOSUB 850
1070    PEN2%=1                         'put pen down and draw line
1080    X0%=X1%
1090    Y0%=Y1%
1100    GOSUB 850
1110 RETURN
1120 REM ****************************
1130 REM Put box from 0,0 to x0%,y0% (as opposite corners)
1140    TX%=X0%
1150    TY%=Y0%
1160    X0%=0
1170    Y0%=0
1180    X1%=TX%
1190    Y1%=0
1200    GOSUB 1040
1210    Y1%=TY%
1220    GOSUB 1040
1230    X1%=0
1240    GOSUB 1040
1250    Y1%=0
1260    GOSUB 1040
1270 RETURN
1280 REM ****************************
1290 REM Encode x0%,y0% and send to plotter
1300    TEMP$=STR$(X0%)                 'get location as string
1310    GOSUB 1740                      'remove blanks from string
1320    A1$=TEMP$+","                   'output string as "X%,Y% "
1330    TEMP$=STR$(Y0%)
```

Program 7.3 (Part 2 of 3). Serial Plotter Driver

```
1340    GOSUB 1740
1350    A1$=A1$+TEMP$+" "
1360    PRINT #3,A1$;
1370    IF LOC(3)=0 THEN RETURN          'check if plotter has returned XOFF
1380    IF LOC(3)=0 THEN 1380 ELSE C$=INPUT$(1,#3)
1390    PRINT "waiting for plotter":IF C$<> CHR$(17) THEN 1380    'wait for XON
1400 RETURN
1410 REM ***************************
1420 REM Plot an array, a%. Xfact is expansion factor in x, yfact is factor in y
1430    X0%=0                            'start at 0,y0%
1440    Y0%=(A%(1)-YMIN)*YFACT
1450    MODE%=0                          'lift pen to go there
1460    PEN2%=0
1470    GOSUB 850                        'position pen
1480    PRINT #3, "D ";                  'put pen down
1490    FOR I=2 TO N%                    'plot rest of points
1500        X0%=(I-1)*XFACT
1510        Y0%=(A%(I)-YMIN)*YFACT
1520        GOSUB 1300
1530    NEXT I
1540    PRINT #3,"U ";                   'lift pen
1550 RETURN
1560 REM ***************************
1570 REM Print string centered at x0%,y0%
1580    DIFF%=PTPERINCH*.06*LEN(MESSG$)*(2^(VAL(SIZE$)-1))/2  'length of string/2
1590    IF VAL(ORIEN$)=1 THEN X0%=X0%-DIFF%    'center string in approp. direct.
1600    IF VAL(ORIEN$)=2 THEN X0%=X0%+DIFF%
1610    IF VAL(ORIEN$)=3 THEN Y0%=Y0%+DIFF%
1620    IF VAL(ORIEN$)=4 THEN Y0%=Y0%-DIFF%
1630    GOSUB 760
1640 RETURN
1650 REM ***************************
1660 REM Put origin at x0,y0%
1670    PEN2%=0                          'lift pen
1680    MODE%=0
1690    GOSUB 850                        'position pen at x0%,y0%
1700    PRINT #3, "O ";                  'make current position the origin
1710 RETURN
1720 REM ***************************
1730 REM Remove leading blank from string (temp$)
1740    IF LEFT$(TEMP$,1)<>" " THEN RETURN
1750    LENGTH%=LEN(TEMP$)
1760    TEMP$=RIGHT$(TEMP$,LENGTH%-1)
1770 RETURN
```

Program   7.3 (Part 3 of 3).   Serial Plotter Driver

tific instruments where serial interfaces are standard, and you will undoubtedly need
to be able to program them at least occasionally.

EXERCISES

1.   Use a breakout box to look at the output from the serial port of the PC. Is it a
      DCE or DTE device? Which would you expect it to be, based upon the dis-
      cussion in this chapter?

2.   Program 7.1 does not display the data that are being sent by the user to the
      device. Modify the program to display the incoming data on one part of the

screen and to print the outgoing data on a separate part of the screen. As a more sophisticated programming task, make provisions for scrolling each section of the screen when it fills.

3. Write a program that will input serial data from a device in your own laboratory and display it on the monitor of your PC. Be sure to read the manual for the device to ascertain all of the proper operating parameters before you begin.

4. Some companies market laboratory interface boards (with A/D, D/A, timer, digital I/O) that are in a separate box from the PC, but which communicate with the PC using a serial interface. What are the advantages and disadvantages of such an approach? (Hint: include in your considerations the speed of the devices and the degree of noise encountered both inside the PC chassis and external to it.)

5. Some companies market devices that will convert GPIB to serial, and vice versa. What would be the advantages and disadvantages of such devices?

8

TIMERS AND COUNTERS

One of the characteristics of almost all data collection programs in a laboratory is that we would like the data collection to occur at specific time intervals. For example, we may wish to collect data from an instrument twice per second. How can we do this?

The most obvious way is to use the clock in the IBM PC. For example, we can use the TIMER function in BASICA, as shown in Program 8.1. With such a program, we can time some event such as the collecting of data with an A/D. Note that you can substitute whatever event you wish in place of line 140 in the program.

However, when we run Program 8.1 as shown, we get a series of times like the following:

Event	Time(Sec)
1	1.039063
2	2.03125
3	3.019531
4	4.011719
5	5.050781
6	6.039063
7	7.03125
8	8.019531
9	9
10	10.05078

137

```
10 REM PROGRAM SYSTEM_CLOCK
20 REM LABORATORY AUTOMATION USING THE IBM PC
30 REM Program to demonstrate the use of TIMER function in BASICA
40 REM *********************************************************************
50 INTERVAL!=1!                          'set interval between events in seconds
60 TOTALPOINTS=10                        'total number of events
70 POINTNUMBER=1                         'starting event number
80 STARTTIME! = TIMER                    'get initial time
90 DELTA! = 0
100    WHILE POINTNUMBER <= TOTALPOINTS  'test for number of events
110        WHILE DELTA! < POINTNUMBER * INTERVAL!  'test for time interval
120            DELTA! = TIMER-STARTTIME!
130        WEND
140        PRINT POINTNUMBER,DELTA!          'our "event"
150        POINTNUMBER = POINTNUMBER + 1
160    WEND
170 END
```

Program 8.1. Timing Using the System Clock.

What happened? Why are not the times all exact seconds? The answer is that the clock accessible to BASIC normally "ticks" 18.2 times per second. This means that the clock ticks once every 0.055 seconds. Hence, we cannot hope to measure events with precision better than 0.055 seconds with Program 8.1.

Although it is possible to change this clock rate, it is much better to avoid modifying the system clock. Hence, almost all general-purpose data acquisition boards now contain a clock. To obtain the necessary precision, these clocks typically have very high rates, usually 100 kHz or more. In many cases, these clocks can also be used to count events as well as time them. It is also usually possible to use the timer to initiate events through direct hardware connections to another section of the board (e.g., using the timer output to initiate A/D conversions). We shall see examples of the more sophisticated uses in Chapters 9, 14 and 15.

8.1 THEORY

The operation of most clock/timers is relatively simple, as shown in Figure 8.1. Although there are many variations of this scheme, the essential process is the counting of the output from an oscillator. Typically, the oscillator operates at a very high frequency, usually in the MHz range. The counter simply records the number of oscillations. There is usually a programmable frequency divider, so that other frequencies than the basic one can be selected, and a load register that is used to hold the initial count for the counter. For example, a 1 MHz oscillator may be used, but with a divider provided to allow rates slower than 1 MHz by a factor of 10, 100, or 1000. The output of the divider then goes to the counter, which may, depending upon the design, count either up, down or in either direction. Most common counters are at least 16-bit counters.

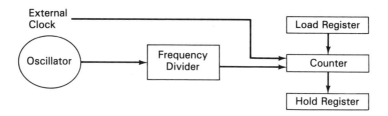

Figure 8.1.   Operation of Timer/Counter. The basic function of the timer/counter is to
count the output from an oscillator. The initial count is obtained from a load
register. The current counter value is accessed from a count register, and in
some systems may be moved to a hold register. On-board control logic provides
frequency division and controls the operation of the timer/counter.

For example, suppose we wish to use the clock to trigger an event every 100
microseconds, i.e. at a rate of 10 kHz, and that we use a down-counter. Assuming an
oscillator frequency of 1 MHz, we would then set the load register to an initial value
of 100. The counting process is initiated by moving the value from the load register
into the counter. After 100 microseconds of counting, the counter would read 0. By
monitoring the counter and waiting until it reaches zero, we could then time an event
to within the precision of the oscillator, namely one microsecond.

Often, the clock can be used to time events both in hardware and in software.
In the former mode, the clock usually generates a signal whenever it reaches zero or
some other predefined count. In some systems, it is even possible to generate a pulse
every time the clock changes. In the latter mode, the user can read the clock in software, and perform some operation whenever the clock reaches zero. Obviously, the
hardware mode is more accurate, because of the delay time in reading the clock using
software.

Although the basic process is simple, there are many complications. For
example, it is undesirable to have the counter change as we read it. This problem is
often avoided by adding another register, typically called the hold register, to which
the counter value is moved upon command. Then the hold register can be read while
the counter continues the counting process. This feature is particularly valuable
when the clock rate of the timer/counter is as fast or faster than the instruction execution speed of the computer.

Another common complication is that often a 16-bit counter is insufficient, particularly when trying to count relatively long time periods with high precision. The
usual method for overcoming this problem is to chain together two or more counters.
Typically, in this mode of operation, whenever the first counter counts down to zero,
the second counter is decremented by one. The second counter is then the one read
by the computer or other device. Each counter typically has its own load and hold
registers.

A third complication is the need for repetitive timing. Usually the clock is being used to time an event that occurs many times. It is inefficient and inaccurate to have to reinitiate the counting process after each event. Ideally, we would like to have the clock keep on counting and just have it notify us whenever the desired time interval is past. This is usually accomplished by utilizing load registers to automatically reinitialize the count. Thus, whenever the counter reaches zero (or some other predefined number), an electrical signal is sent by the board to the user, and then the counter is reloaded with the initial count. In this continuous mode of operation, the counting process continues until explicitly turned off by a command from the computer.

Another option frequently available on most timer/counter systems is the ability to use an external frequency generator instead of the internal oscillator. This allows the timer to be synchronized with some external event, but more importantly allows the timer/counter to be used as a counter. In this mode, we simply monitor the value of the counter to count the number of external events that have occurred, where each external event is measured as a pulse of the appropriate size and duration on the input line to the counter.

In order to offer all of these options, typical timer/counter boards are designed to allow a computer to send commands to them. This is usually done using a command register, which in turn controls the on-board timer logic.

8.2 TIMING DIAGRAMS

Although the timer/counter is inherently a simple device, the existence of many options and the penchant of engineers for jargon often combine to result in a users' manual for the timer/counter that is nearly indecipherable. To help you understand these manuals, let's take a look at several possible clock/timer configurations. We'll do this in part by using **timing diagrams**, or diagrams showing the sequence of events. In the following discussion, the timing diagrams have lines illustrating the oscillator signal and the output from the timer/counter. The diagrams given are for the IBM Data Acquisition and Control Adapter (DACA), but are very similar for most timer/counter systems.

8.2.1 Rate Generator

This is the mode most frequently used to repetitively trigger some event. In this mode, a pulse is generated at regular intervals; the interval is determined by the value in the count register. As shown in Figure 8.2, the frequency of the oscillator used as the frequency source is divided by the value in the count register; e.g., if the oscillator is 1.00 MHz and the count is 4, then the output pulse generated will be at 0.25 MHz. Note that the output typically changes for only one cycle of the oscillator.

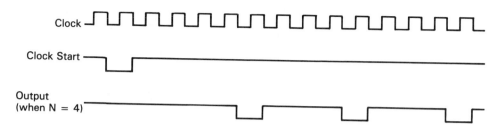

Figure 8.2. Rate Generator. In this mode, the count is initiated by loading the count register with the desired count. This value is loaded into the counter, which then decrements by one with each period of the oscillator. When the count reaches the desired count, the output line is driven low. After one additional count, the output line is driven high again and the count is reset to the initial count from the count register. Note that resetting the count register to another value affects only succeeding counts, not the current count.

8.2.2 Square Wave Rate Generator

This mode is very similar to the rate generator mode, except that the output is a square wave instead of a repetitive pulse. This is accomplished by decreasing the counter value by two instead of one, and having the output change each time the counter reaches the desired count. This is shown in Figure 8.3. Notice that the resultant output is a square wave only as long as the count is even. When the count is odd, in order to make the duration of the square wave be the desired number of counts, a somewhat more complex scheme is used in which the count is alternately decremented by 1 and then by 3 the first time it is decremented; it is then decremented by 2 until zero is reached.

8.2.3 Software-Triggered Strobe

This mode is also similar to the rate generator mode, except that the output is a single pulse, rather than a repetitive series of pulses. When the count is loaded into the counter, it begins counting down until it reaches zero. At that point, a pulse is output that lasts one cycle of the oscillator. This is shown in Figure 8.4.

8.3 PROGRAMMING

The timer/counter is usually relatively easy to program as long as you understand all of the options available. In the following examples, we show how to program the timer/counter on the IBM DACA. The clock on the Data Translation DT2801

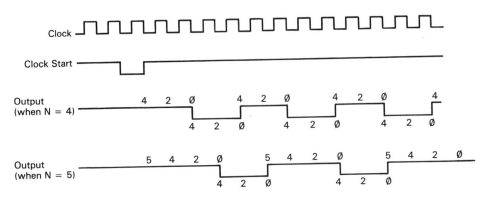

Figure 8.3. Square Wave Rate Generator. In this mode, the count is initiated by loading the count register with the desired count. This value is loaded into the counter, which then decrements by two with each period of the oscillator. When the count reaches the desired value, the output line is driven to the opposite state and the count is reset to the initial count from the count register. For odd counts, the process is slightly different, as described in the text.

board is not directly accessible to the user and can be used only in combination with the A/D or D/A, as shown in Chapter 9.

Table 8.1.   Registers on the IBM DACA used by the Timer/Counter

Register[a]	Name	Address[b]	Function
8	Timer/Counter 0	82E2	Read/load counter 0
9	Timer/Counter 1	92E2	Read/load counter 1
10	Timer/Counter 2	A2E2	Read/load counter 2
11	Timer/Counter Control	B2E2	Control all counters

[a] Only those registers are shown that are used in programming the timer/counter.

[b] The addresses shown are for Adapter 0. Addresses are in hex. If you have more than one DACA card in your IBM PC, then each DACA must have a different adapter number. Because the addresses shown are in hex, you should add 0400 (hex) to the address for each higher numbered adapter. For example, the timer/counter control register of Adapter 1 is referenced at address B6E2 (hex).

Table 8.1 shows the registers on the IBM DACA used by the timer/counter. As for most timer/counter systems, there is a separate register for each of the

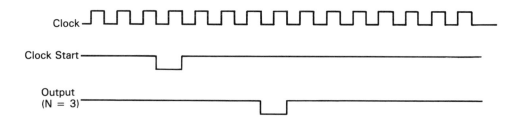

Figure 8.4. Software-Triggered Strobe. In this mode, the count is initiated by loading the count register with the desired count. This value is loaded into the counter, which then decrements by one with each period of the oscillator. When the count reaches zero, the output line is driven to the opposite state for one period of the oscillator.

counters; the DACA has three counters. Counters 0 and 1 are always chained together, such that whenever counter 0 decrements to zero, it resets itself and decrements counter 1. The net effect of this is to have a 32-bit counter. Counter 2 is always used by itself as a 16-bit counter.

In addition, there is a **control register**. This register is used to control the operation of the timer/counter system, and is therefore the most complex. By writing a specific bit pattern to the control register, we can specify which of the many timer/counter options we wish to use. These control patterns are summarized in Table 8.2. Note that each section of this table refers to a set of options, so that you can construct a full control word using all of the bits. For example, suppose you wish to select counter 1, read a 2-byte datum, use the rate generator function, and have the counter report the answer in binary; in this case, you would send the binary value 01110100, which is 74 in hex.

By far the most common use of the timer/counter is as a timer, so let's take a look at timing. We can choose between a 32-bit counter (counters 0 and 1 combined) or a 16-bit counter; the former uses an internal 1.023 MHz clock frequency, while the latter uses a user-provided clock signal. In our program we choose the most common mode of operation, namely 32-bit timing based upon the internal clock. Program 8.2 illustrates how this is done.

First, we must calculate the initial value to be loaded into the two counters. This is based upon the rate of the input to clock 0, which is 1.023 MHz. Hence, we can calculate the number of timer "ticks" between each event by dividing 1.023 MHz by the desired rate. This is done in line 320.

```
10 REM PROGRAM CLOCK1
20 REM LABORATORY AUTOMATION USING THE IBM PC
30 REM Program to time a repetitive event using the DACA timer/counter
40 REM ******************************************************************
50 REM Define all registers on clock/timer
60 BASEADD%=&H2E2                         'change this address if not adapter 0
70 CLOCK0%=&H8000 + BASEADD%              'clock 0 count register
80 CLOCK1%=&H9000 + BASEADD%              'clock 1 count register
90 CLKCONTROL%=&HB000 + BASEADD%          'clock control register
100 REM ********************************
110 REM          Main Program
120   CLS:INPUT "Enter the desired clock rate in Hz ",RATE!
130   INPUT "Enter the desired number of events to be timed ",TOTALPOINTS
140   CLS
150   GOSUB 320                           'start clock
160   DELTA2!=TIMER1%                      'delta2! is previous clock reading
170   GOSUB 520                           'get current clock reading
180   POINTNUMBER=0                       'starting event number
190   STARTTIME=TIMER
200   WHILE POINTNUMBER <= TOTALPOINTS    'test for number of events
210       WHILE (DELTA! <= DELTA2!)       'test clock for rollover
220           DELTA2!=DELTA!
230           GOSUB 520                   'get current reading
240       WEND
250       DELTA2! = DELTA!
260       PRINT POINTNUMBER,TIMER-STARTTIME   'our "event"
270       POINTNUMBER = POINTNUMBER + 1
280   WEND
290 END
300 REM ********************************
310 REM subroutine to set clock to desired rate
320   TICK!=1023000!/RATE!                'convert to ticks of 1.023 MHz clock
330   TIMER1!=INT(TICK!/65536!)+2         'calculate counter values
340   TIMER0!=TICK!/TIMER1!
350   IF TIMER0!>32767 THEN TIMER0%=TIMER0!-65536! ELSE TIMER0%=TIMER0!
360   TIMER1%=TIMER1!
370   CLOCK0LOW%=TIMER0% AND &HFF          'calculate low, high bytes of counters
380   CLOCK0HIGH%=(TIMER0% AND &HFF00)/256
390   CLOCK1LOW%=TIMER1% AND &HFF
400   CLOCK1HIGH%=(TIMER1% AND &HFF00)/256
410   IF CLOCK0HIGH%< 0 THEN CLOCK0HIGH%=256+CLOCK0HIGH%    'can't have negative
420   IF CLOCK1HIGH%< 0 THEN CLOCK1HIGH%=256+CLOCK1HIGH%
430   OUT CLKCONTROL%,&H34                'set to counter 0, rate generator, binary
440   OUT CLOCK0%,CLOCK0LOW%              'output counter 0 setting
450   OUT CLOCK0%+1,CLOCK0HIGH%
460   OUT CLKCONTROL%,&H74                'set to counter 1, rate generator, binary
470   OUT CLOCK1%,CLOCK1LOW%              'output counter 1 setting
480   OUT CLOCK1%+1,CLOCK1HIGH%
490 RETURN
500 REM ********************************
510 REM subroutine to read clock
520   OUT CLKCONTROL%, &H40               'move counter 1 contents to storage reg.
530   J%=INP(CLOCK1%)                     'read low byte
540   K%=INP(CLOCK1%+1)                   'read high byte
550   DELTA!=256!*K% +J%
560 RETURN
```

Program   8.2.   Timing Using the Timer/Counter.

Table 8.2.   Timer/Counter Control Register

Bits	Function Selected
7 6 5 4 3 2 1 0	
0 0 X X X X X X	Select Counter 0
0 1 X X X X X X	Select Counter 1
1 0 X X X X X X	Select Counter 2
1 1 X X X X X X	Illegal
X X 0 0 X X X X	Latch Selected Counter
X X 0 1 X X X X	Read/load Most Significant Byte Only
X X 1 0 X X X X	Read/load Least Significant Byte Only
X X 1 1 X X X X	Read/load Low Byte then High Byte
X X X X 0 0 0 X	Interrupt on Terminal Count
X X X X 0 0 1 X	Not used
X X X X X 1 0 X	Rate Generator
X X X X X 1 1 X	Square Wave Generator
X X X X 1 0 0 X	Software Triggered Strobe
X X X X 1 0 1 X	Not used
X X X X X X X 0	Binary Counter
X X X X X X X 1	BCD Counter

Next, we must calculate appropriate values for each of the two 16-bit counters. Each time counter 0 counts down to zero, it decrements timer 1. Hence, the total number of counts of the 1.023 MHz clock needed is the product of the counts of the two counters; we can therefore use any combination of values that produces the desired count. In Program 8.2, we have used a scheme that produces a small value on counter 1 and a large value on counter 0, but any of several other schemes would work as well. However, we need to be careful to avoid having either counter have a value of 0, so we add a small value (2) to the value of counter 1 in the calculations (line 330). In addition, we must remember that in order to be expressed as signed integers, numbers larger than 32767 must be represented as negative numbers. Hence, in line 350, we test the value to be output to see if it is so large that it should be converted to a negative number.

To send the values to the clock in BASICA, we must send the data as bytes. Hence, lines 370-420 convert the 16-bit values into 8-bit bytes; however, the 8-bit values need to be positive to be correctly represented, so we test them to see if they need to be converted back to positive numbers.

We are finally ready to start the clock. To do so, we must send the appropriate setting to each clock. We first do this by setting up counters 0 and 1 to function as rate generators with a binary output. As we see from Table 8.2, this is done in line 430 by selecting the bit pattern 00110100, which is 34(hex), for: counter 0 used as a rate generator, loading the low byte then the high byte of the count, and counting in binary (instead of binary coded decimal). We then load the low byte, then the high byte into counter 0 in lines 440-450. This process is repeated for timer 1. Note that loading timer 1 with its count restarts the counting process the next time counter 0 reaches zero.

Next, we wish to begin timing events with the timer/counter. To do this, we use two techniques that require some explanation. First, we read the clock by having it "latch the selected counter." This is done with a specific command to the control register, and avoids the problem of having the timer/counter change while it is being read. Thus, the value read in lines 530 and 540 is the latched value that was stored when line 520 was executed.

Second, if the counter is counting very rapidly, we may not read it when the count reaches its lowest value (one if used as a rate generator). In addition, if we just test for the counter reaching the terminal count, our program may be fast enough that it sees the same counter reading (of one) many times. Hence, we use a more general technique, starting in line 190, which is to watch the counter as it decreases, and then watch for it to "rollover", or be reset to a high value.

Actually, in practice we seldom test for the clock in software. Instead, we use the electrical pulse generated when the clock rolls over to physically trigger some event. We shall see an example of this in Chapter 9. Alternatively, we can use the clock to generate interrupts, as shown in Chapter 15.

EXERCISES

1. Program the IBM DACA in each of the three modes we have described. Connect the outputs of timer 0 and timer 1 to a dual-trace oscilloscope and look at the relationship between the two (recall that the timer 0 output is used as the input to timer 1).

2. Write a program to use the counter feature of the data acquisition card. For example, use it to count closures of a switch. (You will want to use a debouncing circuit to ensure only a single signal from each closure.)

3. Run Program 8.2. Why does it still not print values that are all exactly integer multiples of the desired rate? Can you devise a method for printing the time as measured by the clock on the DACA, rather than the PC's clock?

9

COORDINATED DATA COLLECTION AND CONTROL

In the previous chapters we discussed a number of methods for collecting data and controlling instruments. Each of these methods has been considered in isolation; however, it is much more common in the laboratory to use two or more of these methods simultaneously. For example, we may wish to use the D/As to position a device and then take data at that location with an A/D. Or we may use the GPIB to set a programmable power supply and then take a datum with the A/D. Or we may utilize digital I/O to sense switches and then use an A/D that communicates with the computer via an RS-232 interface to acquire data. Or we may wish to use multiple GPIB instruments. Indeed, the list of combinations is almost endless. As usual, the best way for us to understand these combinations is to see some examples.

9.1 COMBINED A/D AND D/A

A simple example of a useful combination is one or two D/As in combination with one or more channels of A/D. Typical examples of this might include using the D/A to output a voltage for a physiological stimulus, and then using the A/D to acquire a voltage from a physiograph. Another example, and the one we illustrate here, might involve using the D/As to position a device in both the X and Y directions and then using the A/D to collect data at that location. This example is illustrated in Program 9.1.

```
10 REM PROGRAM SCAN
20 REM LABORATORY AUTOMATION USING THE IBM PC
30 REM Program to use the two D/A's to generate an X-Y position
40 REM     and then trigger the A/D to collect data at that location.
50 REM Uses the DT2801
60 REM ****************************************************************
70 REM Define all registers
80 DIM XYDATA%(1000)                       'data array
90 BASEADD%=&H2EC                          'base address of board
100 COMMAND%=BASEADD%+1                     'command register
110 STATUS%=BASEADD%+1                      'status register
120 DATUM%=BASEADD%                         'data register
130 REM ********************************
140 REM              Main Program
150   GOSUB 250                            'get user input
160   GOSUB 590                            'do setup
170   FOR I%=1 TO NPOINTS%
180       GOSUB 670                        'output on 2 D/A's
190       GOSUB 390                        'collect and display data
200       PRINT I%,XYDATA%(I%)             'display datum
210   NEXT I%
220 END
230 REM ********************************
240 REM Get parameters from user
250   SCREEN 0:CLS:NPOINTS%=0
260   WHILE ( NPOINTS% < 1 OR NPOINTS% > 1000 )
270     INPUT "How many data do you wish to collect? (1 - 1000) ",NPOINTS%
280   WEND
290   INPUT "What channel of the A/D do you wish to use? ",CHANA2D%
300   INPUT "What A/D gain do you wish to use? (1,2,4,8)";ADGAIN%
310   GAINCODE%=LOG(ADGAIN%)/LOG(2)        'convert gains 1,2,4,8 to 0,1,2,3
320   INPUT "What initial X value do you wish to output? (0-4095) ",X%
330   INPUT "What initial Y value do you wish to output? (0-4095) ",Y%
340   INPUT "What step size in X do you wish? (0-4095) ",XSTEP%
350   INPUT "What step size in Y do you wish? (0-4095) ",YSTEP%
360 RETURN
370 REM ********************************
380 REM Get datum from A/D
390   GOSUB 880
400   OUT COMMAND%,&HC                      'give command READ A/D IMMEDIATE
410   GOSUB 850
420   OUT DATUM%,GAINCODE%                  'set gain
430   GOSUB 850
440   OUT DATUM%,CHANA2D%                   'set channel
450   'get datum, low byte then high byte
460   WAIT STATUS%,&H5                      'low byte ready?  check bits 0,2
470   LOW%=INP(DATUM%)                      'yes, so read in low byte
480   WAIT STATUS%,&H5                      'high byte ready?
490   HIGH%=INP(DATUM%)                     'yes, read it
500   REM convert datum to single word
510   XYDATA%(I%)=(HIGH%*256+LOW%)/ADGAIN%
520   'check status register to see if error occurred
530   GOSUB 880
540   ERRORCHECK%=INP(STATUS%)              'check status register
550   IF (ERRORCHECK% AND &H80) THEN PRINT "Error while reading A/D":STOP
560 RETURN
570 REM ********************************
580 REM Setup for DT2801 board
590   OUT COMMAND%,&HF                      'stop board
600   TEMP%=INP(DATUM%)                     'clear data register
610   REM wait for command to finish, then clear errors
620   GOSUB 880
630   OUT COMMAND%,&H1
640 RETURN
650 REM ********************************
660 REM Output X, Y on D/A
670   GOSUB 880
```

Program 9.1 (Part 1 of 2). Scanning X-Y and Data Collection.

```
680    OUT COMMAND%,&H8                    'give command WRITE DAC IMMEDIATE
690    GOSUB 850
700    OUT DATUM%,2                        'select both channels
710    VALUEOUT%=X%                        'output X, Y values using D/A
720    GOSUB 930
730    VALUEOUT%=Y%
740    GOSUB 930
750    GOSUB 880
760    ERRORCHECK%=INP(STATUS%)
770    IF (ERRORCHECK% AND &H80) THEN PRINT "Error writing to DAC":STOP
780    X%=X%+XSTEP%                        'get new X and Y values
790    Y%=Y%+YSTEP%
800    IF X% > 4095 THEN X%=4095           'make sure within range
810    IF Y% > 4095 THEN Y%=4095
820    RETURN
830    REM *********************************
840    REM Check to see if ready to set a register
850    WAIT STATUS%,&H2,&H2                'wait for bit 1 to be reset
860    RETURN
870    REM check to see if ready to send another command
880    WAIT STATUS%,&H2,&H2                'wait for bit 1 to be reset
890    WAIT STATUS%,&H4                    'wait for bit 2 to be set
900    RETURN
910    REM *********************************
920    REM Output 2 bytes to D/A
930    LOW%=VALUEOUT% AND &HFF             'get low byte
940    HIGH%=(VALUEOUT% AND &HF00)/256     'get high byte (upper 4 bits 0)
950    GOSUB 850
960    OUT DATUM%,LOW%                     'output the datum, low byte first
970    GOSUB 850
980    OUT DATUM%,HIGH%
990    RETURN
```

Program 9.1 (Part 2 of 2).   Scanning X-Y and Data Collection.

In this example, we use two of the functions of the DT2801 card together. The A/D and D/A are programmed much as they were in earlier chapters. Here, we use the two D/As on the DT2801 card so that we can output both an X and a Y voltage. After the D/As have been set, we then acquire a single datum on the A/D.

You should note several things about Program 9.1. First, we are scanning over a single X-Y path. A more common application involves some form of raster scanning (similar to the beam on a television set); this modification is left for you to attempt as an exercise.

Second, before using this program, you need to reread Chapter 2 for its discussion of connecting devices to the D/A. Recall that the D/A can not source much current; hence, a buffer or amplifier circuit should be interposed between the D/A and the device it is driving. However, it provides enough current to drive the A/D, so one way of testing this program is to connect a D/A output to the A/D input.

Finally, recall that the A/D operates in differential mode. Hence, you need to connect the voltage input to the plus and minus inputs of the A/D.

9.2 COMBINED A/D, DIGITAL I/O AND TIMER

A somewhat more sophisticated combination is the use of most of the functions of the DT2801 board: A/D, digital I/O and timer. This is illustrated in Program 9.2, which is based upon the commands summarized in Table 9.1.

Table 9.1.   Registers on the DT2801 used by the Clocked A/D

Register[a]	Name	Address[b]	Meaning of Bits
1	Command	02ED	00001111 for STOP
1	Command	02ED	00000001 for CLEAR ERROR
1	Command	02ED	00000011 for SET CLOCK PERIOD
0	Clock Period	02EC	low byte of period (1st write)
			high byte of period (2nd write)
1	Command	02ED	00001101 for SET A/D PARAMETERS
0	A/D Gain	02EC	bits 1-0=gain code
0	A/D Channel	02EC	bits 3-0=starting channel (1st write)
			bits 3-0=ending channel (2nd write)
0	Conversions	02EC	low byte of count (1st write)
			high byte of count (2nd write)
1	Command	02ED	bits 4-0=01110 for READ A/D
			bit 5=continuous conversions on A/D
			bit 6=enable external clock
			bit 7=enable external trigger
1	A/D Status	02ED	bit 0=datum ready
			bit 2=previous command finished
			bit 3=previous byte was command
			bit 7=composite error
0	A/D Datum	02EC	low byte of datum (1st write)
			high byte of datum (2nd write)

[a] Only those registers and bit meanings are shown that are used in programming the A/D in the "READ A/D" mode.

[b] The addresses shown are for the factory configuration. If you have changed the hardware base address to another setting, the addresses shown should be adjusted accordingly.

Probably the most useful feature of this program is using the timer of the DT2801 board to trigger the A/D. Very often, we wish to collect data at a specific rate. If we use the READ A/D mode of the DT2801, then the A/D will take data only at the intervals we specify. You should compare this to the programs in Chapter

```
10 REM PROGRAM COMBO
20 REM LABORATORY AUTOMATION USING THE IBM PC
30 REM Program to demonstrate the use of A/D, digital I/O, and clock
40 REM It does this by using the A/D to collect data at a rate defined
50 REM     by the clock.  The user starts the data collection using a switch
60 REM     connected to bit 0 of the input port.  When this switch is
70 REM     toggled, a diode attached to bit 0 of the output port is lit.
80 REM     When the desired number of points have been collected, the
90 REM     diode attached to bit 1 of the output is lit and the diode
100 REM    attached to bit 0 is turned off.
110 REM Uses the DT2801
120 REM Note:  The switch should be connected to Port 1, bit 0, the
130 REM    the LEDs to Port 0, bits 0 and 1
140 REM ********************************************************************
150 DIM VALUE%(1000)
160 REM Define all registers, etc.
170 BASEADD%=&H2EC                      'base address of board
180 COMMAND%=BASEADD%+1                 'command register
190 STATUS%=BASEADD%+1                  'status register
200 DATUM%=BASEADD%                     'data register
210 REM ***********************************
220 REM                Main Program
230    GOSUB 340                        'get user input
240    GOSUB 490                        'set up board
250    GOSUB 570                        'set up digital ports
260    GOSUB 740                        'set up clock and A/D
270    PRINT "Please turn on the switch to start data collection"
280    GOSUB 1130                       'wait for toggle of switch
290    GOSUB 1490                       'collect and display data
300    GOSUB 1750                       'change diode to indicate finish
310 END
320 REM ***********************************
330 REM Get user input
340    SCREEN 0:CLS:NPOINTS%=2:TICKS!=0
350    WHILE TICKS! < 3 OR TICKS! > 65535!
360       INPUT "At what rate do you wish to collect data, in Hz(> 6.1)? ",RATE!
370       TICKS!=400000!/RATE!          'assumes 400kHz clock
380    WEND
390    WHILE NPOINTS% < 3
400       INPUT "For how many seconds do you wish to collect data? ",NSECS%
410       NPOINTS%=NSECS% * RATE!
420    WEND
430    INPUT "What channel of the A/D do you wish to use? ",CHANA2D%
440    INPUT "What gain of the A/D do you wish (1,2,4 or 8)? ",ADGAIN%
450    GAINCODE%=LOG(ADGAIN%)/LOG(2)    'convert gains 1,2,4,8 to 0,1,2,3
460 RETURN
470 REM ***********************************
480 REM Set up board
490    OUT COMMAND%,&HF                 'stop board
500    TEMP%=INP(DATUM%)                'clear data register
510    REM wait for command to finish, then clear errors
520    GOSUB 1080
530    OUT COMMAND%,&H1                 'clear the command register
540 RETURN
550 REM ***********************************
560 REM Set up digital ports
570    GOSUB 1080
580    OUT COMMAND%,&H4                 'write SET DIGITAL PORT FOR INPUT command
590    GOSUB 1050
600    OUT DATUM%,1                     'input on port 1
610    GOSUB 1080
620    ERRORCHECK%=INP(STATUS%)         'check for error
630    IF (ERRORCHECK% AND &H80) THEN PRINT "Error writing to Port 1":GOSUB 1810
640    GOSUB 1080
650    OUT COMMAND%,&H5                 'write SET DIGITAL PORT FOR OUTPUT command
660    GOSUB 1050
670    OUT DATUM%,0                     'output on port 0
```

Program 9.2 (Part 1 of 3).   Coordinated Data Collection and Control.

```
 680    GOSUB 1080
 690    ERRORCHECK%=INP(STATUS%)          'check for error
 700    IF (ERRORCHECK% AND &H80) THEN PRINT "Error writing to Port 0":GOSUB 1810
 710 RETURN
 720 REM *************************************
 730 REM Set up clock and A/D
 740    GOSUB 1080
 750    OUT COMMAND%,&H3                  'write SET CLOCK PERIOD command
 760    IF TICKS! > 32767 THEN TICKS%=TICKS!-65536! ELSE TICKS%=TICKS!
 770    CLOCKHIGH%=(TICKS% AND &HFF00)/256
 780    CLOCKLOW%=TICKS% AND &HFF
 790    IF CLOCKHIGH% < 0 THEN CLOCKHIGH%=CLOCKHIGH% +256
 800    GOSUB 1050
 810    OUT DATUM%,CLOCKLOW%              'output low byte
 820    GOSUB 1050
 830    OUT DATUM%,CLOCKHIGH%             'output high byte
 840    GOSUB 1080
 850    ERRORCHECK%=INP(STATUS%)          'check for error
 860    IF (ERRORCHECK% AND &H80) THEN PRINT "Error setting clock":GOSUB 1810
 870    OUT COMMAND%,&HD                  'write SET A/D PARAMETERS command
 880    GOSUB 1050
 890    OUT DATUM%,GAINCODE%              'set gain
 900    GOSUB 1050
 910    OUT DATUM%,CHANA2D%               'set starting a/d channel
 920    GOSUB 1050
 930    OUT DATUM%,CHANA2D%               'set ending a/d channel
 940    HIGHPOINTS%=INT(NPOINTS%/256)     'get low, high bytes of number of points
 950    LOWPOINTS%=NPOINTS% - HIGHPOINTS%*256
 960    GOSUB 1050
 970    OUT DATUM%,LOWPOINTS%
 980    GOSUB 1050
 990    OUT DATUM%,HIGHPOINTS%
1000    ERRORCHECK%=INP(STATUS%)          'check for error
1010    IF (ERRORCHECK% AND &H80) THEN PRINT "Error setting A/D":GOSUB 1810
1020 RETURN
1030 REM *************************************
1040 REM Check to see if ready to set a register
1050    WAIT STATUS%,&H2,&H2              'wait for bit 1 to be reset
1060 RETURN
1070 REM Check to see if ready to send another command
1080    WAIT STATUS%,&H2,&H2              'wait for bit 1 to be reset
1090    WAIT STATUS%,&H4                  'wait for bit 2 to be set
1100 RETURN
1110 REM *************************************
1120 REM Wait for user to toggle switch attached to bit 0, then light diode 0
1130    DEVICE%=0                         'turn off all outputs
1140    GOSUB 1370
1150    GOSUB 1210                        'wait for the switch
1160    DEVICE%=2^0                       'LED at bit 0
1170    GOSUB 1370                        'turn on LED 0
1180 RETURN
1190 REM *************************************
1200 REM Input from switch
1210    LOW%=0
1220    WHILE LOW% = 0                    'wait for switch to be toggled
1230       GOSUB 1080                     'wait for board to be ready
1240       OUT COMMAND%,&H6               'send READ DIO IMMEDIATE command
1250       GOSUB 1050
1260       OUT DATUM%,1                   'input from port 1
1270       WAIT STATUS%,&H5               'check if datum ready
1280       J%=INP(DATUM%)                 'get low and high bytes
1290       LOW%=J% AND 1                  'mask off all but bit 0
1300       GOSUB 1080
1310       ERRORCHECK%=INP(STATUS%)       'check for error
1320       IF (ERRORCHECK% AND &H80) THEN PRINT "Error reading from Port 1":
                   GOSUB 1810
1330    WEND
```

Program 9.2 (Part 2 of 3). Coordinated Data Collection and Control.

```
1340 RETURN
1350 REM *********************************
1360 REM Output on specified bit(s)
1370    GOSUB 1080
1380    OUT COMMAND%,&H7              'WRITE DIGITAL OUTPUT IMMEDIATE command
1390    GOSUB 1050
1400    OUT DATUM%,0                  'output on Port 0
1410    GOSUB 1050
1420    OUT DATUM%,DEVICE%            'output byte
1430    GOSUB 1080
1440    ERRORCHECK%=INP(STATUS%)      'check for error
1450    IF (ERRORCHECK% AND &H80) THEN PRINT "Error writing to Port 0":GOSUB 1810
1460 RETURN
1470 REM *********************************
1480 REM Get data from A/D at rate of 1 point per second
1490    GOSUB 1610                    'start A/D
1500    CLS:PRINT "Beginning data collection...please wait"
1510    FOR I%=1 TO NPOINTS%
1520       GOSUB 1660                 'get datum
1530       VALUE%(I%)=RESULT%
1540    NEXT I%
1550    FOR I%=1 TO NPOINTS%
1560       PRINT VALUE%(I%),
1570    NEXT
1580 RETURN
1590 REM *********************************
1600 REM Start A/D taking data
1610    GOSUB 1080
1620    OUT COMMAND%,&HE              'write READ A/D command
1630 RETURN
1640 REM *********************************
1650 REM Get datum from A/D
1660    WAIT STATUS%,&H5              'check if datum ready
1670    J%=INP(DATUM%)               'get the datum (in 2 bytes)
1680    WAIT STATUS%,&H5              'check if datum ready
1690    K%=INP(DATUM%)
1700    RESULT%=(J%+(256*K%))/ADGAIN%   'convert to an integer, correct for gain
1710    ERRORCHECK%=INP(STATUS%)        'check for error
1720    IF (ERRORCHECK% AND &H80) THEN PRINT "Error reading datum":GOSUB 1810
1730 RETURN
1740 REM *********************************
1750 REM Subroutine to light LED 1 and turn off LED 0
1760    DEVICE%=2^1                   'output on bit 1 of port 0
1770    GOSUB 1370
1780 RETURN
1790 REM *********************************
1800 REM Determine source of error
1810    OUT COMMAND%,&HF              'stop the board
1820    TEMP%=INP(DATUM%)
1830    GOSUB 1080
1840    OUT COMMAND%,&H2              'send READ ERROR REGISTER command
1850    GOSUB 1050
1860    VALUE%=INP(DATUM%)
1870    GOSUB 1050
1880    PRINT "Value of error register low byte is: ";VALUE%
1890    VALUE2%=INP(DATUM%)
1900    PRINT "Value of error register high byte is: ";VALUE2%
1910    IF VALUE2%=4 THEN PRINT "You have tried to collect data at too fast"
1920 STOP
```

Program   9.2 (Part 3 of 3).   Coordinated Data Collection and Control.

3, where we take data as fast as the program runs. Note that the earlier approach results in data taken at some undefined interval. We almost always used timed, rather than untimed data collection because during scientific data collection we wish to have all parameters carefully defined.

In order to understand Program 9.2, we need to discuss three new commands for the DT2801 board. These are the SET A/D PARAMETERS command, the READ A/D command, and the SET INTERNAL CLOCK PERIOD command.

Thus, beginning in line 670 of Program 9.2, we have used the SET CLOCK PERIOD command to choose an appropriate rate for the clock. This is similar to the setting of the timer/counter described in Chapter 8, except that on the DT2801, the output is available only for triggering the A/D and D/A, not for triggering external events. To understand the program, we need to know that the clock rate for the DT2801 is 400 kHz. Hence, to determine the correct count for the clock, we simply divide 400,000 by the desired rate. The resulting count must in general be between 2 and 65,535; however when using the 400 kHz clock, the count must be no less than 50 to avoid going faster than the 13.7 kHz upper limit of the DT 2801 board. In practice, even the 13.7 kHz rate can only be achieved with an assembly language subroutine (see Chapter 14) because BASIC is relatively slow.

An important point to note here is that because the timer rates are set using port I/O, only 8-bit bytes can be sent to the timer at one time. Hence, the timer counts (16 bits) must be broken into low and high bytes. BASICA requires that the value output to a port be between 0 and 255; hence, special care must be taken to ensure that the high bytes are not negative values.

The SET A/D PARAMETERS command allows us to set the gain, channels, and number of conversions on the A/D with a single command sequence. Beginning in line 800, we send the SET A/D PARAMETERS command, followed by the gain code, starting channel, ending channel, and number of conversions (the last number is a 16-bit number, so it is sent low byte first, then high byte). The values must be sent in this order. It is possible to use this command to set up a sequence of channels to be "scanned", but in this example we are using only a single channel; hence, we set the ending channel to be the same as the starting channel.

The READ A/D command is then quite simple: we simply write the appropriate command, check to see that the A/D conversion is finished, and then read the datum into memory, one byte at a time.

The program also uses digital I/O. It does this in two ways. First, a switch must be toggled (flipped to the opposite position) to start the data collection process. This also turns on a light-emitting diode (LED) connected to the digital I/O section of the DT2801. At the end of the data collection process, digital I/O is used to turn the first LED off and another LED on.

A convenient method for demonstrating the use of this and similar programs is to construct a simple device that has a series of switches, LEDs, a voltmeter, and an input for the A/D. The components of this device are shown in Figure 9.1.

Digital Output

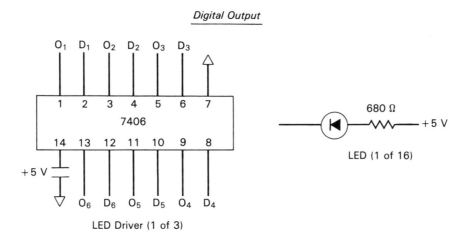

LED Driver (1 of 3)

Digital Input

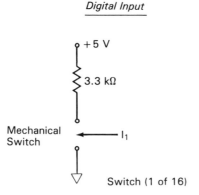

Analog Input

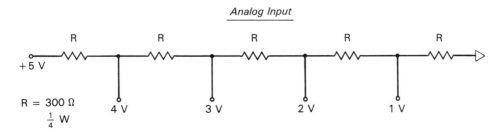

R = 300 Ω
$\frac{1}{4}$ W

Analog Output

Voltmeter Displaying −10 V to +10 V

Power Supply

5 VDC Power Supply, with Power Switch
and LED

Figure   9.1.   Analog and Digital I/O Test System.   This is a convenient circuit for testing digital and analog input and output boards. Program 9.2 can be tested with this circuit, for example.

EXERCISES

1. Rewrite Program 9.1 so that instead of scanning along a single trajectory, the
    D/As are raster scanned. That is, scan X from *a* to *b*. Then increment Y,
    move X back to *a*, and scan X from *a* to *b* again. Repeat this process until you
    have scanned the entire Y range selected by the user.

2. Modify Program 9.1 so that it takes multiple A/D readings at each X-Y
    location and averages them.

3. Modify Program 9.2 to make it display on the 16 LEDs of Figure 9.1 the
    status of the 16 switches. Test for any of the switches being toggled, and begin
    data collection immediately. During data collection, light only the LED that
    corresponds to the switch used to initiate the data collection.

PART TWO

DATA ANALYSIS

10

USER- AND PROGRAMMER-FRIENDLY SOFTWARE

10.1 INTRODUCTION

Now that you have seen how to control instruments and collect data, you may think that you've learned all you need to know and can retire to your computer to begin writing programs. Not so! We first need to talk together about analyzing your data once it has been collected.

Phrased succinctly, the goal of data analysis is to make the data more useful to the scientist or other user. Sometimes this simply means plotting the data so that the user can see it, perhaps even while it is being collected. More often, however, it means taking a large amount of data and reducing it to a few numbers -- for example, reducing a hundred-point region of a graph to a single peak area. And increasingly, it means applying sophisticated algorithms in order to extract information that is not apparent in the raw data -- for example, taking the Fourier transform of the data, deconvolving peaks, or applying colors to enhance desired features.

You will probably find that data analysis is easier, and certainly less intimidating, than data collection and instrument control. This is because you have relatively few worries about hardware compared to the Medusa of cables, amplifiers, power supplies, and switches that you must use in interfacing the instrument.

However, data analysis is in many ways the most challenging part of laboratory automation. In planning the data analysis, we are faced with a variety of questions: What can we do to make the data more useful? How can we make it possible for a user to perceive easily and naturally all of the information contained in the data? How can we manipulate the data without introducing new biases into it? How can we find new relationships in the data? The answers to these questions pose a unique challenge for the scientific programmer -- and certainly requires a considerable understanding of the science involved in the experiment.

There are several keys to successful data analysis programs. Probably the most important is detailed advance planning. Certainly, you should outline the goals you expect to accomplish. It may help to draw out how you would like each of the screen displays (graphs, menus, tables, etc.) to appear. Plan out a list of the variables you will use throughout your program and document the meaning of each of them in a text file that you can easily update.

Second, become an expert about the instrument you are automating. Understand what data are produced and what those data represent. Investigate the theory of the instrument and understand all of the operating parameters of the instrument. Read in the current scientific literature how others analyze the same type of data. Look at commercial systems, if any already exist, for ideas.

The third key to successful data analysis is to avoid "reinventing the transistor." Many of the components of data analysis programs are already available. Don't waste your time redeveloping features that are available in public domain software. Use commercial programs or programming aids when they are appropriate. Talk to others who have done laboratory automation and read the books and journals on the subject before starting your own program.

The last key is to make extensive use of graphics and images. The human eye/brain combination is outstanding at perceiving patterns in data, so give the user as many ways to display the data graphically as possible.

The challenge of data analysis is that virtually every instrument requires a different type of treatment of the data. And as computers become more powerful, we can consider performing types of data analysis that even a few years ago would have been impossible in the laboratory environment. But of course, that is what makes laboratory automation fun!

10.2 DATA STORAGE

Let's begin our discussion of data analysis by discussing some features common to virtually all lab automation programs: data storage, graphing, menus, notebook files and parameter files.

So far, we have avoided discussing a simple feature of virtually all programs: they need to store the data. Recall that the RAM containing your data does not store that data when the computer is turned off. Hence, you will want to store the data more permanently. Normally, this means storing the data on a floppy diskette or hard disk.

10.2.1 Sequential Files

The simplest method of storing data on a disk is to use **sequential files**. If you have been using the Disk Operating System (DOS) on the IBM PC, you are familiar with files. For example, the BASICA program is stored in a file called BASICA.COM. What makes a file "sequential" is that when your program retrieves the data stored in a sequential file, it must start at the beginning of the file and read every datum until it reaches the data of interest. For example, if you have 50 data stored, the computer must physically read back into memory the preceding 49 data before it can reach the 50th datum. By analogy, it is like a novel in which you have to read page 1 before you can read page 2, page 2 before you can read page 3, and so on.

In BASICA, sequential files are created using the

$$\text{OPEN } \textit{filenam} \text{ FOR OUTPUT AS } \#\textit{n}$$

statement, where *filenam* is a string of characters giving the name of the file and *n* is a small integer number.

BASICA allows two similar methods of putting information into a sequential file. The first of these is the

$$\text{PRINT } \#\textit{n}$$

statement, where *n* is the same number used in the OPEN statement. This statement is very similar to the PRINT statement in BASICA, except that it PRINTs to a sequential file. Any of the variants of the PRINT statement that work for printing information on the screen have a corresponding form in the PRINT #*n* statement. For example, you could use

$$\text{PRINT } \#1, \text{ USING } \text{"\#\#\#\#.\#\#"; A!; B!; C!}$$

to save in the sequential file three numbers written with two digits to the right of the decimal point.

The second method of putting information into a sequential file is a statement of the type:

$$\text{WRITE } \#\textit{n}$$
$$\text{e.g., WRITE } \#1, \text{ A!, B\%, C\$}$$

This statement, while less familiar, is preferable to the PRINT #*n* statement for use with sequential files because it makes reading the data out of the file easier: it inserts commas between items and puts quotation marks around strings. You should use it in preference to the PRINT #*n* statement, except where the data must all be written with a specific precision; in that case use the PRINT USING statement.

When you have finished writing all of the desired information into the sequential file, you should use the statement:

$$CLOSE \ \#n$$

where *n* is the same number as used in the OPEN statement. This ensures that all of the information in the file is permanently stored.

To see how all of these commands are used together, let's look at Program 10.1, which has been designed to create a sequential file. Notice that we have kept the line numbering consistent with that used in Program 3.2. By adding lines 195 and 650 through 750 to Program 3.2, in fact, we can store the data generated by Program 3.2 in a sequential file. (This can easily be done by using the MERGE command in BASICA, or any text editor.)

The series of steps you need to follow to access information in a previously-created sequential file is similar to the method used to store the data there. First, you must OPEN the file:

$$OPEN \ filenam \ FOR \ INPUT \ AS \ \#n$$

Notice that a sequential file can be used for either INPUT or OUTPUT, but not both at the same time; it must be CLOSEd before it can be OPENed in the other mode. The file number (*n*) does not need to be the same in the two modes, however.

10.2.1.1 Reading Data from a Sequential File

Data are read from the sequential file with the statement:

$$INPUT \ \#n$$

which reads data from the file back into memory. After all of the data have been read, then the file should be CLOSEd in the same fashion as a file OPENed for OUTPUT.

The simplest and most straightforward method of avoiding problems in using sequential files is to use WRITE statements and INPUT statements that match. For example, if the statement used to store the data in a file is

$$WRITE \ \#1, A\%; B!; C\$$$

then when the file is reopened for input, the data should be read using the statement

$$INPUT \ \#1, A\%, B!, C\$$$

Notice that the variables in the INPUT statement exactly match the variables in the WRITE statement. There should also be the same number of INPUT statements as there were WRITE statements to create the file.

```
10 REM PROGRAM SEQUENTIAL
20 REM LABORATORY AUTOMATION USING THE IBM PC
30 REM Program addendum to Program 3.2 to store data using a sequential file
40 REM Delete lines 10-50 before adding to Program 3.2
50 REM *******************************************************
195    GOSUB 670
650 REM ***************************
660 REM Store data in a sequential file
670    WHILE FILN$=""
680        INPUT "Enter the name of the file in which to store the data. ",FILN$
690    WEND
700    OPEN FILN$ FOR OUTPUT AS #1
710    FOR I%=1 TO NDATA%
720        WRITE #1,RESULT%(I%)
730    NEXT I%
740    CLOSE #1
750 RETURN
```

Program 10.1.   Creating a Sequential File.

10.2.2 Random Files

The second type of file is called a **random file** in BASICA. (It is also called a binary or direct access file in some language manuals.) Data in these files can be stored or retrieved in any order. Thus, for example, the 100th datum may be retrieved first, then the 39th datum, and so forth. This is analogous to a book, such as a cookbook, where we can read page 105 without first reading all of the pages 1 to 104. It is thus particularly suitable for large sets of data where you probably will only be retrieving a small fraction of the data at any given time.

Before showing how to use random files, we need to discuss three concepts: **records**, **random file buffers**, and **binary data storage**. The way in which BASICA knows where to store a given piece of data is that all data are stored in areas of fixed size on the disk known as **records.** Because the records are fixed size, BASICA can easily compute where a particular record is on the disk by using the **record number** for that record and the size of the record. The formula is:

$$\text{starting byte number} = (\,(m\text{-}1)\times s\,) +1$$

where m is the record number and s is the size of the record in bytes. For example, the 11th record of a file having a record length of 128 will start at byte $(\,(\,(11\text{-}1)\times 128)\,+1\,)=1281$ of the file.

A somewhat more difficult concept is that of a **random file buffer.** For random files, data storage is a two-step process. In the first step, data are stored in a special area of memory known as a random file buffer. When all of the desired information has been placed in the buffer, the entire contents of the buffer are moved onto the disk. This is done to reduce the number of times information is written to the disk. (Actually, the data may not be moved to the disk immediately, but may be further buffered by DOS in order to reduce the number of disk accesses. This is controlled by the BUFFERS statement in the CONFIG.SYS file. However, we have no direct control over this subsequent buffering in BASICA, and so we do not concern ourselves with it.)

The third concept used with random files is **binary data storage**. Unlike sequential files, where all data are stored in ASCII, random files require that all data be stored in binary, exactly as they are stored in ROM. Thus, integers are stored in 2 bytes, single precision floating point numbers in 4 bytes, and double precision values in 8 bytes. Strings are stored just as they are in sequential files. Binary data storage is used because it is generally more efficient than storage of ASCII values, as we shall see.

The first step in using random files is to open the file:

$$\text{OPEN } \textit{filenam} \text{ AS } \#n \text{ LEN}=yy$$

where *filenam* is the name of the file, *n* is the file number and *yy* is the length of a data record in bytes. Notice that the OPEN statement for random files is similar to that for sequential files, except that there is no INPUT or OUTPUT and the record length must be specified. Thus, one of the advantages of the random file format is that it can be read from or written to at any time. The record length is also the length of the random file buffer.

If you attempt to open a file with a record length of greater than 128 bytes using BASICA, you will get an error message. To avoid this, simply use the /S option when starting BASICA. For example, if the largest record size your program will use is 300 bytes, then start BASICA by typing:

$$\text{BASICA /S:300}$$

The second step in using random files is to describe how data are to be placed in the buffer (and ultimately in the file). This is done with the FIELD statement, which is of the form:

$$\text{FIELD } \#n, \text{ } m1 \text{ AS } a\$, \text{ } m2 \text{ AS } b\$, \text{ } m3 \text{ AS } c\$ \text{ .....}$$

where *n* is the file number used in the OPEN statement and *m1* is the length of string variable *a$* in bytes, *m2* is the length of *b$*, etc. The a$, b$, c$, and so forth are the **field variables** representing the data to be stored, where a field variable is a string variable that stores the binary representation of a number or, if the variable is a string, stores the string itself. The sum of m1 + m2 + m3 + ... must be no more than the record length.

It should be emphasized that the FIELD statement does nothing more than define how the data are to be ordered in the buffer. To actually move the data into the record in BASIC requires the LSET statement, which is of the form:

$$\text{LSET } a\$= q\$$$

where *a$* is one of the variables defined in the FIELD statement and *q$* is the string variable to be stored. RSET can be used in place of LSET if the strings are to be right-justified instead of left-justified; for most applications it does not matter which is used, however.

This can be very confusing until we realize that *all variables must be stored as strings when being transferred to the random file buffer*. Three special functions in BASICA, namely, MKI$, MKS$, and MKD$, are available for converting integers, single-precision floating point, and double-precision floating point variables, respectively, to strings; the STR$ of BASICA should *not* be used. For example, to move an integer, I%, a single precision variable, X!, and a 10-byte string, S$, into the random file buffer, the following statements could be used:

```
10  FIELD #1, 2 AS A$, 4 AS B$, 10 AS C$
20  LSET A$= MKI$(I%)
30  LSET B$= MKS$(X!)
40  LSET C$= S$
```

Notice that S$ must be moved into the buffer with the LSET statement even though it is already a string.

Once we have written all of the desired information for the current record into the buffer, it is very simple to move the data from the buffer into the disk file. This is done with a statement of the form:

$$PUT \#n, j$$

where *n* is the file number defined in the OPEN statement and *j* is the record number. All of the data in the buffer are moved by this one statement.

We can now write a program segment that performs the same functions as Program 10.1, except that it uses a random file. Program 10.2 could also be added to the end of our data collection program (Program 3.2, on page 46) without change. Note that the program is designed to write records containing 12 integer data per record; hence, the record length is 24 bytes (line 680). This program also assumes that the first record contains a count of the number of data (lines 700-710).

10.2.2.1 Reading Data from a Random File

Data can be retrieved from the random file in a similar fashion. The OPEN and FIELD statements are identical to those used for storing data in a random file. Again, data are read from the file in a two step process. The statement used to read the data from the file into the random file buffer is:

$$GET \#n, m$$

which reads record *m* from file number *n*. There is no equivalent to the LSET statement to read the data from the buffer into memory; however, numerical data must be converted back from string form using the CVI, CVS, and CVD which perform the

```
10 REM PROGRAM RANDOM_FILE_CREATE
20 REM LABORATORY AUTOMATION USING THE IBM PC
30 REM Addendum to Program 3.2 to save data using a random file
40 REM NOTE:  delete lines 10-50 before adding to Program 3.2
50 REM ******************************************************
55 DIM A$(12)
195    gosub 670
650 REM *************************
660 REM Save data in a random file
670    INPUT "Enter the name of the file in which data are to be stored. ",FILN$
680    OPEN FILN$ AS #1 LEN=24
690    FIELD #1, 2 AS A$(1), 2 AS A$(2), 2 AS A$(3), 2 AS A$(4), 2 AS A$(5),
       2 AS A$(6), 2 AS A$(7), 2 AS A$(8), 2 AS A$(9), 2 AS A$(10), 2 AS A$(11),
       2 AS A$(12)
700    LSET A$(1)=MKI$(NDATA%)                'store number of data in file
710    PUT #1,1
720    K%=1                                   'K% is record number
730    FOR I%=1 TO NDATA% STEP 12
740       FOR J%=0 TO 11                      'move data to random file buffer
750          IF I% +J% <= NDATA% THEN LSET A$(J%+1)=MKI$(RESULT%(I%+J%))
760       NEXT J%
770       K%=K%+1
780       PUT #1,K%                           'move data from buffer to disk
790    NEXT I%
800    CLOSE #1
810 RETURN
```

Program 10.2.   Creating a Random File.

opposite functions of MKI$, MKS$, and MKD$, respectively. Program 10.3 illustrates the process of reading back data from the file created in Program 10.2.

You may find one additional hint about random files useful. If you need to store a large number of data in each record, you can use multiple FIELD statements. For example:

```
10  FIELD #3, 2 AS A$(1), 2 AS A$(2), 2 AS A$(3)
20  FIELD #3, 6 AS DUM$, 2 AS A$(4), 2 AS A$(5), 2 AS A$(6)
30  FOR I=1 TO 6
40     LSET A$(I)=C%(I)
50  NEXT I
60  PUT #3,1
```

would store 6 integer data in a single record. DUM$ is a "dummy" variable used to show that the variable A$(4) starts at the seventh byte of the record. You can use as many FIELD statements as necessary to allocate space for the data you are storing.

10.2.3 Comparison of Sequential and Random Files

Sequential and random files each have their own advantages for scientific applications. Sequential files are easier to understand and program than are random files. In addition, a sequential file is stored in standard ASCII format, much like a text

```
10  REM PROGRAM RANDOM_FILE_READ
20  REM LABORATORY AUTOMATION USING THE IBM PC
30  REM Program to retrieve data stored in a random file
40  REM       created with Program 10.2
50  REM ********************************************
60  DIM RESULT%(1000),A$(12)
70     INPUT "Enter the name of the file containing the data. ",FILN$
80     OPEN FILN$ AS #1 LEN=24
90     FIELD #1, 2 AS A$(1), 2 AS A$(2), 2 AS A$(3), 2 AS A$(4), 2 AS A$(5),
       2 AS A$(6), 2 AS A$(7), 2 AS A$(8), 2 AS A$(9), 2 AS A$(10), 2 AS A$(11),
       2 AS A$(12)
100    REM get number of data stored in file
110    GET #1,1
120    NDATA%=CVI(A$(1))
130    K%=1
140    PRINT "Retrieving data...."
150    FOR I%=1 TO NDATA% STEP 12
160       K%=K%+1
170       GET #1,K%                          'move data to random file buffer
180       FOR J%=0 TO 11                      'move data from buffer to memory
190       N%=I%+J%
200             IF N% <= NDATA% THEN RESULT%(N%)=CVI(A$(J%+1))
210       NEXT J%
220    NEXT I%
230    CLOSE #1
240    FOR I%=1 TO NDATA%
250       PRINT I%,RESULT%(I%)
260    NEXT I%
270 END
```

Program 10.3.   Reading Data from a Random File.

file. Hence, we can type it out on the screen using the DOS command TYPE, e.g.,
TYPE MYDATA.DAT. Or we can edit it with a standard text editor.

Data stored in such a random file cannot be viewed using the TYPE or PRINT
statements in DOS. However, random files have two advantages over sequential
files. First, random files typically occupy less disk space because they store the
binary values for the numbers instead of the ASCII codes for each digit of the
number. Thus, for example, the integer number 2048 would take four bytes to store
in a sequential file (one byte for each digit) while it would take only 2 bytes to store
in a random file (because BASICA uses only 2 bytes to store any integer in binary).
Usually, sequential files also contain delimiters (e.g., commas, quotation marks,
spaces) while random files do not; this makes sequential files even less efficient.

Second, the individual values in a random file can be retrieved in the same
amount of time, regardless of their position in the file. In contrast, data at the end of
a sequential file require a longer time to retrieve than those at the beginning. The
time savings can thus be quite significant if the file contains many data and we only
wish to examine a few data at the end of the file.

As a rough rule of thumb, you might wish to consider random files for storing
more than 1000 data; for smaller amounts of data, sequential files are often more
convenient. If you store data in a random file, however, you will probably want to
include in your program the ability to display a section of the raw data on the screen

or even to store it in a sequential file, both for testing purposes during program design and for utilization by the end-user of the program.

10.3 PLOTTING

Almost all of the laboratory automation data analysis programs you will write will include a wide variety of graphical representations of your data. Plots are the scientist's "picture that tells a thousand words" (or more) of data. Fortunately, one of the tasks made easy by BASICA and many other languages is plotting.

10.3.1 Using BASIC

Suppose, for example, that we wish to plot a set of 1000 data points that we have collected from an instrument. We can break the process of plotting the points into three steps:

1.   Scaling the data.
2.   Plotting the data.
3.   Creating axes and graph labels.

These steps are illustrated in Program 10.4, which demonstrates several of the plotting features available in BASICA.

   Scaling is used to make the data fit on the graphics screen; normally, you will want to have the graph fill the entire screen. This is particularly easy to do using BASICA. To scale the data you must first find the minimum and maximum values in the data (lines 550 to 680 in Program 10.4). Once you know the minimum and maximum values of the data, you can use the WINDOW statement (line 730) to scale the data between these extremes automatically. In this case, we simply define a window such that the data completely fill it.

   The VIEW statement (line 720) selects the portion of the screen we wish to use for the graph. Since our screen is 640 x 200, we define the graph boundaries such that we leave an 80 x 200 area to the left of the graph, a 40 x 640 area below the graph, and a small space above the graph for axis labels and the graph title. We then simply use the LINE statement (line 770) to plot each of the data. A similar process is used to put tick marks on each of the axes.

   The most difficult portion of plotting using BASICA is labeling the axes and tick marks, because different units of distance are used for graphics and text. The best advice we can give you is to use a piece of graph paper to sketch a sample plot first and carefully figure out the conversions between the two systems of units. Pay particular attention to the ordinate (distance from the x-axis), because BASICA starts with the origin at the upper left corner, rather than the lower left-hand corner as you might expect. One of the functions of the VIEW statement is to convert the

```
10 REM PROGRAM GRAPHING_USING_BASIC
20 REM LABORATORY AUTOMATION USING THE IBM PC
30 REM Program to read either a one or two-dimensional array
40 REM        from a sequential file and plot it on the
50 REM        high resolution (640 x 200) display, with axes.
60 REM ********************************************************
70 DEFINT I-N
80 DIM X(1000),Y(1000)
90 CLS:SCREEN 0:KEY OFF
100 REM *********************
110 REM              Main program
120   GOSUB 180                       'get data from file
130   GOSUB 500                       'scale data and plot
140   GOSUB 830                       'label axes
150 END
160 REM *********************
170 REM Get data from sequential file
180   INPUT "Enter the name of the data file ",FILN$
190   ON ERROR GOTO 1100              'Go to error routine if can't open file
200   OPEN FILN$ FOR INPUT AS #1
210   ON ERROR GOTO 0                 'Disable error checking
220   INPUT "Enter the number of points to be plotted ",NPTS
230   INPUT "Does the data file contain only y-values? ",YVAL$
240   IF YVAL$="y" THEN YVAL$="Y"
250   IF YVAL$="Y" THEN GOSUB 330 ELSE GOSUB 430 'get data
260   INPUT "Enter the graph title ",TITLE$
270   INPUT "Enter the x-axis label ",XLABEL$
280   INPUT "Enter the y-axis label ",YLABEL$
290   CLS
300 RETURN
310 REM *********************
320 REM Subroutine to read only y-values from file
330   FOR I=1 TO NPTS
340        INPUT #1,Y(I)
350        X(I)=I                     'x-axis value = point number
360   NEXT I
370   PRINT "The x values range from 1 to ";NPTS
380   PRINT "You may wish to scale these values."
390   INPUT "Enter the scaling factor desired; enter 1.0 for no scaling ",XSCALE
400 RETURN
410 REM *********************
420 REM Subroutine to read both x and y-values from file
430   FOR I=1 TO NPTS
440        INPUT #1, X(I),Y(I)
450   NEXT I
460   XSCALE=1                        'only scale data if no x-values read in
470 RETURN
480 REM *********************
490 REM Scale data and plot
500   IF YVAL$="Y" THEN GOSUB 550 ELSE GOSUB 630 'scale data
510   GOSUB 710                       'plot data
520 RETURN
530 REM *********************
540 REM find minimum and maximum of y-array only
550   YMIN=Y(1):YMAX=Y(1)            'set initial values
560   FOR I=1 TO NPTS
570        IF Y(I) > YMAX THEN YMAX=Y(I) ELSE IF Y(I) < YMIN THEN YMIN=Y(I)
580   NEXT I
590   XMIN=1:XMAX=NPTS
600 RETURN
610 REM *********************
620 REM find minimum and maximum of both x and y arrays
630   YMIN=Y(1):YMAX=Y(1):XMIN=X(1):XMAX=X(1)
640   FOR I=1 TO NPTS
650        IF Y(I) >YMAX THEN YMAX=Y(I) ELSE IF Y(I) < YMIN THEN YMIN=Y(I)
660        IF X(I) >XMAX THEN XMAX=X(I) ELSE IF X(I) < XMIN THEN XMIN=X(I)
670   NEXT I
```

Program 10.4 (Part 1 of 2).   Plotting with BASIC.

```
 680 RETURN
 690 REM **********************
 700 REM plot data
 710   SCREEN 2                             'set in high resolution graphics mode
 720   VIEW (80,12)-(600,172)              'define viewport
 730   WINDOW (XMIN,YMIN)-(XMAX,YMAX)       'scale data to fit entire viewport
 740   LINE (XMIN,YMIN)-(XMAX,YMAX),,B      'put box around graph
 750   PSET (X(1),Y(1)),1                   'Plot first point
 760   FOR I=2 TO NPTS                      'Draw line connecting rest of points
 770        LINE -(X(I),Y(I)),1
 780   NEXT I
 790 RETURN
 800 REM **********************
 810 REM Put tick marks on axis and label
 820 REM calculate number of divisions of x-axis to use
 830   VIEW (80,172)-(600,180)              'redefine viewport for x-axis tick marks
 840   DELTA=(XMAX-XMIN)/5
 850   FOR I=0 TO 5
 860        X2=I*DELTA+XMIN
 870        LINE (X2,YMIN)-(X2,YMAX)  'draw tick mark
 880        LOCATE 24,I*13+6
 890        PRINT USING "##.##^^^^";X2*XSCALE; 'label tick mark
 900   NEXT I
 910   REM calculate number of divisions of y-axis to use
 920   VIEW (72,12)-(80,172)
 930   DELTA=(YMAX-YMIN)/5
 940   FOR I=0 TO 5
 950        Y2=I*DELTA+YMIN
 960        LINE(XMIN,Y2)-(XMAX,Y2)         'draw tick mark
 970        LOCATE (5-I)*4+2,1              'text position goes from top to bottom
 980        PRINT USING "##.##^^^^";Y2
 990   NEXT I
1000   X2=40-(LEN(XLABEL$)/2)              'label x-axis
1010   LOCATE 25,X2
1020   PRINT XLABEL$;
1030   LOCATE 1,1                          'label y-axis
1040   PRINT YLABEL$;
1050   VIEW                                'return to entire screen as viewport
1060   X2=40-(LEN(TITLE$)/2)               'add centered title
1070   LOCATE 1,X2;
1080   PRINT TITLE$;
1090 RETURN
1100 REM **********************
1110 REM Error on opening sequential file
1120   PRINT "That file was not found. "
1130   PRINT "Either type another file name"
1140   PRINT " or type a return to see the list of data files"
1150   INPUT "File name? ",FILN$
1160   IF FILN$="" THEN CLS:FILES "*.DAT":INPUT "File name? ",FILN$
1170 RESUME
```

Program 10.4 (Part 2 of 2).   Plotting with BASIC.

graphics units so that the origin is in the lower left-hand corner. However, text coordinates (using the LOCATE statement) always begin at the upper left-hand corner.

Particularly for graphs made in real time (as the data are being collected), you should try to optimize the performance of the graphing routine. Remember that integer operations proceed much more rapidly than floating point operations. Hence, if your data are all integers, you should convert Program 10.4 to make the data (X and Y arrays) integer. If the integer data represent floating point numbers in your instrument (e.g., if the A/D values of 0 to 4095 represent 0 to 2.5 Absorbance units in a spectroscopy system), simply label the axes appropriately rather than multiply all of the data by the conversion factor.

10.3.2 Using the Virtual Device Interface

While BASICA and some other programming languages provide direct support for screen graphics, considerably more difficulty arises when you wish to plot data on a plotter, color printer, or screens other than the standard Color Graphics screen that BASICA expects. Also, some other languages do not support graphics directly. Because almost every scientist wants to do plots on one or more of these other graphics devices, it is very common to need to write special graphics subroutines each time a new device is added to your system. Because each device has different methods of sending and receiving data, different scaling factors, different color options, and so forth, you can end up spending an inordinate amount of time writing graphics programs.

For this reason, there have been numerous efforts made to develop a standard graphics language for all devices. Unfortunately, programs written in these standard languages on a microcomputer often end up being too slow for normal scientific or engineering usage. However, we encourage you to investigate such systems for use in your own laboratory.

As an example of such systems, one of the standard graphics systems which we use in many applications is the Virtual Device Interface (VDI) system developed by Graphics Software Systems and available in a variety of forms including the IBM Graphics Development Toolkit. Although this is not a truly universal system, it is reasonably fast and supports a wide variety of graphics devices from a number of manufacturers. The major advantage of this system is that virtually no new BASIC code needs to be written to output to a new device. Instead, all device-dependent code is contained in **device drivers**. VDI device drivers are programs loaded at boot-up time by CONFIG.SYS that *convert generic graphics statements sent by your program into device-specific instructions*. For example, if your program sends a command that says "draw a line," then the VDI device driver converts this into the specific instructions needed to tell your plotter or printer or screen to draw a line at the desired location. By comparison, using BASIC alone you would have to write different commands to draw a line on each of the devices you are using, and you would need to have a detailed understanding of the programming of each device.

Many manufacturers of graphics devices in fact now supply a VDI device driver with their hardware to ensure ease of use. The same device drivers can also be used with a variety of programming languages.

As an example, let's look at Program 10.5. In this program any one of three different devices (or workstations, in VDI terminology) can be opened: a graphics screen, a plotter, or a graphics printer. Notice that the user simply selects one of the three devices; *the remaining plotting code does not use anything that is device-specific*. In the VDI system, all devices are assumed to have coordinates that extend from 0 to 32767 in each direction; the device drivers perform the conversion to the units actually used by each device.

If you look at Program 10.5 more carefully, you may discover another advantage of VDI. There is no reference to a specific plotter or printer. Instead, three generic devices, PLOTTER, PRINTER, and DISPLAY, are mentioned. How then does the program know, for example, that we are using an EGA display instead of a CGA display, say, or which manufacturer's printer we are using?

The association between the generic device name in the program (e.g., DISPLAY) and the physical device is a two-step process. First, we must load a device driver by describing it in a DEVICE statement in CONFIG.SYS. The second step is to use several commands, usually placed in AUTOEXEC.BAT, to associate the generic names with the specific device drivers and to initialize the device drivers. We use the SET command to perform the former task.

Thus, for example, we might insert lines such as the following into the CONFIG.SYS file in the root directory of our hard disk or on the diskette we use to "boot" the system:

```
DEVICE=VDIPRGRA.SYS /R
DEVICE=VDIDY006.SYS /R
DEVICE=VDI.SYS
```

At boot-up time, these VDI device drivers are loaded into memory. In this case, we have specified device drivers for the IBM Graphics Printer, the IBM CGA display used in high resolution mode (640 x 200), and the VDI generic software (which must always be loaded last). You would of course use the device drivers for the hardware in the your own system. The /*R* (be sure there are two spaces in front of the slash!) specifies that the drivers remain resident in memory (it is also possible to have several drivers share the same memory space).

Then, in our AUTOEXEC.BAT file, we would place the following commands:

```
SET DISPLAY=VDIDY006
SET PRINTER=VDIPRGRA
INIT__VDI
```

These associate the specific device with the generic names, and initialize all of the loaded VDI device drivers, respectively. You can write the program to support a wide variety of devices for different users without even knowing in advance what graphics devices they will be using. Then each user can select the appropriate device drivers for the particular devices they are currently using.

We use VDI or similar interfaces whenever possible in our own programs. The programs where we do not use it are those in which extremely high speed plotting is desired or when a VDI device driver does not exist for the graphics device we are using. We recommend using it or an equivalent system in virtually all other cases.

```
10 REM PROGRAM GRAPHING_USING_VDI
20 REM LABORATORY AUTOMATION USING THE IBM PC
30 REM Program to read either a one or two-dimensional array
40 REM      from a sequential file and plot it on a graphics device.
50 REM      Uses IBM Graphics Development Toolkit commands.
60 REM May be used with BASIC compiler only;
70 REM      will not work with BASIC Interpreter
80 REM Compile as         BASCOM PROG105/X;
90 REM    Link as         LINK PROG105,,,BASVDI+BASRUN20
100 REM *********************************************************
110 OPTION BASE 1                     'use this for arrays sent to VDI
120 CLEAR ,,3000                      'increase available stack size for VDI use
130 DIM X(1000),Y(1000),WORKIN%(19),WORKOUT%(66),XYPLOT%(4),DEVICE$(3)
140 CLS:SCREEN 0:KEY OFF
150 REM **********************
160 REM            Main Program
170    GOSUB 250                      'get data from file
180    GOSUB 570                      'initialize VDI
190    GOSUB 810                      'scale data and plot
200    GOSUB 1220                     'label axes
210    Y$=INKEY$: IF Y$="" THEN 210   'wait for user to type any key
220 END
230 REM **********************
240 REM Get data from sequential file
250    INPUT "Enter the name of the data file ",FILN$
260    ON ERROR GOTO 1860             'go to error routine if can't open file
270    OPEN FILN$ FOR INPUT AS #1
280    ON ERROR GOTO 0                'disable error checking
290    INPUT "Enter the number of points to be plotted ",NPTS
300    INPUT "Does the data file contain only y-values? ",YVAL$
310    IF YVAL$="y" THEN YVAL$="Y"
320    IF YVAL$="Y" THEN GOSUB 400 ELSE GOSUB 500 'get data
330    INPUT "Enter the graph title ",TITLE$
340    INPUT "Enter the x-axis label ",XLABEL$
350    INPUT "Enter the y-axis label ",YLABEL$
360    CLS
370 RETURN
380 REM **********************
390 REM Read only y-values from file
400    FOR I=1 TO NPTS
410         INPUT #1,Y(I)
420         X(I)=I                    'x-axis value = point number
430    NEXT I
440    PRINT "The x values range from 1 to ";NPTS
450    PRINT "You may wish to scale these values."
460    INPUT "Enter the scaling factor desired; enter 1.0 for no scaling ",XSCALE
470 RETURN
480 REM **********************
490 REM Read both x and y-values from file
500    FOR I=1 TO NPTS
510         INPUT #1, X(I),Y(I)
520    NEXT I
530    XSCALE=1                       'only scale data if no x-values read in
540 RETURN
550 REM **********************
560 REM Initialize VDI
570    WHILE (GRDEVICE < 1 OR GRDEVICE > 3 )
580         PRINT "Please select a graphics device:"
590         INPUT "1=screen, 2=plotter, 3=printer",GRDEVICE
600    WEND
610    DEVICE$(1)="DISPLAY ":DEVICE$(2)="PLOTTER ":DEVICE$(3)="PRINTER "
620    FOR I=1 TO 11                  'get default values for VDI
630         READ WORKIN%(I)
640    NEXT I
650    REM workin%(1)=do not preserve aspect ratio
660    REM workin%(2-10)= use default values for other parameters
670    REM workin%(11)= do not display device-dependent prompts
```

Program 10.5 (Part 1 of 3).   Plotting using Graphics Development Toolkit.

```
680    DATA 0,1,1,1,1,1,1,1,1,0
690    FOR I=1 TO 8                        'convert device name to ASCII codes
700        WORKIN%(I+11)=ASC(MID$(DEVICE$(GRDEVICE),I,1))
710    NEXT I
720    REM open VDI workstation
730    CALL VOPNWK(WORKIN%(1),DEVHANDLE%,WORKOUT%(1),VERROR%)
740    IF VERROR% >-1 THEN RETURN
750      PRINT "Error opening VDI workstation"
760      PRINT "Check both CONFIG.SYS and AUTOEXEC.BAT for proper installation"
770      PRINT "      of VDI device drivers."
780 STOP
790 REM **********************
800 REM Scale data and plot
810    IF YVAL$="Y" THEN GOSUB 860 ELSE GOSUB 940 'scale data
820    GOSUB 1000                        'plot data
830 RETURN
840 REM **********************
850 REM Find minimum and maximum of y-array only
860    YMIN=Y(1):YMAX=Y(1)              'set initial values
870    FOR I=1 TO NPTS
880        IF Y(I) > YMAX THEN YMAX=Y(I) ELSE IF Y(I) < YMIN THEN YMIN=Y(I)
890    NEXT I
900    XMIN=1:XMAX=NPTS
910 RETURN
920 REM **********************
930 REM Find minimum and maximum of both x and y arrays
940    YMIN=Y(1):YMAX=Y(1):XMIN=X(1):XMAX=X(1)
950    FOR I=1 TO NPTS
960        IF Y(I) >YMAX THEN YMAX=Y(I) ELSE IF Y(I) < YMIN THEN YMIN=Y(I)
970        IF X(I) >XMAX THEN XMAX=X(I) ELSE IF X(I) < XMIN THEN XMIN=X(I)
980    NEXT I
990    RETURN
1000 REM **********************
1010 REM Plot data
1020    CALL VCLRWK(DEVHANDLE%,VERROR%)      'clear workstation
1030    REM define factors to scale data in both dimensions
1040    XFACT=.7 * WORKOUT%(52) / (XMAX-XMIN)     'scale x and y axes to fill
1050    YFACT=.7 * WORKOUT%(53) / (YMAX-YMIN)     '   70% of plot area
1060    XOFF=WORKOUT%(52)*.25             'leave room for labels
1070    YOFF=WORKOUT%(53)*.14
1080    DEF FNXSCALE%(A!)=XOFF + (A!-XMIN) * XFACT
1090    DEF FNYSCALE%(A!)=YOFF + (A!-YMIN) * YFACT
1100    XYPLOT%(1)=FNXSCALE%(X(1))
1110    XYPLOT%(2)=FNYSCALE%(Y(1))
1120    FOR I=2 TO NPTS                   'scale data to fit graphics device and plot
1130        XYPLOT%(3)=FNXSCALE%(X(I))
1140        XYPLOT%(4)=FNYSCALE%(Y(I))
1150        CALL VPLINE(DEVHANDLE%,2,XYPLOT%(1),VERROR%)
1160        XYPLOT%(1)=XYPLOT%(3)         'use second point as first point next time
1170        XYPLOT%(2)=XYPLOT%(4)
1180    NEXT I
1190 RETURN
1200 REM **********************
1210 REM Put tick marks on axis and label
1220    XYPLOT%(1)=FNXSCALE%(XMIN)        'draw box around graph
1230    XYPLOT%(2)=FNYSCALE%(YMIN)
1240    XYPLOT%(3)=FNXSCALE%(XMAX)
1250    XYPLOT%(4)=XYPLOT%(2)
1260    CALL VPLINE(DEVHANDLE%,2,XYPLOT%(1),VERROR%)
1270    XYPLOT%(1)=XYPLOT%(3)
1280    XYPLOT%(4)=FNYSCALE%(YMAX)
1290    CALL VPLINE(DEVHANDLE%,2,XYPLOT%(1),VERROR%)
1300    XYPLOT%(2)=XYPLOT%(4)
1310    XYPLOT%(3)=FNXSCALE%(XMIN)
1320    CALL VPLINE(DEVHANDLE%,2,XYPLOT%(1),VERROR%)
1330    XYPLOT%(1)=XYPLOT%(3)
```

Program 10.5 (Part 2 of 3). Plotting using Graphics Development Toolkit.

```
1340    XYPLOT%(4)=FNYSCALE%(YMIN)
1350    CALL VPLINE(DEVHANDLE%,2,XYPLOT%(1),VERROR%)
1360    DELTA=(XMAX-XMIN)/5              'plot and label tick marks on X axis
1370    XYPLOT%(2)=FNYSCALE%(YMIN)
1380    XYPLOT%(4)=XYPLOT%(2)-1000
1390    REM set text alignment to top center
1400    CALL VSTALN(DEVHANDLE%,1,2,HOROUT%,VEROUT%,VERROR%)
1410    REM set character height to approximately 5% of screen size
1420    HEIGHT%=.05*WORKOUT%(52)
1430    CALL VSTHGT(DEVHANDLE%,HEIGHT%,XHEIGHT%,XWIDTH%,CWIDTH%,CHEIGHT%,VERROR%)
1440    FOR I=0 TO 5
1450        X2=I*DELTA+XMIN
1460        XYPLOT%(1)=FNXSCALE%(X2)
1470        XYPLOT%(3)=XYPLOT%(1)
1480        CALL VPLINE(DEVHANDLE%,2,XYPLOT%(1),VERROR%) 'draw tick mark
1490        LABEL$=STR$(X2*XSCALE)    'get tick label
1500        CALL VGTEXT(DEVHANDLE%,XYPLOT%(1),XYPLOT%(4)-300,LABEL$,VERROR%)
1510    NEXT I
1520    XYPLOT%(1)=FNXSCALE%(.5*(XMAX+XMIN))      'put centered X axis label
1530    XYPLOT%(2)=FNYSCALE%(YMIN)-1300-CHEIGHT%-200
1540    CALL VGTEXT(DEVHANDLE%,XYPLOT%(1),XYPLOT%(2),XLABEL$,VERROR%)
1550    DELTA=(YMAX-YMIN)/5                       'plot and label y-axis tick marks
1560    REM set text alignment to middle right
1570    CALL VSTALN(DEVHANDLE%,2,1,HOROUT%,VERTOUT%,VERROR%)
1580    XYPLOT%(3)=FNXSCALE%(XMIN)
1590    XYPLOT%(1)=XYPLOT%(3)-1000
1600    FOR I=0 TO 5
1610        Y2=I*DELTA+YMIN
1620        XYPLOT%(2)=FNYSCALE%(Y2)
1630        XYPLOT%(4)=XYPLOT%(2)
1640        CALL VPLINE(DEVHANDLE%,2,XYPLOT%(1),VERROR%) 'draw tick mark
1650        LABEL$=STR$(Y2)
1660        CALL VGTEXT(DEVHANDLE%,XYPLOT%(1)-300,XYPLOT%(2),LABEL$,VERROR%)
1670    NEXT I
1680    XYPLOT%(1)=.05*WORKOUT%(52)  'put centered Y axis label
1690    XYPLOT%(2)=FNYSCALE%((YMAX+YMIN)/2)
1700    CALL VSTROT(DEVHANDLE%,900,VERROR%)    'rotate axis to vertical
1710    REM display label
1720    CALL VGTEXT(DEVHANDLE%,XYPLOT%(1),XYPLOT%(2),YLABEL$,VERROR%)
1730    REM add graph title
1740    REM set text alignment to bottom center
1750    CALL VSTALN(DEVHANDLE%,1,0,HOROUT%,VEROUT%,VERROR%)
1760    REM set character height to approximately 8% of screen size
1770    HEIGHT%=.08*WORKOUT%(52)
1780    CALL VSTHGT(DEVHANDLE%,HEIGHT%,XHEIGHT%,XWIDTH%,CWIDTH%,CHEIGHT%,VERROR%)
1790    CALL VSTROT(DEVHANDLE%,0,VERROR%)    'rotate axis label to horizontal
1800    XYPLOT%(1)=FNXSCALE%((XMIN+XMAX)/2)
1810    XYPLOT%(2)=.9*WORKOUT%(53)
1820    CALL VGTEXT(DEVHANDLE%,XYPLOT%(1),XYPLOT%(2),TITLE$,VERROR%)
1830    REM for plotter or printer, plot is not done until workstation is cleared
1840    IF GRDEVICE > 1 THEN CALL VCLRWK(DEVHANDLE%,VERROR%)
1850 RETURN
1860 REM *********************
1870 REM Error on opening sequential file
1880    PRINT "That file was not found. "
1890    PRINT "Existing files are:"
1900    FILES "*.DAT"               'display list of data files
1910    INPUT "New file name? ",FILN$
1920 RESUME
```

Program 10.5 (Part 3 of 3).   Plotting using Graphics Development Toolkit.

10.3.3 Pseudo-Three-Dimensional Plots

Another general plotting technique that we have found useful, particularly in applications involving multiple detectors or data collected repetitively over fixed time intervals, is pseudo-three-dimensional plotting. In this technique, data are made to appear three-dimensional by displacing each set of data some distance from the previous set. The displacement may be in any of several directions, but is commonly above and to the right of the previous data set.

A very common example of this type of problem is scanning spectrometers (UV, IR, and mass spectrometers often do this). They produce a complete "spectrum" every few seconds or minutes, where a spectrum is a plot of detector response as a function of the detector energy being scanned. In many applications, the user wishes to use such a device to monitor a dynamic event; for example, use the spectrometer to monitor the effluent from a chromatograph, or to monitor a chemical reaction or physical change in a sample. In such cases, it is useful to be able to plot the spectra as a function of time. In this example, we might use the displacement to give the appearance of time. That is, we add both an X and Y displacement to each spectrum, where the displacements are proportional to the time elapsed. Figure 10.1 shows an example of such an effect.

To make the image more realistic, a **hidden line** technique is used to avoid plotting data which in a real three-dimensional image would be obscured by a larger set of data in front. This effect is also illustrated in Figure 10.1.

One method for implementing pseudo-three-dimensional plotting with hidden line elimination uses the following steps. The following discussion assumes that you have a series of data sets to plot, with each data set containing an equal number of values (Z) at equally spaced X intervals. Each data set represents a different value of Y (where Y is usually time).

1. Create a "plot array" that initially contains all zeros. Make the array as long as the longest array to be plotted plus the longest X-axis offset to be added to any plot. This array is the array to be plotted.

2. Read in the first data set into a "data array," which now contains *n* Z values. The first data set will be the one that appears to be closest to you when you view the finished plot.

3. Copy the data array into the plot array; i.e., copy each Z value to the corresponding point in the current plot array.

4. Plot the first n points of the plot array. In effect, you are simply plotting the n points in the first data set.

5. Read another set of data into the data array. To each value in the data array, add an offset, ΔZ, which is proportional to the difference in Y values between the first data set and this data set; i.e.,

    $$\Delta Z = k * \Delta Y$$

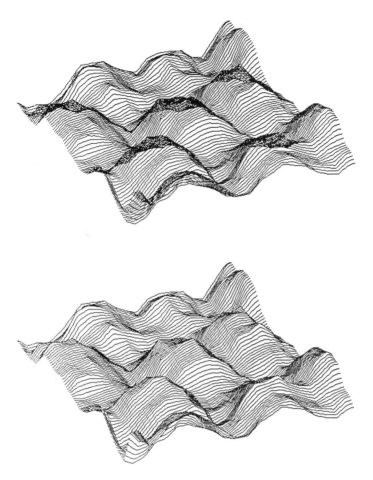

Figure 10.1. Pseudo-Three-Dimensional Plots. **Top**Pseudo-three-dimensional plot with no hidden line elimination. **Bottom**Same, except with hidden line elimination. Notice that the three-dimensional effect is achieved by plotting each line slightly above and to the left of the previous line, starting at the front of the image.

where **k** is a constant that can be based upon the size of the display area, the number of data sets, and the maximum Z value to be plotted. If Y is time, then ΔZ will be proportional to the time elapsed.

6. Calculate a ΔX in a fashion similar to that of the ΔZ. This ΔX is the X offset, and proportional to ΔY.

7. Beginning at point ΔX in the plot array and point 1 in the data array, compare each pair of points. If the Z value in the data array is larger than the Z value in the corresponding point in the plot array, then replace the point in the point

array with the point in the data array. If the data array point is smaller, then
leave the plot array unchanged.

8. Plot N points in the plot array, beginning at point ΔX.

9. Repeat steps 4 to 6 until all data sets have been plotted.

A program to plot such data is given in Program 10.6.

If you perform such plots frequently, you may also wish to read the literature
on filtering data in two dimensions and on techniques for implementing these types of
plots more efficiently or with more sophisticated algorithms. You may also wish to
read Chapter 12 for information about presenting such data sets as gray-scale
images.

10.4 MENUS

It is obvious that a laboratory automation program is going to include quite a few
graphs. What may not be so obvious is that an even more widely-used feature of
such programs is **menus**. By menus, we mean portions of the program that do one or
more of the following:

- Give the user a choice of available options.
- Obtain numeric entries from the user.
- Display "help" information.
- Display error or other information.

As we will see, the latter two types of displays are not traditional menu items, but
you should consider them as such nonetheless.

In a typical lab automation program, much of the program consists of gathering information and choices from the user. For example, you may want to ask the
user, "How many points per second should be collected?" Or, "Do you want to
collect data or plot previously collected data?" Or, "Which of the following data
files do you wish to analyze?" Because a high proportion of a typical lab automation
program is typically spent asking such questions, it will considerably shorten your
task if you develop a clear-cut system for designing your menus. In addition,
deciding in advance on a menu system will make your program look more professional. It will certainly decrease the amount of time you spend programming.

10.4.1 Using BASICA

One way of achieving consistent menus is to use a menu-building subroutine written
in BASICA. An example of a program that uses such a menu subroutine is
Program 10.7.

```
10 REM PROGRAM 3-D
20 REM LABORATORY AUTOMATION USING THE IBM PC
30 REM Program to plot successive data sets in pseudo-three-dimensional plot
40 REM      using simple hidden-line elimination technique
50 REM ************************************************************************
60 DEFINT A-Z
70 DIM D(3600),B(3600)
80 CLS:KEY OFF
90 MAXINT=32767                       'maximum integer value allowed
100 REM               Main program
110    GOSUB 190                      'open data file and read
120    GOSUB 250                      'get user input
130    GOSUB 380                      'plot data
140    CLOSE #1
150    Y$=INKEY$: IF Y$="" THEN 150   'wait for user to strike key
160 END
170 REM *************************************
180 REM Open data file
190    INPUT "Enter the file name of the data to be displayed ",FILN$
200    OPEN FILN$ FOR INPUT AS #1
210    INPUT #1,TOTSETS,STARTX,ENDX,STEPX
220 RETURN
230 REM *************************************
240 REM Get user input
250    PRINT "There were ";TOTSETS;" total data sets collected."
260    PRINT "The x-values ranged from ";STARTX;" to ";ENDX; "in steps of ";STEPX
270    INPUT "Enter the minimum Z value to plot ",ZMIN
280    INPUT "Enter the maximum Z value to plot ",ZMAX
290    INPUT "Enter the x-axis label ",XLABEL$
300    INPUT "Enter the y-axis label ",YLABEL$
310    INPUT "Enter the z-axis label ",ZLABEL$
320    INPUT "Enter the graph title ",TITLE$
330    INPUT "Enter the x-axis offset between successive plots ",XOFFSET
340    INPUT "Enter the z-axis offset between successive plots ",ZOFFSET
350 RETURN
360 REM *************************************
370 REM Main plotting routine
380    LOCATE 20,1:PRINT "Type the 'A' key to abort the plotting process"
390    FOR H=1 TO 3600:B(H)=ZMIN:NEXT H   'initialize array
400    'calculate step size in x-direction
410    XOFFSET=50/TOTSETS
420    IF XOFFSET < 1 THEN XOFFSET=1
430    'step size in z-direction = XOFFSET * ZOFFSET
440    ZOFFSET=(ZMAX-ZMIN)/200
450    'adjust for offsets added during plotting
460    ZMAX=ZMAX+XOFFSET*ZOFFSET*(TOTSETS-1)
470    XMIN=1
480    XMAX=XOFFSET*(TOTSETS-1)+(ENDX-STARTX+1)
490    GOSUB 850                      'set up graph mode
500    WINDOW (XMIN,ZMIN)-(XMAX,ZMAX)  'set window for all plots
510    XSTART!=STARTX
520    XEND!=ENDX
530    GOSUB 990                      'box and label
540    XMAX=ENDX-STARTX+1
550    T=0
560    FOR NSPECT=1 TO TOTSETS
570        GOSUB 790 'get one data set
580        'Plot first point
590        IF (D(1)+T*OFFSET>B(1+T))THEN PSET (1+T,D(1)+T*OFFSET),0 ELSE
           PSET (1+T,B(1+T)),0
600        'Plot remaining points offset from old points
610        FOR K=1 TO ENDX-STARTX+1 STEP STEPX
620            M=K+T                      'x-value to plot=x-value plus x-offset
630            N=D(K)                     'get z-value then check if offscale
640            IF N+NSPECT*ZOFFSET! > MAXINT THEN N=MAXINT-NSPECT*ZOFFSET
650            'check if current value is "hidden" by previous value
```

Program 10.6 (Part 1 of 3). Pseudo-Three-Dimensional Plot with Hidden Line Elimination.

```
660                'if not, update B array with new value; otherwise, use
670                'previous value stored in B array
680        IF N+NSPECT*ZOFFSET > B(M) THEN B(M)=N+NSPECT*ZOFFSET
690      NEXT K
700      GOSUB 920 'plot B array without resetting window size
710      T=T+XOFFSET
720      XMIN=XMIN+XOFFSET
730      XMAX=XMAX+XOFFSET
740      Y$=INKEY$:IF Y$="a" OR Y$="A" THEN 760 'abort plot?
750    NEXT NSPECT
760 RETURN
770 REM **************************************
780 REM Read one data set
790    FOR I=STARTX TO ENDX STEP STEPX
800        INPUT #1,D(I)
810    NEXT I
820 RETURN
830 REM **************************************
840 REM Set to high resolution screen
850    SCREEN 2:CLS
860    VIEW (35,20)-(635,175)
870    WINDOW (0,-100)-(1000,8000)
880 RETURN
890 REM **************************************
900 REM Graph an array (B) with lines connecting points
910    WINDOW (XMIN,ZMIN)-(XMAX,ZMAX)
920    PSET (XMIN,B(XMIN)),0
930    FOR I=XMIN TO XMAX
940        LINE -(I,B(I))
950    NEXT I
960 RETURN
970 REM **************************************
980 REM Label 3-D axes
990    DEF FND1(X2)=X2/639*80  'convert view units to character units on x-axis
1000   'convert window units to character units on x-axis
1010   DEF FNXDIST(X)=(X-XMIN)/(XMAX-XMIN)*(FND1(635)-FND1(35))+FND1(35)
1020   DEF FNYDIST(Y)=(Y-ZMIN)/(ZMAX-ZMIN)*(20/199*25-175/199*25)+(175/199*25)
1030   ZMAX2=XOFFSET*ZOFFSET*(TOTSETS-1)+ZMIN 'length of y axis (in y-direct)
1040   ZMAX3=ZMAX-ZMAX2                'top of z axis
1050   XMAX2=ENDX-STARTX+2             'location of x,y axis intersect (x-value)
1060   VIEW(35,20)-(635,175)          'view graph area
1070   WINDOW (XMIN,ZMIN)-(XMAX,ZMAX)
1080   'draw three axes
1090   LINE (XMIN,ZMIN)-(XMIN,ZMAX3)  'z axis
1100   LINE (XMAX2,ZMIN)-(XMAX,ZMAX2)  'y axis
1110   LINE (XMIN,ZMIN)-(XMAX2,ZMIN)  'x axis
1120   'draw z axis ticks and labels
1130   VIEW (30,20)-(35,175)
1140   J=-1
1150   FOR I=ZMIN TO ZMAX3 STEP (ZMAX3-ZMIN)/4
1160       LINE(XMIN,I)-(XMAX,I)
1170       LOCATE FNYDIST(I),1        'print tick label
1180       J=J+1
1190       PRINT USING "####";ZMIN+J/4*(ZMAX3-ZMIN)
1200   NEXT I
1210   LOCATE 5,1:PRINT ZLABEL$;
1220   'x axis ticks and labels
1230   VIEW (35,175)-(635,180)
1240   FOR I=XMIN TO XMAX2 STEP (XMAX2-XMIN)/3
1250       LINE (I,ZMIN)-(I,ZMAX)
1260   NEXT I
1270   LOCATE 25,(FNXDIST(XMAX2)-FNXDIST(XMIN))/2-7:PRINT XLABEL$;
1280   X1=FNXDIST(XMIN)
1290   X2=FNXDIST(XMAX2)
1300   J=-1
```

**Program 10.6 (Part 2 of 3). Pseudo-Three-Dimensional Plot with Hidden Line Elimi-
nation.**

```
1310    FOR I=X1 TO X2 STEP (X2-X1)/3  'label tick marks
1320        IF (I > 78) THEN GOTO 1380
1330        LOCATE 24,I
1340        J=J+1
1350        PRINT USING "####";XSTART!+J/3!*(XEND!-XSTART!);
1360    NEXT I
1370    'y axis ticks and labels
1380    VIEW (35,20)-(635,175)
1390    YSTART!=1
1400    YEND!=TOTSETS
1410    J=-1
1420    FOR I=XMAX2 TO XMAX STEP (XMAX-XMAX2)/3
1430        J=J+1
1440        Y=(ZMAX2-ZMIN)/3!*J+ZMIN
1450        LINE (I,Y)-(I+7,Y)
1460        IF (FNXDIST(I+10)>76) THEN GOTO 1500
1470        LOCATE FNYDIST(Y)+1,FNXDIST(I+10)+1
1480        PRINT USING "####";YSTART!+(J/3!)*(YEND!-YSTART!);
1490    NEXT I
1500    LOCATE 22,70:PRINT YLABEL$;
1510    LOCATE 1,40-LEN(TITLE$)/2:PRINT TITLE$
1520    VIEW(35,20)-(635,175)
1530 RETURN
```

Program 10.6 (Part 3 of 3).   Pseudo-Three-Dimensional Plot with Hidden Line Elimination.

Notice that you can very easily use Program 10.7 to build a wide variety of menus that require numeric or alphanumeric input. For each menu, you must add data lines like those in lines 1200 to 1230 to display the questions you need to have answered by the user. The subroutine in lines 670 to 830 then takes care of gathering the user's answers. An example of a menu built with this program is shown in Figure 10.2.

If you find yourself designing a number of such menus, you might also wish to consider reading in the menu information from a sequential file instead of using DATA statements.

Program 10.7 also illustrates the use of function keys as part of a menu. BASICA allows you to define and display up to 10 function keys at the bottom of the screen; a function key is one of the 10 keys marked F1 through F10 on most keyboards. Typing these keys followed by the return key at any time during the program will cause it to jump to a subroutine you have defined. A particularly important key is the HELP key. You can use this key to display on-line help about anything the user might wish to know when using the program. A good general rule-of-thumb is that *every menu should have a HELP key and that you should provide enough information so that the user does not need to consult a manual to use your program.*

Finally, notice that Program 10.7 displays a question when the user has completed all of the entries (lines 1280 to 1320); you can use a similar technique to display short informational entries about each highlighted item in the menu or to display error information.

```
10 REM PROGRAM MENU
20 REM LABORATORY AUTOMATION USING THE IBM PC
30 REM Program to use a menu to get numeric or alphanumeric entries
40 REM ********************************************************************
50 DEFINT I-N
60 DIM LOCATNUM(10),LOCATALPH(10),XNUMANS(10)
70 DIM NUMERICS$(10),ALPHANUM$(10),PFKEY$(10),XALPHA$(10)
80 REM              Main program
90    GOSUB 230                         'read in screen information
100   ON KEY(1) GOSUB 1060              'enable help key
110   ON KEY(3) GOSUB 670               'enable numeric key
120   ON KEY(5) GOSUB 860               'enable alphanumeric key
130   ON KEY(10) GOSUB 1160             'enable end key
140   KEY(1) ON: KEY(3) ON: KEY(5) ON: KEY(10) ON
150   KEY ON                            'display keys
160   GOSUB 460                         'display screen and get values
170   GOTO 170                          'wait for user to type a key
180   FOR I=1 TO 10:KEY(I) OFF:NEXT 'turn off keys
190   CLS:KEY OFF
200 END
210 REM ******************************
220 REM get screen information
230   READ NNUM,NALPHA                  'get number of numeric, alphanumeric entries
240   FOR I=1 TO NNUM                   'get screen locations of numeric entries
250       READ LOCATNUM(I)
260   NEXT I
270   FOR I=1 TO NNUM
280       READ NUMERICS$(I)
290   NEXT I
300   FOR I=1 TO NALPHA                 'get screen locations alphanumeric entries
310       READ LOCATALPH(I)
320   NEXT I
330   FOR I=1 TO NALPHA                 'get questions for alphanumeric entries
340       READ ALPHANUM$(I)
350   NEXT I
360   READ TITLE$                       'get screen title
370   FOR I=1 TO 10                     'get function keys
380       READ PFKEY$(I)
390   NEXT I
400   FOR I=1 TO 10                     'define PF keys
410       KEY I,PFKEY$(I)
420   NEXT I
430 RETURN
440 REM ******************************
450 REM Display screen
460   CLS:SCREEN 0:WIDTH 80             'clear screen
470   LOCATE 1,1:PRINT DATE$;" ";TIME$;
480   LOCATE 1,40-(LEN(TITLE$)/2):PRINT TITLE$
490   LOCATE 2,1:FOR I=1 TO 79:PRINT CHR$(205);:NEXT I 'draw double line
500   LOCATE 24,1:FOR I=1 TO 79:PRINT CHR$(205);:NEXT I
510   LOCATE 23,20:COLOR 7,0:PRINT "Type the F1 key for help.":COLOR 7,0
520   FOR I=1 TO NNUM                   'print current values of numeric entries
530       LOCATE LOCATNUM(I),1
540       PRINT NUMERICS$(I)
550       LOCATE LOCATNUM(I),LEN(NUMERICS$(I))+2
560       PRINT XNUMANS(I)
570   NEXT I
580   FOR I=1 TO NALPHA                 'print current values of alphanumerics
590       LOCATE LOCATALPH(I),1
600       PRINT ALPHANUM$(I)
610       LOCATE LOCATALPH(I)+1,1
620       PRINT XALPHA$(I)
630   NEXT I
640 RETURN
650 REM ******************************
660 REM Get user's numeric entries
670   Y$="Y"
```

Program 10.7 (Part 1 of 3). General-Purpose Menu.

```
680    WHILE Y$ = "Y" OR Y$="y"
690       FOR I=1 TO NNUM
700          L1=LOCATNUM(I):L2=LEN(NUMERICS$(I))+2
710          LOCATE L1,L2
720          A$=STR$(XNUMANS(I))
730          COLOR 0,7:PRINT A$;   'print current value in reverse video
740          LOCATE L1,L2
750          INPUT "",Y$              'get new value
760          IF Y$<>"" THEN XNUMANS(I)=VAL(Y$)  'check if retain old value
770          COLOR 7,0:LOCATE L1,L2:PRINT SPACE$(LEN(A$));
780          LOCATE L1,L2
790          PRINT XNUMANS(I);
800       NEXT I
810       GOSUB 1270               'check if entries OK
820    WEND
830 RETURN
840 REM ****************************
850 REM Get user's alphanumeric entries
860    y$="Y"
865    WHILE Y$="y" OR Y$="Y"
870       FOR I=1 TO NALPHA
880          L1=LOCATALPH(I)+1
890          LOCATE L1,1
900          COLOR 0,7:PRINT SPACE$(79); 'highlight space for entry
910          LOCATE L1,1
920          PRINT XALPHA$(I);    'print previous value
930          LOCATE L1,1
940          INPUT "",Y$           'wait for input
950          IF Y$<>"" THEN  XALPHA$(I)=Y$    'check for RETURN key
960          COLOR 7,0:LOCATE L1,1
970          PRINT SPACE$(80);    'clear off reverse video
980          LOCATE L1,1
990          COLOR 7,0:PRINT XALPHA$(I); 'and reprint in normal color
1000      NEXT I
1010      GOSUB 1280               'check if entries are OK
1020   WEND
1030 RETURN
1040 REM ****************************
1050 REM User has typed function key #1 = HELP
1060   COLOR 7,0:CLS:KEY OFF
1070   LOCATE 10,1:PRINT "Type F3 to change numeric entries"
1080   LOCATE 12,1:PRINT "Type F5 to change the alphanumeric entries"
1090   LOCATE 14,1:PRINT "Type F10 to exit"
1100   LOCATE 20,1:INPUT "Type the return key to continue",RK$
1110   GOSUB 460
1120   KEY ON
1130 RETURN
1140 REM ****************************
1150 REM User has typed function key #10=EXIT
1160    COLOR 7,0:CLS
1170 END
1180 REM ****************************
1190 REM Data statements-- insert yours here to get correct values
1200 DATA 2,2,5,6
1210 DATA "Numeric entry one?","Numeric entry two?"
1220 DATA 8,10
1230 DATA "Alphanumeric entry one?","Alphanumeric entry two?"
1240 DATA "Title of menu"
1250 DATA "HELP","    ","NUMB","    ","ALPH","    ","    ","    ","    ","EXIT"
```

Program 10.7 (Part 2 of 3).   General-Purpose Menu.

A somewhat simpler program illustrating one method of providing the user a series of items to choose among (a **selection menu**) is illustrated in Program 10.8.

Note that with a selection menu, the user uses the arrow keys to move the highlighted area and then types the return (enter) key to indicate the desired selection. Of course, you may wish to combine this type of menu with that illustrated in

```
1260 REM *****************************
1270 REM check if user thinks entries are OK
1280   COLOR 0,7
1290   LOCATE 21,1:INPUT "Do you wish to change any answer? <Y or N> ",Y$
1300   COLOR 7,0
1310   LOCATE 21,1:PRINT SPACE$(79);
1320 RETURN
```

Program 10.7 (Part 3 of 3). General-Purpose Menu.

Program 10.7. Other possibilities include arranging the choices horizontally rather than vertically and letting the user type the first letter of the choice instead of selecting the item with the arrow keys.

10.4.2 Using Menu-Generating Software

For the occasional programmer, using BASICA directly to build menus is the easiest approach. However, if you are planning to write a very large program or a number of programs, it may be worth your time to investigate one of the programs commercially available for designing such menus. These programs are generally referred to as screen-designers or panel-designers. The advantages of these programs include the following:

■ Generally they allow you to design the menu directly on the screen instead of having to use LOCATE statements to position everything.

■ They are language independent, so you can use the same menu if you rewrite your program in another language.

■ They work on more than one type of display (e.g., both Color Graphics and Monochrome).

■ They store the menus on the disk instead of consuming space in your program.

■ They encourage the development of menus with consistent appearance.

■ They provide a number of sophisticated features (e.g., scrolling, selection menus, "pop-up" menus, easy use of screen colors) that are difficult to provide using BASIC alone.

The major disadvantage of these programs is that they are generally rather complicated for first-time users.

Thus, the major advantage of menu-display programs is that they allow you to modify the appearance of a screen without having to modify your program. Furthermore, you can see exactly what the menu screen will look like as you design it.

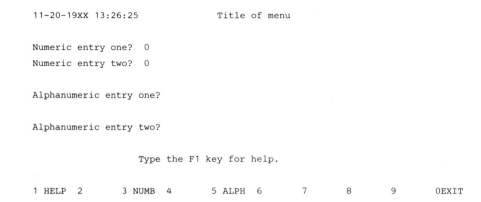

```
11-20-19XX 13:26:25              Title of menu

Numeric entry one?  0
Numeric entry two?  0

Alphanumeric entry one?

Alphanumeric entry two?

              Type the F1 key for help.

1 HELP  2        3 NUMB  4       5 ALPH  6        7        8        9        0EXIT
```

Figure 10.2.   Sample Menu. This menu was generated using Program 10.7. When the
program is run, each entry area is highlighted as the user enters information,
and then returned to the normal color when the user types the return key.

Overall, we use commercial menu-generating software when writing very large
programs or programs that will be distributed to a variety of users.   For small,
single-user programs we generally prefer using BASICA directly.

10.5 NOTEBOOK ENTRIES

As a final suggestion for a simple addition to your programs, we would recommend
that you add the ability to display a complete set of laboratory notebook entries. We
strongly recommend that every set of data collected in your lab have notebook entries
as a permanent part of the data file; *every data file should include an "electronic
notebook."* Notebook entries should include a complete description of the sample, all
operating parameters for your instrument, and a space for special remarks by the
operator; in short, everything that will go into the scientist's notebook.

   You should make it possible to recall this information during the data analysis
portion of the program as well as at the end of data collection. You might wish to use
a menu to collect this information and then provide a function key to display it at any
time during the data analysis process, for example. You should also give the user the
option of printing a copy of it on a printer.

   Too often, programmers neglect to provide the user the opportunity to enter
notebook information and too often the user will try to bypass this portion of the
program. However, it is simply good laboratory practice to have all notebook information permanently associated with the data. Before laboratory computers were
available, all data and comments went into the lab notebook; now, both should go
into a permanent disk file as well. You can help the users by providing an extensive

```
10 REM PROGRAM SELECTION_MENU
20 REM LABORATORY AUTOMATION USING THE IBM PC
30 REM Program to create a selection menu
40 REM       User selects only one of several choices
50 REM ***********************************************************************
60 DEFINT I-N
70 DIM M$(100)
80 M$(1)="USER CHOICE ONE":    M$(2)="USER CHOICE TWO"
90 M$(3)="USER CHOICE THREE":  M$(4)="USER CHOICE FOUR" : M$(5)="EXIT"
100 NCHOICE=5                          'number of choices for user
110 CHOICE=0
120 REM **********************
130 REM               Main program
140    WHILE CHOICE < 5
150       GOSUB 290                     'call subroutine which gets choice
160       LOCATE 10,1
170       ON CHOICE GOSUB 220,230,240,250,260
180    WEND
190 END
200 REM **********************
210 REM User subroutines -- put your routines in place of these
220    PRINT "User selected choice one"  :GOSUB 610: RETURN
230    PRINT "User selected choice two"  :GOSUB 610: RETURN
240    PRINT "User selected choice three":GOSUB 610: RETURN
250    PRINT "User selected choice four" :GOSUB 610: RETURN
260 RETURN
270 REM **********************
280 REM general purpose menu choice subroutine
290    SCREEN 0:CLS:WIDTH 80            'make sure is in 80-column text mode
300    FOR IC=1 TO NCHOICE              'print choices on screen
310       PRINT M$(IC)
320    NEXT IC
330    LOCATE NCHOICE+2,1               'print help message below choices
340    PRINT "Use arrow keys to move cursor; use <ENTER> key to make choice."
350    CHOICE=1                         'position cursor on first choice
360    WHILE Y$ <> CHR$(13)             'has user typed the <ENTER> key?
370       LOCATE CHOICE,1,0             'with cursor blink turned off
380       COLOR 0,7:PRINT M$(CHOICE);   'print current choice in reverse video
390       COLOR 7,0                     'return to original color
400       Y$=INKEY$:IF Y$="" THEN 400   'wait for user to type some key
410       LOCATE CHOICE,1,1             'return line to normal color
420       PRINT M$(CHOICE)
430       'arrow keys generate 2 characters; a 0 and another
440       'if an arrow key, then go to new line and print it reverse video
450       IF Y$=CHR$(0)+CHR$(80) THEN GOSUB 520  'cursor down key
460       IF Y$=CHR$(0)+CHR$(72) THEN GOSUB 580  'cursor up key
480    WEND
490    Y$=""                           'clear <ENTER> key as choice
500 RETURN
510 REM **********************
520 REM Down arrow key
530    CHOICE=CHOICE+1                  'go to next choice on list
540    IF CHOICE > NCHOICE THEN CHOICE=1   'trying to move past end of list?
550 RETURN
560 REM **********************
570 REM Up arrow key
580    CHOICE=CHOICE-1
590    IF CHOICE < 1 THEN CHOICE=NCHOICE   'trying to move before beginning?
600 RETURN
610 REM **********************
620 REM Pause for 1 second so user can see response
630    START=TIMER
640    IF START > 86398! THEN START=0   'avoid problems at midnight
```

Program 10.8 (Part 1 of 2).   Selection Menu.

space for notebook entries and by requiring entries not be left blank. To avoid the
inevitable protests from users about having to complete long notebook entries, you

```
650    WHILE CURRENT-START < 1
660        CURRENT=TIMER
670    WEND
680 RETURN
```

Program 10.8 (Part 2 of 2).   Selection Menu.

should store on the disk a file containing all of the answers most recently entered into the notebook. This will allow you to display the old entries and permit the user to change only those that are different for the current sample. Usually only a few entries will need to be changed for each sample. Program 10.7 can be very easily modified to do this for you.

EXERCISES

1.   Write a program that will read the data from the file created in Program 10.1 and display the data on the screen in tabular form.

2.   Write a program that will read the data set from the file created in Program 10.2 and either display it or store it in a sequential file.

3.   Modify Program 10.4 so that it reads the data from random access files of the type created by Program 10.2.

4.   Write a program that will plot data as they are collected by an A/D or other input device of your choosing. Give the user a choice of how to scale the data on both the X and Y axes.

5.   Take Program 9.1 and modify it so that it has menus of the type described in this chapter. Provide context-sensitive help (i.e., if the user asks for help when the cursor is positioned upon a particular command, give information about that command).

6.   Take Program 9.1 and modify it to store parameter and notebook information collected from the user. Store all of the data collected, including parameters and notebook information, in a random access file.

11

PEAK DETECTION

One of the general problems encountered in a many areas of science and engineering is the detection and measurement of **peaks**, or local extremes in continuous data. Examples of this type of data include spectroscopic and chromatographic data of all types, physiological data such as electrocardiograms, and the results of a wide variety of physical measurements in the engineering laboratory. The general problem with this type of data is to locate the peaks in the data, often in the presence of substantial noise, and, in many cases, to accurately measure properties of the peaks such as area, asymmetry, and location of the peak apex. Hence, you may find it useful to examine some aspects of peak detection and analysis.

Before proceeding, we should distinguish between two somewhat different problems of peak detection. The first is **real-time peak detection**; that is, we analyze the peak data as it is received at the computer. In this mode, the computer has access to only a few data points at a time, and therefore must make decisions about the peaks based upon incomplete data. This mode of operation is the one used by almost all commercially-available peak integration systems (integrators) and some microcomputer systems, where there is insufficient memory to store every point collected during the analysis.

The second mode of peak detection is **off-line**, or **post-run** peak analysis. In this mode, we typically have the entire data set to examine, and presumably can make more sophisticated decisions about the peaks. It is this mode of peak detection that we shall primarily consider here, although many of the same concepts can be applied to real-time peak detection as well.

In general, we can subdivide the problem of peak detection into five discrete steps:

1. Detection of the peak apex.
2. Detection of the start of the peak.
3. Detection of the end of the peak.
4. Detection of the baseline.
5. Dealing with overlapping peaks.

Let's examine each of these steps in turn. Before doing so, however, we should put in a general disclaimer: there is no one best method for peak detection that will fit all circumstances. You should read the following for some ideas about peak detection, but then modify the ideas to best match the data you are collecting in your own laboratory.

11.1 PEAK APEX

Locating the apex of each peak is often the fist step of peak detection. Typically, we begin by searching for local maxima in our data set. (In some cases, the peaks we wish to view are actually local minima, but such cases can be treated analogously.) In the simplest case, all we need do is test each data point to see if is higher than both the preceding and succeeding data point in the set of data.

The difficulty with this simple approach is that it almost never works because of noise in the data. For example, as shown in Figure 11.1, baseline noise around the peak also contains local maxima. Hence, a more sophisticated algorithm for peak apex detection will use one or a combination of criteria for eliminating noise. Commonly-used criteria include the following:

- *Define a minimum peak height*. The apex must be at least this distance above the baseline before being considered a true peak.
- *Define a minimum peak width*. The apex must occur at least this distance from other peaks to be considered a peak. If two local maxima are closer than this distance, only the higher maximum is considered to be a peak.
- *Highest point between start and end of a peak*. In systems that first find the peak start and then look for the peak apex, the apex is usually just the highest point before the peak ends.

The minimum peak height criterion is easy to implement, although you must have some knowledge of what the baseline is. The minimum peak width criterion requires no knowledge of the baseline, but may require some care in implementation to consider cases of more than two local maxima that are close together. It may also

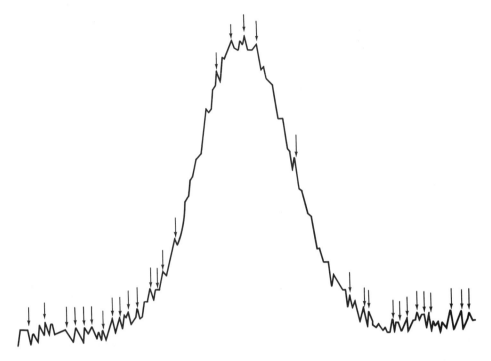

Figure 11.1. Detecting the Peak Apex. If we use only a local maximum criterion for peak detection, then "peaks" will be detected at all of the locations marked with an arrow. A more sophisticated algorithm will detect only the one major peak.

cause problems in cases where the peak width changes markedly during the analysis (e.g., in isothermal gas chromatography).

In many cases, a more subtle problem arises; for example, if the top of the peak is relatively flat, or even is literally flat over some distance. In this case, we may need to add an algorithm for determining the middle of a flat region. In other cases, we may need a very accurate determination of the location of the apex, which may be particularly difficult for small peaks in the presence of large amounts of noise. In some such circumstances, it may be appropriate to assume a particular peak shape and locate the apex by either fitting an ideal peak to it, or by extrapolating the sides of the peak at the inflection point to form a triangle.

Many of the problems in detecting the peak apex (and start and end, for that matter) arise from the noise present in the signal. Hence, you should examine several typical data sets for the noise level present. If your data have even moderate amounts of noise, then you should attempt to eliminate that noise using the techniques described in Chapter 4. Begin first, as usual, with trying to eliminate sources of noise. Then use hardware (usually lowpass) filters to reduce any remaining noise, and, finally, apply software filters (e.g., the Savitzky-Golay filters) to the data. Do all of these things before attempting to analyze the peaks and you will find the process of peak detection much easier.

11.2 PEAK START

The method used to detect the peak start depends upon whether the peak apex has already been located. Most real-time peak detection systems detect the peak start before detecting the peak apex, and hence much of the literature concerns methods for detecting peak starts prior to detecting the peak apex. We shall examine such methods first, and then describe a technique useful if the peak apex has already been located.

Most of the methods for detecting the start of a peak are in some manner based upon a slope criterion; the more sophisticated systems allow the slope criterion to be chosen by the user. The most obvious technique of this type is simply to look for any point that is higher than the previous point. The difficulty with this approach is at least two-fold. First, there may be substantial noise present, in which case a point may be higher just because of random noise. Second, in many systems, there may a background that increases during at least part of the sample analysis (e.g., gradient elution liquid chromatography and temperature-programmed gas chromatography).

Hence, a more sophisticated approach is required. One approach that may occur to you is to require that the peak rise a certain distance above the baseline before it is deemed to have started. However, this technique is still rather susceptible to noise problems, and, more importantly, may underestimate the width of the peak. This is because many systems produce peaks that are (to a first approximation only) Gaussian in shape. This means that the slope of the peak at the true start of the peak is very low, and hence the peak may not rise the minimum distance until quite some time after its true start, as shown in Figure 11.2.

Hence, most peak detection systems use some form of measurement of the slope of a number of sequential points. One such technique that can be quite successful is to require that each of **n** points be higher than the first point by at least some amount; that is, that the slope increase between each of several successive pairs of points. An even better technique, because it is less susceptible to noise problems, is to require an increasing slope between a point and each of several points after the point being tested. This technique thus requires that:

$$s_1 > s_2 > s_3 > s_4 > \text{etc.}$$

where s_n is the slope between point x_0 and point x_n. This is illustrated in Figure 11.3. The number of points considered in this criterion may be fixed or varied by the user, depending upon the application.

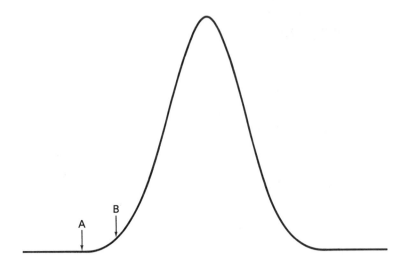

Figure 11.2.   Peak Start Detection with Minimum Height Criterion.   Because a Gaussian-shaped peak has a small slope at its beginning, at point A, the peak may not rise above the minimum height criterion until point B. Hence, the minimum height criterion is a poor choice for systems that measure peak area or other parameters that depend upon knowing the location where the peak starts.

11.2.1 Peak Start or End Detection when the Apex is Known

The problem of peak start detection is somewhat different when the location of the peak apex is already known. In this case, the peak start and end can be located in virtually identical fashions by starting at the peak apex and moving outwards until the slope between points decreases below a certain value for more than a certain number of points. This approach is illustrated in Figure 11.4.

11.3 PEAK END

In essence, the problem of detecting the peak end in off-line systems is virtually identical to that of detecting the peak start. In real-time systems, usually a "peak detected" flag is set in software that signals that some slope criterion should be applied in looking for where the peak ends. Of course, if another peak start is found, then the previous peak is considered to have ended, so real-time systems usually check both for peak end and the start of another peak simultaneously.

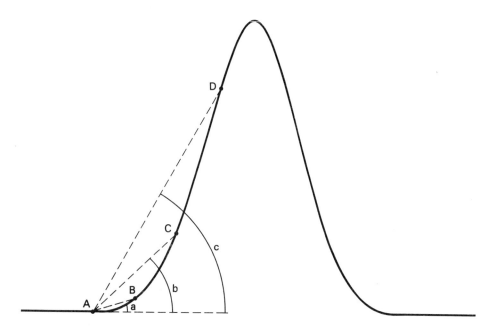

Figure 11.3.   Increasing Slope Criterion for Peak Start.   Point A is considered to be the
start of the peak by this criterion if the slopes between point A and points B, C,
and D are increasing; that is, if angle a < angle b < angle c.

11.4 BACKGROUND BASELINE

The **background baseline**, or more simply just the **baseline**, of a data set can be
defined as the values that would be recorded in there were no components in the
sample being analyzed. It is used in a variety of calculations; for example, the peak
height is typically calculated between the baseline and the peak apex. Hence, it is
important to determine it accurately. Unfortunately, this is not always easy to do.

The simplest method, at least in principle, is to run the instrument under the
desired conditions, except with no sample added. Then run the instrument under
identical conditions. Subtract the "blank" from the sample run, and the result will
be a data set with the baseline at zero. (In fact, some types of instruments allow this
to be done by physically measuring the sample and the blank at the same time and
electronically subtracting the blank from the sample.) The difficulty with this technique is that many instruments are not entirely reproducible from run to run, and in
addition, it may be difficult to prepare a blank that contains all of the non-sample
components (the "matrix") in proportions equal to those in the sample.

Hence, under many circumstances, it is desirable to *estimate* the baseline from
the data set itself. In real-time systems, this is often done by drawing a line between
the peak start and peak end, as shown in Figure 11.5. As shown in that figure,

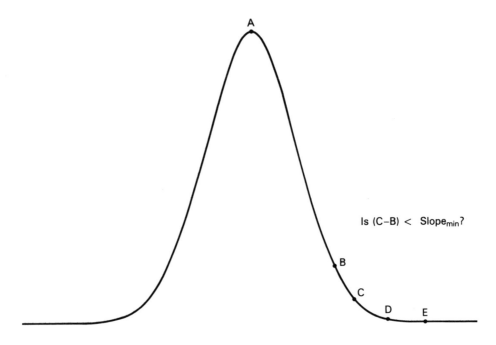

Is $(C-B) <$ Slope$_{min}$?

Figure 11.4. Peak Start when Peak Apex is Known. If the location of the apex is known, then both the peak start and peak apex can be located by starting at the apex, point A, and testing the slope between successive points, B and C, until the slope becomes smaller than some predefined value. Immunity from problems with noise can be increased by considering the slope between B and D, B and E, etc.

however, this approach can be inaccurate in the presence of either substantial amounts of noise or overlapping peaks.

In off-line systems, we can use a somewhat more sophisticated approach. One possibility, which we shall illustrate later, is to consider the baseline to consist of all points that are not part of a peak. In this case, we can then fit a polynomial or a series of line segments or a spline curve to the baseline in order to estimate its value under all of the peaks. This approach is shown in Figure 11.6.

11.5 OVERLAPPING PEAKS

In our discussions so far, we have tacitly assumed that the peaks we are trying to detect are well-resolved from one another. If they are not, then the problem of analyzing the data becomes considerably more complex. How do we know if peaks are well-resolved? If one peak starts before or just as the previous peak ends, then the

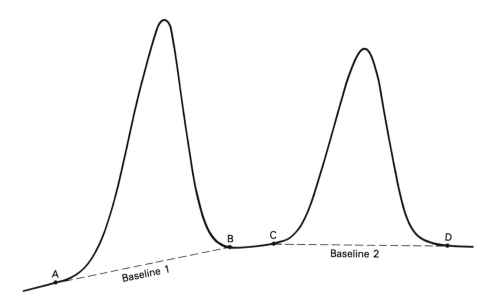

Figure 11.5.   Estimation of Baseline (Method 1).   The simplest technique for estimating the baseline is to draw a straight line between the start and end of each peak.

peaks are said to **overlap,** or to be poorly resolved. In the extreme cases, two or more peaks may even completely overlap.

The most satisfactory method for analyzing such peaks is to simply change instrumental conditions so that the peaks do not overlap. If this is not possible, it may then be possible to separate the peaks using mathematical techniques. Although a number of such techniques have been published, few are sufficiently robust to be used in routine practice except in cases where the peaks are only mildly overlapped, the number of peaks is known in advance, the shape of the peaks is known, and the system is relatively free from noise. Since these conditions are seldom met in practice, we need to develop methods for dealing with peaks when they are overlapped and cannot be separated mathematically with any high degree of accuracy.

Generally, two methods are used in commercial peak detection systems; neither of these methods is intellectually very satisfying, but are nonetheless widely accepted. The first of these is the **perpendicular drop method.** In this method, the minimum between two partially overlapped peaks is considered the end of the first peak and the beginning of the second, as shown in Figure 11.7. Obviously, this method ignores the true shape of the peak. However, it is a reasonably accurate method for apportioning peak area between two peaks *if the two peaks are approximately the same size* and area already almost fully resolved.

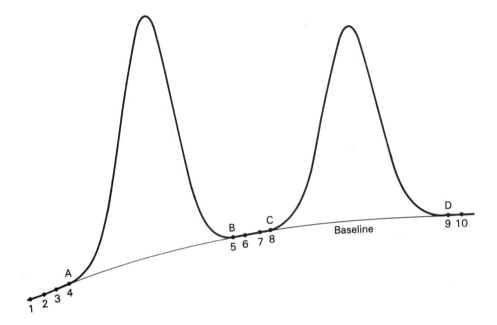

Figure 11.6. **Estimation of Baseline (Method 2).** In this method, a second-order curve is drawn through those data points (numbered 1-10) that are not part of peaks. The baseline is then estimated for the points of the peak, as shown by the solid line.

If one of the peaks is significantly larger than the other, then a more accurate method for apportioning peak area is the **tangent-skim method**, shown in Figure 11.8. In this method, an attempt is made to approximate the shape of the larger peak; that is, to estimate its boundary underneath the smaller of the peaks.

In cases where the overlapped peaks do not fit either of these two cases (equal size or large disparity in size), neither technique works well, and the results should be viewed with some caution. Again, the best advice is to try to analyze the sample under somewhat different conditions that separate the two peaks of interest.

Several studies have shown that considerable peak overlap is likely to occur even in fairly simple systems. Hence, there is substantial work going on in a number of laboratories to find better methods of mathematically resolving peaks.

11.6 QUANTITATIVE ANALYSIS

In many cases, we wish to obtain some measure of the amount of a substance present in our sample. In most such measurements, there is a direct linear relationship between the amount of substance present and the size of the peak representing the

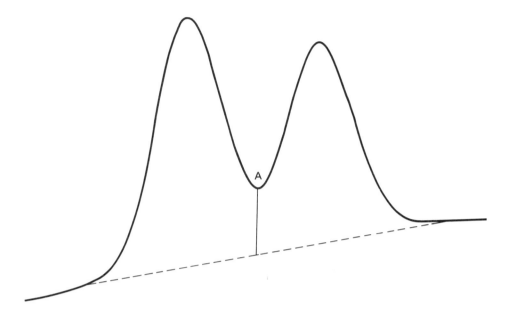

Figure 11.7. Perpendicular Drop Method. If two overlapped peaks are approximately the same size, an approximate division of the area between the two peaks can be obtained by dropping a straight line from the minimum between the two peaks (point A in this example).

substance in our data set. Hence, by measuring the size of the peak, and knowing the response of our system to different concentrations of the substance of interest, we can perform quantitative analysis of the sample.

Several caveats are required, however. First, such quantitative analysis is highly sensitive to system noise, peak overlap, and a variety of instrumental problems. Quantitative analysis is difficult to do!

The second problem is that it is not always obvious what to use as a measure of the size of the peak. In a simple, noise-free, non-overlapping peak system, either peak height above the baseline, or peak area, can be used as measures of the size of the peak, and hence the concentration of substance present. However, in many cases, either peak area or peak height, or both may be affected by having less than an ideal separation of the component peaks. In such cases, you are well advised to test the accuracy of determinations using both peak height and peak area, and to investigate cases where the two methods do not agree.

Peak height is usually simply measured as the distance between the peak apex and the baseline. Peak area is slightly more complex, at least from a theoretical point of view, because we are using digital data to measure the peak area. Most peak integration systems simply sum the distances from the baseline to the data points

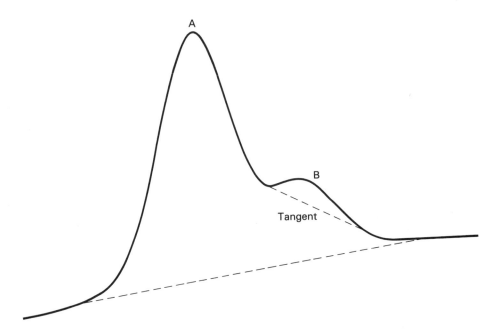

Figure 11.8. Tangent-Skim Method. If one of two overlapped peaks is significantly larger than the other, then an approximate division of the area between the two peaks can be obtained by drawing a tangent to the larger peak at the point where the smaller starts. This tangent is the dotted line in this figure, and divides Peak A from Peak B.

within the peak. Technically, it would be more accurate to sum the areas of the irregular tetrahedrons formed by connecting each pair of points with the baseline below it; however, as long as the peaks contain sufficiently large number of points (a common recommendation is 30 or more points per peak), the summing of the data points is sufficiently accurate. Obviously, in any case, the accuracy of the determination of the peak area depends significantly upon the accuracy of the determination of the peak start, peak end, and the baseline.

11.7 PROGRAM FOR IMPLEMENTING PEAK DETECTION

As we have mentioned, it is very difficult to write a single peak detection program that will cover all possible circumstances. Hence, we will simply show you a program that is based upon one we have used in our laboratories. This program was designed to detect and quantify peaks in complex samples with many overlapping peaks. It is a **post-run program** that is designed to accomplish the following steps:

1.   Read in a set of data.

2.   Get user input on the portion of the data to be analyzed.

3.   Read a file of parameters that are used to determine peak detection and peak integration limits.

4.   Find the apex of each peak.

5.   Find the start and end of each peak.

6.   Refine the estimates of the peak start and end.

7.   Determine the baseline.

8.   Integrate the peaks using the perpendicular drop method.

These steps are shown in Program 11.1.

You will quickly notice that this is the longest program in this textbook. To help you see the flow of the program, we have written each of the steps as one or more subroutines. These subroutines are all called from the main program, beginning in line 180.

The program begins by reading in the data. We have illustrated doing this in lines 360 to 460 using an ASCII file; of course, you should modify this routine to match your own data format.

The first unusual portion of this program is the use of a parameters file, shown in lines 750 to 870. These are parameters that can be changed to customize the program to data of the type normally encountered in your system. However, we have assumed that most of your data files are similar, so that the parameters normally will be the same for all of your samples. A sample set of parameters is included on the diskette.

There are 10 parameters input by the program. They are:

1.   XNOISE!, which is the approximate noise level of the data, expressed as the standard deviation of the noise. If the actual number is known, this should be substituted for the value read in from the parameters file.

2.   NDOWN%, which is the number of points that will be searched on each side of a putative apex to determine if it is the highest point.

3.   SEC! is the minimum number of seconds between two peak apexes. Peaks closer than this are merged together and considered a single peak.

4.   NBASE%, which is the number of points that must be increasing before a peak is considered to have started.

5.   MINHT%, the minimum peak height.

6.   SLOPFC!, which is a factor used in determining whether a peak has actually started or ended.

7.   IDIFF%, which is the minimum number of points between the end of one peak and the start of a subsequent peak. Peaks closer than this are considered unresolved.

8.   XMINAR!, the minimum allowed peak area.

```
10 REM PROGRAM INTEGRATE
20 REM LABORATORY AUTOMATION USING THE IBM PC
30 REM Integrates peak data (e.g., chromatography data).
40 REM Note:  requires use of a graphics adapter and monitor
50 REM ******************************************************************
60   DEFINT I-N
70   DIM XRAY%(640),ARRAY%(3600),LDATA%(3600),APEX%(250,3),AREA!(250)
80   'L is number of points in data array
90   'M counts number of maxima (potential peaks)
100  'ARRAY% contains data points
110  'LDATA% contains baseline points
120  'APEX%(M,1) stores position of peak apex
130  'APEX%(M,2) stores position of start of peak
140  'APEX%(M,3) stores position of end of peak
150  'AREA contains peak areas
160 REM *****************************
170 REM              Main Program
180    GOSUB 360                       'open data file
190    GOSUB 490                       'get information from user
200    GOSUB 750                       'get integration parameters
210    GOSUB 900                       'get data
220    GOSUB 1010                      'find local maxima
230    GOSUB 1180                      'check for flat-topped peaks
240    GOSUB 1520                      'find peak start, end
250    GOSUB 1630                      'check for peaks that are too close together
260    GOSUB 1800                      'get better estimate of start, end of peaks
270    GOSUB 2050                      'check for minima close to one another
280    GOSUB 2210                      'determine baseline
290    GOSUB 2640                      'display peaks on screen
300    GOSUB 2780                      'subtract baseline from data
310    GOSUB 2860                      'integrate peaks using perpendicular drop
320    GOSUB 3200                      'store results in disk file
330 END
340 REM *****************************
350 REM Open data file
360    GOSUB 4770                      'display standard screen border
370    INPUT "What is the name of the data file to be analyzed? ",FILNAM1$
380    OPEN FILNAM1$ FOR INPUT AS #1
390    INPUT #1,L                      'number of data points in file
400    INPUT #1,IPOINT!                'number of data points / second
410    INPUT #1,STARTTIM!              'time corresponding to first point
420    FOR I=1 TO L                    'get the data
430        INPUT #1,ARRAY%(I)
440    NEXT I
450    CLOSE #1
460 RETURN
470 REM *****************************
480 REM Get information from user
490    Y1$="y"
500    WHILE Y1$ <> "n" AND Y1$ <> "N"
510        GOSUB 4770                  'display standard screen border
520        'get start,end point of run to integrate
530        PRINT "Number of points/second =          ";IPOINT!
540        PRINT "Starting at time =                 ";STARTTIM!
550        ISTART%=1:IEND%=L
560        PRINT "Number of data points in memory is ";L
570        COLOR 7,4,1:PRINT
580        PRINT "Desired starting point? ":LOCATE 10,44:PRINT ISTART%
590        PRINT "Desired ending point? ":LOCATE 11,44:PRINT IEND%
600        LOCATE 10,45,1: INPUT "",T$
610        IF LEN(T$)>0 THEN ISTART%=VAL(T$):LOCATE 10,45:PRINT ISTART%
620        LOCATE 11,45,1: INPUT "",T$
630        IF LEN(T$)>0 THEN IEND%=VAL(T$):LOCATE 11,45:PRINT IEND%
640        INPUT "Do you wish to see plots during integration?Y or N> ",Y$
650        LOCATE 18,1
660        INPUT "Do you wish to change any of the above entries?<Y or N> ",Y1$
670    WEND
```

Program 11.1 (Part 1 of 8).   Peak Detection.

```
680     IF Y$="y" OR Y$="Y" THEN IGOPLOT=1 ELSE IGOPLOT=0
690     CLS:SCREEN 1:COLOR 0,0:VIEW (35,15)-(317,185)
700     LOCATE 10,1:PRINT "Please wait while I find the peaks";
710     WINDOW (1,1)-(32767,32767)
720 RETURN
730 REM ****************************
740 REM Get file containing integration parameters
750     OPEN "AREA.CO1" FOR INPUT AS #3
760     INPUT #3,XNOISE!             'approximate noise level (std. dev.) of data
770     INPUT #3,NDOWN%              'no. pts. going down before consider an apex
780     INPUT #3,SEC!                'no. sec. between peaks to be a shoulder
790     INPUT #3,NBASE%              'no. points checked for start of peak
800     INPUT #3,MINHT%              'minimum peak height
810     INPUT #3,SLOPFC!             'time above SD of noise for peak start or end
820     INPUT #3,IDIFF%              'no. pts. between peaks before are "resolved"
830     INPUT #3,XMINAR!             'minimum allowed peak area
840     INPUT #3,OFFSCALE%           'values above this are considered off-scale
850     INPUT #3,TOPDIFF%            'difference at top of peak to call noise
860     CLOSE #3
870 RETURN
880 REM ****************************
890 REM Get data array of interest
900     M=0                         'M counts number of peaks found
910     IF ISTART%=1 THEN L=IEND%-ISTART%+1:RETURN
920     K=0
930     FOR I=ISTART% TO IEND%      'delete all points from time=0 to time=istart
940           K=K+1
950           ARRAY%(K)=ARRAY%(I)
960     NEXT I
970     L=K
980 RETURN
990 REM ****************************
1000 REM Find local maxima
1010    FOR I=NDOWN%+1 TO L
1020          XI1=ARRAY%(I)
1030          'check for local maxima, but watch for flat-topped peaks
1040          FOR J=-1*NDOWN% TO -1
1050             IF(ARRAY%(J+I)>XI1)THEN GOTO 1120
1060          NEXT J
1070          FOR J=1 TO NDOWN%
1080             IF(ARRAY%(J+I)>=XI1)THEN GOTO 1120
1090          NEXT J
1100          M=M+1
1110          APEX%(M,1)=I
1120    NEXT I
1130    IF M = 0 THEN PRINT "No peaks were found":INPUT "Type <return>",Y$:STOP
1140 RETURN
1150 REM ****************************
1160 REM   check for roughly flat-topped peaks so not count noise on top as
1170 REM   > 1 peak apex
1180    IF(M <=1) THEN GOTO 1290
1190    FOR  I=2 TO M
1200          'check if peaks are close together, approx. same size
1210          IF(APEX%(I,1)-APEX%(I-1,1)-1.5*NDOWN% > 0) THEN GOTO 1280
1220          XI1=ARRAY%(APEX%(I-1,1))
1230          XI2=ARRAY%(APEX%(I,1))
1240          I3=ABS(XI2-XI1)
1250          IF(I3 > TOPDIFF%) THEN GOTO 1280
1260          IF(XI1 > XI2)THEN APEX%(I,1)=0       'if so, flag smaller peak
1270          IF(XI2 >= XI1)THEN APEX%(I-1,1)=0
1280    NEXT I
1290    GOSUB 4330                  'delete peaks marked as noise
1300    FOR  I=1 TO M               'check for flat peaks
1310          I2=APEX%(I,1)
1320     IF(ARRAY%(I2-1)<>ARRAY%(I2) AND ARRAY%(I2)<>ARRAY%(I2+1)) THEN GOTO 1440
1330          I3=I2
1340          I4=I2
```

Program 11.1 (Part 2 of 8).   Peak Detection.

```
1350              FOR J=I2-1 TO 1 STEP -1
1360                  IF(ARRAY%(J) <> ARRAY%(I2))THEN GOTO  1390
1370                  I3=J
1380              NEXT J
1390              FOR  J=I2+1 TO L
1400                  IF(ARRAY%(J) <> ARRAY%(I2))THEN GOTO 1430
1410                  I4=J
1420              NEXT J
1430              APEX%(I,1)=(I3+I4*1!)/2  'put apex at middle of flat region
1440      NEXT I
1450 RETURN
1460 REM ****************************
1470 REM Find estimate of peak start, end
1480 REM Find minimum between each pair of peaks.  For each peak, apex%(i,1)
1490 REM stores the location of the peak maximum, apex%(i,2) stores position
1500 REM of the minimum to the left of the peak, apex%(i,3) stores
1510 REM the position of the minimum to the right of the peak
1520   IF (M > 1) THEN GOSUB 4470  'find minimum between each pair of points
1530   FOR I=1 TO M              'eliminate noise peaks
1540         FOR J=2 TO 3
1550            IF(ARRAY%(APEX%(I,1))-ARRAY%(APEX%(I,J))-MINHT%>0)THEN GOTO 1580
1560         NEXT J
1570         APEX%(I,1)=0         'flag peak to be eliminated
1580   NEXT I
1590   GOSUB 4330                 'delete flagged noise peaks
1600 RETURN
1610 REM ****************************
1620 REM Check if maxima are too close together
1630   IPTCHK=SEC!*IPOINT! 'IPTCHK is no. points occuring in SEC! seconds
1640   IF (M<=1) THEN RETURN
1650   FOR I=2 TO M
1660         I1=APEX%(I-1,1)
1670         I2=APEX%(I,1)
1680         IF(I2-I1-IPTCHK > 0)THEN GOTO 1710  'are peaks far enough apart?
1690         'peaks are too close, so delete the smaller one
1700         IF ARRAY%(APEX%(I-1,1)) > ARRAY%(APEX%(I,1)) THEN APEX%(I,1)=0
                 ELSE APEX%(I-1,1)=0
1710   NEXT I
1720   GOSUB 4330                 'eliminate shoulders
1730   IF (M > 1) THEN GOSUB 4470  'find new minima between peaks
1740 RETURN
1750 REM ****************************
1760 REM  Starting at local minimum and going toward peak apex, calculate
1770 REM  whether have any point more than a slope factor x sd of baseline
1780 REM  noise above the rest in a small local region--if not, move start
1790 REM  or end over one nearer to apex
1800   FOR I=1 TO M              'search for peak start, end
1810         LEFT%=APEX%(I,2)
1820         IRIGHT%=APEX%(I,3)
1830         FOR J=LEFT% TO APEX%(I,1)   'find new left-side minimum
1840              APEX%(I,2)=J
1850              FOR J2=J+1 TO J+NBASE%
1860            IF(ARRAY%(J2) > ((SLOPFC!*XNOISE!)+ARRAY%(J)))THEN GOTO 1890
1870              NEXT J2
1880         NEXT J
1890         FOR J=IRIGHT% TO APEX%(I,1) STEP -1 'find new right-side minimum
1900              APEX%(I,3)=J
1910              FOR J2=J-1 TO J-NBASE% STEP -1
1920            IF(ARRAY%(J2) > ((SLOPFC!*XNOISE!)+ARRAY%(J)))THEN GOTO 1980
1930              NEXT J2
1940         NEXT J
1950         'flag peak if too small in width
1960         IF (APEX%(I,3)-APEX%(I,2) < 2) THEN APEX%(I,1)=0
1970    IF (APEX%(I,2)=APEX%(I,1)) OR (APEX%(I,3)=APEX%(I,1)) THEN APEX%(I,1)=0
1980   NEXT I
1990   'last peak too short?
2000   IF APEX%(M,3)-APEX%(M,1)< .5*(APEX%(M,1)-APEX%(M,2)) THEN APEX%(M,1)=0
```

Program 11.1 (Part 3 of 8). Peak Detection.

```
2010    GOSUB 4330                      'delete peaks as necessary
2020  RETURN
2030  REM ****************************
2040  REM If minima are nearby, make coincide
2050    IF (M <= 1) THEN RETURN
2060    FOR I=2 TO M
2070        IF APEX%(I,2)=APEX%(I-1,3) THEN GOTO 2170 'SEE IF PEAKS OVERLAP
2080        'peaks are too far apart
2090        IF(APEX%(I,2)-APEX%(I-1,3) > IDIFF%*IPOINT!)THEN GOTO 2170
2100        'peaks are "nearby" so make peaks join at lowest point
2110        LEFT%=APEX%(I-1,3):RIGHT%=APEX%(I,2):IMID%=LEFT%:JMID=ARRAY%(IMID%)
2120        FOR J=LEFT% TO RIGHT%  'find lowest point
2130            IF ARRAY%(J) < JMID THEN JMID=ARRAY%(J):IMID%=J
2140        NEXT J
2150        APEX%(I,2)=IMID%
2160        APEX%(I-1,3)=IMID%
2170    NEXT I
2180  RETURN
2190  REM ****************************
2200  REM Determine the baseline
2210    'get all of points into LDATA array (which is baseline array)
2220    FOR I=1 TO L
2230            LDATA%(I)=ARRAY%(I)
2240    NEXT I
2250    'now determine baseline points under each peak.  Assumes that if
2260    'several peaks fused together, pick baseline with lowest slope
2270    APEX%(M + 1, 2) = 0           'so last peak is treated accurately
2280    FOR I=1 TO M                  'any peaks overlap?
2290        IF APEX%(I,3) = APEX%(I+1,2)THEN GOTO 2400
2300        LEFT%=APEX%(I,2)          'no overlap, so interpolate baseline between
2310        RIGHT%=APEX%(I,3)         '  peak start and peak end
2320        SLOPE!=(ARRAY%(RIGHT%)-ARRAY%(LEFT%))/(RIGHT%-LEFT%)
2330        YBASE=ARRAY%(LEFT%)
2340        FOR J=LEFT% +1 TO RIGHT%-1
2350            YBASE=YBASE+SLOPE!
2360            LDATA%(J)=YBASE
2370        NEXT J
2380        GOTO 2600
2390        'peaks do overlap, so look for lowest slope of set of fused peaks
2400        K=I                       'keep track of starting peak of set
2410        KK=K 'KK is peak that is the last one in the set of fused peaks
2420        YSLOPE!=30000             'initial guess of slope
2430        FOR J=K TO M-1            'look for more overlapped peaks
2440            IF APEX%(J,3) < APEX%(J+1,2) THEN GOTO 2470
2450            KK=J+1
2460        NEXT J
2470        LEFT%=APEX%(K,2)          'left side of starting peak
2480        FOR J=K TO KK             'look for lowest slope
2490            RIGHT%=APEX%(J,3)
2500            SLOPE!=(ARRAY%(RIGHT%)-ARRAY%(LEFT%))/(RIGHT%-LEFT%)
2510            'found a lower slope
2520            IF SLOPE! < YSLOPE! THEN YSLOPE!=SLOPE!:XRIGHT!=RIGHT%:XSTOP=J
2530        NEXT J
2540        YLEFT%=ARRAY%(LEFT%)
2550        FOR J=LEFT%+1 TO XRIGHT!-1    'interpolate baseline
2560            YLEFT%=YLEFT%+YSLOPE!
2570            LDATA%(J)=YLEFT%
2580        NEXT J
2590        I=XSTOP 'reset I so skips over peaks already used in baseline
2600    NEXT I
2610  RETURN
2620  REM ****************************
2630  REM Display integrated data on screen in groups of 240 seconds
2640    IX=0
2650    N=IPOINT!*60!
2660    IBEG%=240*IPOINT!
2670    IF(IBEG% >L)THEN IBEG%=L
```

Program 11.1 (Part 4 of 8). Peak Detection.

```
2680    FOR J=IBEG% TO L STEP IBEG%
2690          IX=J/N
2700          GOSUB 3450                      'graph data
2710          LN=J                            'set up graph label
2720    NEXT J
2730    'do for last minute if not already plotted
2740    IF LN < L THEN J=L:GOSUB 3450   'graph data
2750 RETURN
2760 REM ****************************
2770 REM Subtract baseline from data
2780    FOR I=1 TO L
2790          LDATA%(I)=ARRAY%(I)-LDATA%(I)
2800          IF(LDATA%(I) < 0) THEN LDATA%(I)=0
2810    NEXT I
2820 RETURN
2830 REM ****************************
2840 REM  Integrate each baseline-subtracted peak using
2850 REM  perpendicular drop method
2860    GOSUB 4770:COLOR 7,4,1
2870    LOCATE 5,1
2880    INPUT "Enter the name for the output data file ",FILNAM2$
2890    OPEN FILNAM2$ FOR OUTPUT AS #2
2900    'calculate area of each peak for M peaks
2910    FOR I=1 TO M
2920          IST=APEX%(I,2)
2930          IEND%=APEX%(I,3)
2940          AREA!(I)=0
2950          FOR J=IST TO IEND%
2960                AREA!(I)=AREA!(I)+LDATA%(J)
2970          NEXT J
2980    NEXT I
2990    'get rid of peaks with less than minimum area
3000    COLOR 7,1,1:PRINT "Eliminating peaks with areas below ",XMINAR!
3010    J=0
3020    FOR I=1 TO M
3030          IF(AREA!(I) < XMINAR!)THEN GOTO 3090
3040          J=J+1
3050          FOR K=1 TO 3
3060                APEX%(J,K)=APEX%(I,K)
3070                AREA!(J)=AREA!(I)
3080          NEXT K
3090    NEXT I
3100    'don't include off-scale peaks in area sum
3110    SUM!=0
3120    FOR I=1 TO M
3130          IF(ARRAY%(APEX%(I,1)) = OFFSCALE%)THEN GOTO 3150
3140          SUM!=SUM!+AREA!(I)
3150    NEXT I
3160    M=J
3170 RETURN
3180 REM ****************************
3190 REM Store results in file
3200    PRINT #2," For the ";M;" peaks in file ";FILNAM1$;" the peak values are:"
3210    PRINT #2," Pk. no.  Ret. time    Area          Relative area"
3220    IFLAG=0
3230    FOR I=1 TO M
3240          RELAR=AREA!(I)/SUM!*100!
3250          RT!=(1!*APEX%(I,1)+ISTART%-2)/(60!*IPOINT!)+STARTTIM!/60!
3260          ICODE$=" "
3270          'check for off-scale peaks
3280          IF(ARRAY%(APEX%(I,1)) = OFFSCALE%)THEN ICODE$="OS":IFLAG=1
3290          PRINT #2,USING "###";I;:PRINT #2,"        ";
3300          PRINT #2,USING "###.##";RT!;:PRINT #2, "        ";
3310          PRINT #2,USING "#########.";AREA!(I);:PRINT #2,"      ";
3320          PRINT #2,USING "#######.####";RELAR;:PRINT #2,"   ";
3330          PRINT #2,ICODE$
3340    NEXT I
```

Program 11.1 (Part 5 of 8).   Peak Detection.

```
3350    PRINT #2," Variables from default file were:"
3360    PRINT #2,XNOISE!;NDOWN%;SEC!;NBASE%;MINHT%;SLOPFC!;IDIFF%;XMINAR!
3370    IF IFLAG<>1 THEN GOTO 3400
3380         PRINT #2,"NOTE: OS means peak was off-scale."
3390         PRINT #2,"Its area should not be considered to be accurate."
3400    CLOSE #2
3410 RETURN
3420 REM  ****************************
3430 REM Plot partial areas during integration
3440    'plot 320 points at a time
3450    IF IGOPLOT=0 THEN RETURN
3460    I2=0
3470    VIEW:CLS:VIEW (35,15)-(317,185)
3480    J2=J-319                        'J2 is starting point
3490    IF(J2 < 1)THEN J2=1
3500    J3=J2+319                       'J3 is ending point
3510    IF(J3 > L)THEN J3=L
3520    XMAX!=0
3530    XMIN!=OFFSCALE%
3540    IT=(J3-J2+1)*2                  'IT is no. of points to plot
3550    LF!=32766/(J3-J2+1)             'LF is x-axis scaling factor
3560    FOR I3=J2 TO J3                 'find min, max
3570         IF(ARRAY%(I3) > XMAX!)THEN XMAX!=ARRAY%(I3)
3580         IF(LDATA%(I3) < XMIN!)THEN XMIN!=LDATA%(I3)
3590    NEXT I3
3600    FACT2!=28767!/(XMAX!-XMIN!)
3610    LOCATE 1,1:PRINT "Min= ";XMIN!,"max= ";XMAX!
3620    ICOL%=1
3630    FOR I3=J2 TO J3                 'plot data
3640         I2=I2+2
3650         XRAY%(I2-1)=I2/2*LF!
3660         XRAY%(I2)=(ARRAY%(I3)-XMIN!)*FACT2!+4000
3670    NEXT I3
3680    GOSUB 4290                      'plot xray array
3690    I2=0
3700    FOR I3=J2 TO J3                 'plot baseline (same x-values as data)
3710         I2=I2+2
3720         XRAY%(I2)=(LDATA%(I3)-XMIN!)*FACT2!+4000
3730    NEXT I3
3740    GOSUB 4290                      'plot xray array
3750    IP=0
3760    'now fill in peak areas for each integrated peak
3770    LINE(1,4000)-(32767,32767),1,B
3780    ICOL%=1
3790    FOR I2=1 TO M
3800         CENT%=APEX%(I2,1)
3810         IF CENT%-J2 < 2 OR CENT%-J2 >318 THEN GOTO 3850 'check if on screen
3820         I3=(CENT%-J2)*LF!+1
3830         I4=(  ( LDATA%(CENT%)+1!*ARRAY%(CENT%) )/2-XMIN!  )*FACT2!+4000
3840         PAINT(I3,I4),1,1
3850    NEXT I2
3860    'display lines at 1.0 min. intervals on screen
3870    'compute where first line at integer no. of minutes is = J4
3880    'convert spectrum number to time
3890    X4=(((J2+ISTART%-2)/IPOINT!)+STARTTIM!)/60!
3900    J4=INT(X4)  'truncate
3910    IF(X4-J4 > .01)THEN J4=J4+1  'make sure is on screen
3920    X4=J4
3930    'convert time to spectrum number
3940    J44=(X4*60!-STARTTIM!)*IPOINT!-ISTART%+2
3950    'convert spectrum no. to time
3960    X4=(((J3+ISTART%-2)/IPOINT!)+STARTTIM!)/60!
3970    J5=INT(X4) 'truncate
3980    XRAY%(2)=4000
3990    XRAY%(4)=32767
4000    FOR I6=J4 TO J5                 'draw lines at one-minute intervals
4010         XRAY%(1)=(J44-J2+1)*LF!+1 'see line 5 below for calc j44
```

Program 11.1 (Part 6 of 8).   Peak Detection.

```
4020          XRAY%(3)=XRAY%(1)
4030          IT=4:GOSUB 4290 'PLOT 2 POINTS
4040          'print time label on each line
4050          'compensate for view port starting at 35 and window width of 32767
4060          YPOS%=(XRAY%(1)+4067!)/36834!*40-1:IF YPOS% >37 THEN GOTO 4090
4070          LOCATE 23,YPOS%:PRINT I6;
4080          J44=J44+60*IPOINT!      'add one minute's worth of points
4090     NEXT I6
4100     I4=0
4110     ICOL%=2
4120     'put in marks indicating integration limits of peak
4130     FOR I2=1 TO M
4140        FOR I3=2 TO 3                   'check peak end, start to see if on-screen
4150           IF(APEX%(I2,I3) < J2 OR APEX%(I2,I3) > J3) THEN GOTO 4220
4160           XRAY%(1)=(APEX%(I2,I3)-J2)*LF!+1
4170           XRAY%(3)=XRAY%(1)
4180           J4=APEX%(I2,I3)
4190           XRAY%(2)=(ARRAY%(J4)-XMIN!)*FACT2!+5500
4200           XRAY%(4)=XRAY%(2)-1500
4210           IT=4:GOSUB 4290          'line between 2 points
4220        NEXT I3
4230     NEXT I2
4240     LOCATE 24,16:PRINT "MINUTES";:LOCATE 1,1
4250     Y$=INKEY$: IF Y$="" THEN 4250           'halt so user can see graph
4260 RETURN
4270 REM ****************************
4280 REM Graph an array (xray%) with lines connecting points
4290    FOR I9=3 TO IT-1 STEP 2
4300        LINE (XRAY%(I9-2),XRAY%(I9-1))-(XRAY%(I9),XRAY%(I9+1)),ICOL%
4310    NEXT I9
4320 RETURN
4330 REM ****************************
4340 REM Delete noise peaks
4350    J=0
4360    FOR  I=1 TO M
4370        IF(APEX%(I,1) <= 0)THEN GOTO 4420
4380        J=J+1
4390        FOR K=1 TO 3
4400            APEX%(J,K)=APEX%(I,K)
4410        NEXT K
4420    NEXT I
4430    M=J
4440 RETURN
4450 REM ****************************
4460 REM Find minimum between each pair of peaks
4470    FOR  I=2 TO M
4480        XMIN!=1E+10
4490        LEFT%=APEX%(I-1,1)
4500        IRIGHT%=APEX%(I,1)
4510        FOR J=LEFT% TO IRIGHT%
4520            IF(ARRAY%(J)>XMIN!) THEN GOTO 4550
4530            XMIN!=ARRAY%(J)
4540            JMIN%=J
4550        NEXT J
4560        APEX%(I-1,3)=JMIN%
4570        APEX%(I,2)=JMIN%
4580    NEXT I
4590    XMIN!=1E+10
4600    FOR I=1 TO APEX%(1,1)          'find left minimum of first peak
4610        IF(ARRAY%(I)>=XMIN!)THEN GOTO 4640
4620        XMIN!=ARRAY%(I)
4630        JMIN%=I
4640    NEXT I
4650    APEX%(1,2)=JMIN%
4660    XMIN!=1E+10
4670    FOR I=APEX%(M,1) TO L          'find right minimum of last peak
4680        IF(ARRAY%(I)>XMIN!)THEN GOTO 4710
```

Program 11.1 (Part 7 of 8).   Peak Detection.

```
4690            XMIN!=ARRAY%(I)
4700            JMIN%=I
4710     NEXT I
4720     APEX%(M,3)=JMIN%
4730     IF(JMIN%<=APEX%(M,1))THEN M=M-1   'avoid ending peak at apex
4740 RETURN
4750 REM ******************************
4760 REM  Standard screen display
4770    KEY OFF:SCREEN 0:WIDTH 80:COLOR 7,1,1:CLS:DL$=STRING$(79,205)
4780    LOCATE 1,1:PRINT DATE$ :LOCATE 1,25:COLOR 4,7,1
4790    PRINT "PEAK INTEGRATION PROGRAM":LOCATE 1,70:COLOR 7,1,1:PRINT TIME$
4800    LOCATE 2,1:PRINT DL$;
4810    LOCATE 25,1:PRINT DL$;
4820    LOCATE 5,1
4830 RETURN
```

Program 11.1 (Part 8 of 8).   Peak Detection.

9.   OFFSCALE%, the maximum allowed A/D reading; values at or above this level are considered to be off-scale.

10.   TOPDIFF% is a factor used only for peaks that are suspected of being large, flat-topped peaks. If two nearby "peaks" differ in intensity by less than this amount, then the two peaks are considered to be in actuality a large, single, flat-topped peak.

Next, in lines 900 to 980, we read in only those points that are of interest, based upon what the user has selected as the beginning and ending points of data. Then in lines 1010 through 1140 we find all local maxima. Notice that we can reduce the effect of noise in the data by searching through more points; that is, by increasing the NDOWN% parameter. The higher the value of NDOWN%, the less likely that a point will be considered an apex when it is just a high value produced by noise. However, we do not wish to have NDOWN% be larger than the half-width of the smallest width peak we expect to encounter.

Next we check for peaks that are approximately flat-topped (lines 1180 to 1450). We do this because we wish to avoid designating the top of a relatively large, flat peak as more than one peak (because of noise). This also allows us to handle off-scale peaks, which will usually be flat-topped.

The next, and very important, step is to try to find where each peak begins and ends. First, we start by eliminating any peaks that are too small, using a minimum height criterion (lines 1520 to 1600). We will eliminate peaks using several different criteria, so we call a special subroutine to perform this task (lines 4350 to 4440). Then, in lines 1630 to 1740, we check to see if we have detected two peak apexes that are very close together (closer than SEC! seconds). If they are too close (i.e., unresolved), then we merge them together by eliminating the peak with the smaller intensity. We then call a subroutine (lines 4470 to 4740) that finds the minimum between each pair of peak apexes. This minimum is (temporarily) considered to be the end of the first peak of each pair and the end of the last peak of each pair. We use an array (APEX%) to store the apex, start, and end of each peak, in that order.

Of course, the minimum between each pair of peaks is not a very accurate measure of where the peak starts or stops. Hence, beginning at the local minimum, we use a slope test to tell whether the peak has truly started or stopped (lines 1800 to 2020). This test is applied repetitively until we find a series of points, each of which is enough above the initial point, thus satisfying our slope test. We then eliminate all peaks with trivial widths (less than two points between the apex and start or end of the peak). We also check the last peak to make certain that the peak has not been cut off substantially by the end of the data set.

The next step (lines 2050 to 2180) is to check for peaks that almost overlap; that is, the end of one peak is near the start of next peak. If they are very close, the program simply puts the common point at the lowest point between the end of the first peak and the start of the next peak. This is done in order to better estimate the start and end of the adjoining peaks, and to get a better estimate of the baseline.

We now are ready to determine the baseline (lines 2210-2610). The algorithm used is simple. If the peak is fully resolved from its neighbors, then the baseline is simply the straight line drawn from the beginning of the peak to its end. However, if a group of peaks is unresolved from one another, it draws the baseline from the start of the first peak to the end of whichever other peak in the group results in the lowest slope baseline. Thus, in effect, we draw the minimum slope baseline across each set of peaks. This is illustrated in Figure 11.9.

If the user requested it, we now display the data, including the calculated baseline, on the graphics screen of the PC (lines 2640 to 2750). We then subtract the baseline from all of the data (lines 2780 to 2820), and integrate peaks using the perpendicular drop method (lines 2860 to 3170). The results are stored in a file (lines 3200 to 3410) and the program then is finished.

EXERCISES

1. Program 11.1 uses a parameter file to determine appropriate parameters for peak width, noise levels, etc. Write a series of subroutines that will examine the data file and attempt to estimate these parameters from the data.

2. Add to Program 11.1 a routine for tangent skimming.

3. Try modifying Program 11.1 to use a different method for finding the baseline; for example, try one of the methods discussed earlier in the chapter. Which method gives the best results for your data?

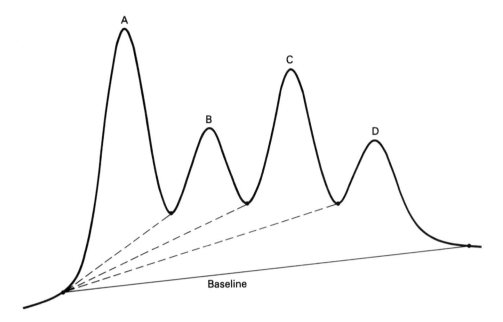

Figure 11.9.   Baseline Selection for Unresolved Peaks.   Peaks A to D are unresolved.
Hence, the baseline for peak A is calculated by drawing lines to the end of each
of the other unresolved peaks (dotted lines).   The line with the lowest slope
(solid line) is considered to be the actual baseline.

12

DIGITAL IMAGE PROCESSING

One of the most exciting new areas for the microcomputer is that of digital image processing. Once restricted to mainframe systems because of the need for large amounts of memory and specialized displays, image processing is now possible on machines such as the IBM PC/AT, where up to 16 Mbytes of memory and high resolution displays are available.

In general, digital image processing seeks to digitize data of three dimensional objects and present a meaningful representation of the objects on a two-dimensional surface (e.g., representing an orange on a TV screen). The resulting image is then manipulated in a variety of ways either to make it more "realistic" or to increase the amount of useful information that can be obtained from the image. In some circumstances, the computer analysis of the data may even result in a simple numerical output, such as a count of the number of objects present, or a yes/no decision on the reliability of a manufactured part.

In general, images are utilized to present a large amount of information to the user in a small physical space, making use of the fact that the human eye/brain combination can very easily analyze information presented as images. Indeed, the human brain is still far better at analyzing patterns in image data than are computers; the major advantage of computers is that they do not tire of looking at the same types of images over long periods of time.

Images tend to have some special properties that make them easier to analyze. In the first place, there is a great amount of non-interesting information in a typical image. Hence, our job is usually to emphasize or extract a small portion of the information. Second, much of the information in a typical image is correlated; i.e., each point is related to its neighbors, and often to many other points in the image. Images exhibiting periodicity are particularly easy to analyze.

Basically, the imaging process can be divided into five steps: image acquisition, image storage, image display, image manipulation, and image analysis. There are entire shelves of books on each of these topics; our discussion is therefore necessarily limited to an introductory level.

12.1 IMAGE CAPTURE

Image data can be acquired in a large number of ways. The most obvious technique is to use a television camera attached to a **frame digitizer**. The camera acquires an analog signal, which is then digitized by the frame digitizer, which is in essence a high-speed A/D converter. This arrangement is shown in Figure 12.1.

In addition to using a camera to acquire an image, however, you can use almost any of the data acquisition techniques discussed earlier in this book. For example, you might use a GPIB-connected instrument that acquires three channels of information to be plotted against one another, or an A/D that collects data over time. The resulting data are subsequently stored and treated in the same manner as images acquired using a camera.

Because images are essentially just data acquired by a A/D, much of our discussion in Chapter 3 applies here as well. However, because of the increased complexity of the data, we must introduce several new concepts. For example, the **resolution** of the resulting images is measured in two ways. The **brightness resolution** is simply the resolution of the A/D. The typical brightness resolution for most frame digitizers for scientific applications is 8 bits. Thus, images can be represented with 256 (2^8) levels of gray; on a scale of 0 to 255, 0 is black and 255 is white.

You may recall that in Chapter 3 we said that 8 bits was generally inadequate resolution for most A/Ds used for scientific purposes, so you may be surprised to find 8-bit frame digitizers as the standard. However, 256 levels of gray is approximately the limit of resolution of most TV-type monitors and has thus become standard. In practice, this resolution is quite adequate for most applications, and many applications can use systems with even less resolution. In any case, 256 gray levels is more than the human eye can readily distinguish and allows the use of high-speed A/Ds.

The **spatial resolution** refers to the number of samples made by the A/D in traveling across the image in either the horizontal or vertical direction. The larger the number of individual picture elements, or **pixels** in the image, the more closely it approximates the continuous image seen by the human eye. Most frame digitizers for scientific use display data on a TV screen, which has a relatively limited spatial resolution. Thus frame digitizers produce images that are typically 512 pixels x 512 lines. Most digitize images at rates of up to 30 frames (images) per second; some

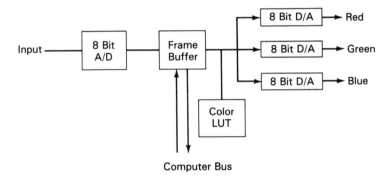

Figure 12.1. Frame Digitizer. The components of a typical frame digitizer consist of a high-speed 8-bit A/D to digitize the signal, memory (usually called a frame buffer), a high-speed 8-bit D/As to display the image in three colors, and on-board logic for controlling the digitizer.

models have fixed rates, while other models can be synchronized with some external event.

12.2 IMAGE STORAGE

A quick calculation will show you that a typical 512 x 512 x 8-bit frame digitizer produces images that occupy 0.25 Mbyte. Hence, at 30 such frames per second, you would produce over 2 Gbytes of information per hour. It is thus beyond the permanent storage capacity of even the largest mainframe computer system to store all of the collected data from an imaging system for any substantial period of time.

Hence, two alternatives are commonly used. One method is simply to record the images on videotape. Unfortunately, this often results in some degradation in the quality of the image; high quality recording systems are generally also of high cost. A second alternative is to store only some of the data. This latter approach means either selectively storing only certain frames, or applying data analysis techniques to select only certain features of the data for storage. Techniques are also available to compress the data before storage. The selected data is then stored on regular magnetic or optical disks.

12.3 IMAGE DISPLAY

Because of the techniques used to acquire the images, most images are displayed on what is essentially a common TV monitor as they are acquired. During processing of the digital images, however, the images can be displayed on almost any graphics monitor. For some scientific applications which require higher spatial resolution, special-purpose monitors are now available. Displays of 0.75 Mpel or 1.0 Mpel are becoming quite common on PCs, and even higher resolution displays are available. Most of these higher resolution displays are driven by special-purpose adapter cards that also provide useful image processing functions of the type described below.

Most displays expect data to be in one of several standard formats. There are detailed electronic specifications for each of these formats that are of interest to designers, but the user primarily needs to understand them only in general terms.

The original black-and-white format used by television displays is the EIA RS-170 specification produced by the Electronic Industries Association. In essence, this format has 485 lines of image; the lines are displayed at a rate of 15.75 kHz. This format uses **interlacing** to reduce flicker; i.e., it displays all even-numbered lines first, then returns to the top of the image and displays all odd-numbered lines. The horizontal resolution in a standard TV display is approximately 380 pixels; however, because the signal is an analog rather than digital one, this resolution is determined by the quality of the analog circuitry of the monitor rather than by the specification itself.

There are several variants of the RS-170 specification. For example, some monitors are said to accept **RGB** signals. These are monitors that in effect use three RS-170 signals, one each for red, green, and blue, to produce colored images. In this case, each of the three images must be treated separately when applying image processing techniques. The RS-343 specification gives yet another format for black-and-white images that is often used with high-resolution displays because it allows a vertical resolution of between 675 and 1023 lines. The **NTSC** specification promulgated by the National Television System Committee is a modification of the RS-170 standard that adds color information to the RS-170 signals. Many of the modern high-resolution displays support more than one of these formats.

12.3.1 Printing Images

Unfortunately, the printing of an image is often less easy than the displaying of the image. This is because printers basically only have two gray-levels: white and black. Shading in the original image must therefore be approximated by varying the density of dots; the more dense the dots, the blacker that area of the image appears. Hence, in order to print pleasing images, the printer must be able to print several hundred dots per inch or more.

There are many techniques for changing a gray-level image to a printed image. We will examine one such method to give you a feeling for the type of processing required. This method is referred to as the **Floyd-Steinberg error diffusion** technique.[7]

Suppose that we wish to print an image that contains 256 levels of gray on a printer. Since we can only print dots on the printer, we can view each location on the

---

[7]   This technique is described in more detail in, for example, D. F. Rogers, **Procedural Elements for Computer Graphics**, McGraw-Hill, 1985.

printed page as a 0 (black) or a 1 (white). The question is then how can we convert from the 256 levels of gray to a binary image.

One technique for doing this is to simply assign all values between 0 and 127 as a 0 in the printed image, and all values between 128 and 255 as 1. However, this considerably misrepresents the value of some pixels in the original image; e.g., a value of 100 is represented the same as a value of 2. Hence, the Floyd-Steinberg technique uses a method of **error diffusion** to distribute some of the error to neighboring dots in the image.

We can easily understand this technique if we consider the printed dots to be either 0 (black) or 255 (white) instead of 0 and 1, respectively. At a given pixel in the input image, a dot (black) is output on the printer if the pixel intensity is 127 or less. The location is left blank (white) for input values of 128 to 255. Thus, except for the cases where the input pixel is 0 or 255, there is some error, because the inputs are 0 through 255, whereas the outputs are only either 0 or 255. This error, or difference between the input and output values, is distributed (diffused) to the three neighboring pixels on the right, below, and diagonally to the right and below of the current input pixel.

Thus, if the error at the current pixel at location X,Y is:

$$\Delta I_{X,Y} = \text{input value - printed value}$$

then the value stored at location X+1,Y is:

$$I_{X+1,Y} = I_{X+1,Y} + 0.375 * \Delta I_{X,Y}$$

the value stored for location X,Y+1 is:

$$I_{X,Y+1} = I_{X,Y+1} + 0.375 * \Delta I_{X,Y}$$

and the value stored for location X+1,Y+1 is:

$$I_{X,Y+1} = I_{X,Y+1} + 0.25 * \Delta I_{X,Y}$$

One advantage of this technique is that it requires storing very little of the image in RAM. Each line of data from the image is processed left to right, and the rows from top to bottom; i.e., in the same directions they are printed. Thus, only the current line and the next line need be held in memory.

Let's see an example. Suppose that the input image contains three values of 100 in the first row. In this case, the first value is less than 128, so it is printed as a black dot (0). The error is 100 - 0 = 100, so we add 38 to the next pixel, 38 to the pixel below it, and 25 to the pixel below and to the right of it. Hence, the second pixel is now 138. This is greater than 127, so the printed area is left white (255). The error is this location is thus 138 - 255 = -117. This error is distributed as -44 to the point to its right, -44 to the point below it, and -29 to the point below it and to the right.   Hence, the third point in this row is now 100 - 44 = 56 and is thus printed as a black dot (0).

Program 12.1 shows how we might implement this algorithm using BASICA. Note again that this program only requires that we have two rows of data in memory at one time; this feature is particularly important because most images are too large to conveniently hold entirely in a microcomputer's RAM.

In Program 12.1 we have made several assumptions. First, we have assumed the data to be in a sequential file (although you probably will want to use a random file because it is faster). Second, we have assumed that the data are all in the range of 0 to 255. Third, we have assumed that you want to interpolate the data to fill the screen more completely in the horizontal (X) direction (lines 470 to 540); this will give a better appearance. Finally, we have assumed you will plot the data on a CGA (640 x 200) screen (line 80). The program can easily be changed if these assumptions are not true for your system. In particular, you will almost certainly wish to use a higher resolution monitor than the CGA.

In line 250 of this program, we have output the results onto the screen as dots to simulate the output to a printer. You can of course use the usual print-screen option (typing Shift-PrtSc) to print this on your standard graphics printer, assuming you have typed GRAPHICS first (from DOS). However, you will need to modify the program to output dots directly to your printer, to take advantage of its higher resolution. Unfortunately, there is no standard method for sending image data to a printer. You will need to consult the printer manual to see how this is done for your printer. Usually it involves sending some specific code sequence to put the printer in all-points-addressable graphics mode, and then computing a sequence of bytes to be sent for each line of the image to be printed.

If you run this program with your printer, you will notice one additional feature needs to be added to it: the printed image needs to be enlarged. We will discuss two methods for enlarging (zooming) the image below.

12.4 IMAGE MANIPULATION

Once we have acquired an image, there are many kinds of things we can do to process the image that require no knowledge of the contents of the image. We shall refer to these types of context-independent processing as **image manipulation** to distinguish them from **image analysis**, which requires some knowledge of the contents of the image. There are many such techniques; we'll look at some of the most common ones.

12.4.1 Zoom

One of the most common things we wish to do with an image is to enlarge some or all of the image; this process is generally referred to as image **zoom**. Many of the more common image display systems perform this function in hardware, so that it can be very rapid.

```
10 REM Program F-S_error_diffusion
20 REM LABORATORY AUTOMATION USING THE IBM PC
30 REM Program to perform Floyd-Steinberg error diffusion on an image
40 REM Assumes display will be on graphics screen in 640 x 200 mode
50 REM Requires a pair of image lines to be in memory
60 REM ***********************************************************
70 DIM IMAGE%(2,1000)
80 SCREENWIDTH%=640               'screen resolution
90 SCREEN 2:KEY OFF:CLS           'display on high-resolution screen
100 REM  \            Main program
110    GOSUB 330                  'get user input
120    CLS:WINDOW (0,0)-(PIXELS%*(2^XEXPAND%),ROWS%) 'fill entire screen
130    I% = 1
140    GOSUB 410                  'get first row of data
150    GOSUB 580                  'move from second row to first row of IMAGE%
160    FOR I% = 2 TO ROWS% + 1
170        Y$=INKEY$: IF Y$="Q" OR Y$="q" THEN END
180        IF I% < = ROWS% THEN GOSUB 410    'get the next row of data
190        FOR J%= 1 TO PIXELS2%
200            IF IMAGE%(1,J%) < 128 THEN DOT% =0 ELSE DOT% = 255
210            DOTERROR% = IMAGE%(1,J%)-DOT%     'calculate "error"
220            IMAGE%(2,J%) = IMAGE%(2,J%)+.375 * DOTERROR%    'diffuse error
230            IMAGE%(1,J%+1) = IMAGE%(1,J%+1)+.375 * DOTERROR%   'to 3 adjoining
240            IMAGE%(2,J%+1) = IMAGE%(2,J%+1) + .25 * DOTERROR%  'locations
250            IF DOT% =255 THEN PSET (J%,I%-1)
260        NEXT J%
270        GOSUB 580                    'move row #2 to row #1 of IMAGE% array
280    NEXT I%
290    Y$=INKEY$: IF Y$="" THEN 290      'Type any key to stop
300 END
310 REM *****************************
320 REM Get user input
330    INPUT "Enter the name of the data file ",FILN$
340    OPEN FILN$ FOR INPUT AS #1
350    INPUT "Enter the number of pixels per row in your image ",PIXELS%
360    XEXPAND%=LOG(SCREENWIDTH%/PIXELS%)\LOG(2)     'expand to fill screen
370    INPUT "Enter the number of rows in your image ",ROWS%
380 RETURN
390 REM *****************************
400 REM Read in a line of data
410    FOR K% = 1 TO PIXELS%
420        INPUT #1,IMAGE%(2,K%)
430    NEXT K%
440    PIXELS2%=PIXELS%
450    IF XEXPAND%=0 THEN RETURN
460    'if possible, interpolate image to try to fill screen in X direction
470    IMAGE%(2,0)=0
480    FOR J% = 1 TO XEXPAND%
490        FOR K%=PIXELS2% TO 1 STEP -1
500            IMAGE%(2,K%*2)=IMAGE%(2,K%)
510            IMAGE%(2,K%*2-1)=(IMAGE%(2,K%)+IMAGE%(2,K%-1))/2
520        NEXT K%
530        PIXELS2%=2*PIXELS2%
540    NEXT J%
550 RETURN
560 REM *****************************
570 REM Move row 2 to row 1 of IMAGE%
580    FOR K% = 1 TO PIXELS2%
590        IMAGE%(1,K%)=IMAGE%(2,K%)
600    NEXT K%
610 RETURN
```

Program 12.1.   Floyd-Steinberg Error Diffusion.

There are two commonly-used types of zoom: **pixel replication zoom** and **interpolation zoom**. Let's look at each of these in turn.

12.4.1.1 Pixel Replication Zoom

Pixel replication zoom is the easiest zoom to implement in hardware. In this type of zoom, as the name implies, the image is enlarged by replicating every pixel in the image. Thus, for example, if we have the pixels in a 2 x 2 section of an image with values as follows:

$$10\ 20$$
$$30\ 40$$

then the image after replication would contain a 4 x 4 section that would be as follows:

$$10\ 10\ 20\ 20$$
$$10\ 10\ 20\ 20$$
$$30\ 30\ 40\ 40$$
$$30\ 30\ 40\ 40$$

Notice that the replication results in an image that is twice as large as the starting image in both dimensions. The replication can be repeated as many times as necessary to provide the desired degree of enlargement; the resulting image will always be some power of two larger than the starting image.

12.4.1.2 Interpolation Zoom

The interpolation zoom, by contrast, can enlarge an image by any desired factor. It does so by interpolating the values between the pixels. Thus, if we start out with a 2 x 2 image segment:

$$10\ 20$$
$$30\ 40$$

and perform an interpolation zoom by a factor of 2, we would produce:

$$10\ 13\ 17\ 20$$
$$17\ 20\ 24\ 27$$
$$23\ 26\ 30\ 33$$
$$30\ 33\ 37\ 40$$

Actually, another type of interpolation zoom, **midpoint interpolation**, is much faster and is therefore more commonly used. In this technique, the image is always enlarged by some power of two and the values between each pair of pixels in the original image are then calculated as a simple average. The reason this is an attractive alternative is that division and multiplication are slow processes compared to addition and subtraction, and floating point operations are slow compared to integer operations. The average of a pair of points can be calculated by adding the two points together and dividing by the integer 2; some computer languages permit bit shifting operations, in which case a divide-by-two can be accomplished even more rapidly by shifting the bits one bit to the right.

One method of implementing a midpoint interpolation zoom is illustrated in Program 12.2. Notice that interpolation results in an enlarged image that is one pixel less than twice the size of the original image. Hence, to keep the image exactly twice as large, we use extrapolation at two of the four edges of the image; however, you may prefer to simply set those edges equal to the background color (e.g., to zero). Figure 12.2 shows the same image enlarged by both a pixel replication zoom and an interpolation zoom. Note that the output of Program 12.2 can be displayed by Program 12.1.

We begin Program 12.2 by getting parameters form the user (lines 450 to 570). We then read in a line (row) of data (lines 600 to 690). The reading is done such that two lines of data are always in memory, but each line is read only once (because RAM to RAM data transfers are fast, while disk to RAM transfers are slow). Each pair of rows is then interpolated to the desired size, first in the X direction (lines 130 to 270) and then in the Y direction (lines 280 to 390).

12.4.1.3 Memory Restrictions

Before leaving the topic of zooming an image, we should point out a feature that somewhat complicates this process, whether we are performing a pixel replication or interpolation zoom. This special case occurs when the entire image is in display memory and hence can be treated more efficiently. You will notice in the previous examples that we have treated the image as though it is in memory. For most displays used with the IBM PC, the data are indeed in a section of memory accessible to both the CPU and the card controlling the display. However, because DOS is restricted to 640 Kbyte segments and yet many displays are capable of displaying color images of 1 Mpel or more, we must have some means of addressing more than 640K of memory. Hence, many display controller adapters use some form of bank switching (where a software switch controls the mapping of many banks of display memory into a single bank of memory in the 640 K space addressable by DOS) so that 1 Mbyte or more of image can be accessed. Hence, when you write your own routines for zoom (or any other process that acts directly on the displayed image), you may need to perform bank switching operations as well.

12.4.2 Pan and Scroll

Pan and scroll refer to moving an image in the horizontal and vertical directions, respectively. Often these are operations that are performed by the display hardware. They almost always involve the entire display, rather than just one of several images that may be on the display. In essence, pan and scroll are operations that involve

```
10   REM Program Interpolation_Zoom
20   REM LABORATORY AUTOMATION USING THE IBM PC
30   REM Program to perform a midpoint interpolation zoom of an image
40   REM Note: changing IMAGE! to IMAGE% will increase speed, decrease accuracy
50   REM **********************************************************
60   DIM IMAGE!(16,500),NEWLINE%(2,500)
70   REM                 Main program
80      GOSUB 450                                'get user input
90      LINE1%=0
100     GOSUB 600                                'read in a line of data (line 0)
110     FOR I% = 1 TO ROWS%-1
120        GOSUB 600                             'get two rows of data
130        COUNT%=PIXELS%                        'counts no. of pixels in current row
140        FOR Z%=1 TO ZOOM%                     'for each zoom factor, zoom in X
150           M%=COUNT%*2
160           IMAGE!(1,0)=2*IMAGE!(1,1)-IMAGE!(1,2)'extrapolate first end point
170           IMAGE!(2,0)=2*IMAGE!(2,1)-IMAGE!(2,2)
180           FOR J%=COUNT% TO 1 STEP -1         'perform midpoint interpolation
190              IMAGE!(1,M%)=IMAGE!(1,J%)
200              IMAGE!(2,M%)=IMAGE!(2,J%)
210              M%=M%-1
220              IMAGE!(1,M%)=(IMAGE!(1,J%)+IMAGE!(1,J%-1))/2
230              IMAGE!(2,M%)=(IMAGE!(2,J%)+IMAGE!(2,J%-1))/2
240              M%=M%-1
250           NEXT J%
260           COUNT%=COUNT%*2
270        NEXT Z%
280        COUNT%=PIXELS%*2^(ZOOM%)
290        FOR Z%=1 TO ZOOM%                     'now interpolate in y direction
300           FOR Q%=1 TO COUNT%                 'extrapolate zeroth row
310              IMAGE!(0,Q%)=2*IMAGE!(1,Q%)-IMAGE!(2,Q%)
320              FOR Y%=2^Z% TO 1 STEP -1
330                 M%=Y%+Y%
340                 IMAGE!(M%,Q%)=IMAGE!(Y%,Q%)
350                 M%=M%-1
360                 IMAGE!(M%,Q%)=(IMAGE!(Y%,Q%)+IMAGE!(Y%-1,Q%))/2
370              NEXT Y%
380           NEXT Q%
390        NEXT Z%
400        GOSUB 720                             'print out results
410     NEXT I%
420  END
430  REM ********************
440  REM Get user input
450     INPUT "Enter the name of the data file ",FILN$
460     OPEN FILN$ FOR INPUT AS #1
470     INPUT "Enter the name of the output file ",FILN2$
480     OPEN FILN2$ FOR OUTPUT AS #2
490     INPUT "Enter the number of pixels per row in your image ",PIXELS%
500     INPUT "Enter the number of rows in your image ",ROWS%
510     INPUT "Enter the zoom factor (must be 2, 4, 8, or 16) ",ZOOMMAG%
520     ZOOM% = INT (LOG(ZOOMMAG%)/LOG(2))       'get log base 2 of zoommag%
530     WHILE (2^ZOOM%)*PIXELS% > 500            'too much zoom?
540        ZOOM%=ZOOM%-1
550     WEND
560     IF ZOOM% > 4 THEN ZOOM% = 4
570  RETURN
580  REM ********************
590  REM Read in a line of data
600     FOR K% = 1 TO PIXELS%                    'read one line of data
610        INPUT #1,NEWLINE%(LINE1%,K%)
620     NEXT K%
630     LINE2% = LINE1% XOR 1
640     FOR K% = 1 TO PIXELS%                    'put pair of lines in IMAGE! array
650        IMAGE!(1,K%)=NEWLINE%(LINE2%,K%)
660        IMAGE!(2,K%)=NEWLINE%(LINE1%,K%)
670     NEXT K%
```

Program 12.2 (Part 1 of 2).   Interpolation Zoom.

```
 680    LINE1% = LINE1% XOR 1                    'switch to other line next time
 690 RETURN
 700 REM ********************
 710 REM Print results to sequential file
 720    LOCATE 10,1:PRINT "Row ";I%
 730    IF I%=1 THEN P%=1 ELSE P%=2∧ZOOM%+1
 740    FOR QQ%=P% TO 2∧(ZOOM%+1)
 750          FOR Q%=1 TO COUNT%
 760                PRINT #2, INT(IMAGE!(QQ%,Q%));
 770          NEXT Q%
 780          PRINT #2,""
 790    NEXT QQ%
 800 RETURN
```

Program 12.2 (Part 2 of 2).   Interpolation Zoom.

movement of data from one place in memory to another, and hence are generally very rapid and trivial to program.

12.4.3 Histogram Enhancement

When we look at a gray-scale image, we often notice that the balance of shades of gray is not acceptable. For example, we may have an image that has too little contrast, or an image where all of the information of interest is in the dark areas of the image.

The technique often used to assess the balance of gray tones in the image is **histogramming**. In this technique, we set up as many **bins**, or locations in an array, as we have values on the gray scale; for example, this would be 256 bins for a typical 8-bit image. We then simply read each pixel of the gray-scale image and increment the count in the bin corresponding to the pixel intensity until we have examined all of the pixels (e.g., for a pixel with intensity of 114, we increment the count in bin 114 by 1). A simple program to do this is illustrated in Program 12.3. Note that the bin incrementing occurs in line 150. Figure 12.3 shows a typical image and its histogram.

Once we have the histogram calculated, we can use it to perform any of several operations that improve the distribution of pixel values; i.e., we perform **histogram enhancement**.

12.4.3.1 Contrast Enhancement

As shown in Figure 12.4, a low-contrast image has a histogram with almost all of its pixel values clustered in one region (usually the middle) of the histogram. A high contrast image, on the other hand, tends to have most of its pixels with either very high or very low values. Hence, we can increase the contrast of an image by moving some of the pixel values toward the two extremes. One of the easiest methods for doing this is to let the user choose a minimum and maximum value, and then stretch

Figure 12.2.   Image Zoom. The image on the top has been enlarged using a midpoint interpolation zoom. The image on the bottom has been enlarged the same amount using a pixel replication zoom. Notice the difference in spatial resolution of the two images.

```
10 REM Program Histogram
20 REM LABORATORY AUTOMATION USING THE IBM PC
30 REM Program to calculate a histogram of an image
40 REM *************************************************************
50 DIM IMAGE%(1000),HIST%(255)
60 SCREEN 0:CLS
70 REM                    Main program
80     GOSUB 230                           'get user input
90     GOSUB 300                           'zero histogram array
100    REM Get the input line and calculate its histogram
110    FOR I% = 1 TO ROWS%
120        GOSUB 360                       'get a row of data
130        FOR J% = 1 TO PIXELS%
140            K%=IMAGE%(J%)               'calculate bin number
150            HIST%(K%)=HIST%(K%)+1       'increment count in that bin
160        NEXT J%
170    NEXT I%
180    GOSUB 420                           'print the histogram results
190    INPUT "",Y$                         'type <return> to finish
200 END
210 REM *****************************
220 REM Get user input
230    INPUT "Enter the name of the file containing the data ",FILN$
240    OPEN FILN$ FOR INPUT AS #1
250    INPUT "Enter the number of pixels per row in your image ",PIXELS%
260    INPUT "Enter the number of rows in your image ",ROWS%
270 RETURN
280 REM *****************************
290 REM Zero array containing histogram
300    FOR I=0 TO 255
310        HIST%(I)=0
320    NEXT I
330 RETURN
340 REM *****************************
350 REM Read in a line of data
360    FOR K% = 1 TO PIXELS%
370        INPUT #1,IMAGE%(K%)
380    NEXT K%
390 RETURN
400 REM *****************************
410 REM Plot results on screen
420    SCREEN 2:CLS:KEY OFF                'high resolution graphics
430    MAXHT%=0
440    FOR I%=0 TO 255                     'get maximum histogram value
450        IF HIST%(I%) > MAXHT% THEN MAXHT%=HIST%(I%)
460    NEXT I%
470    VIEW (80,12)-(600,172)              'define viewport
480    WINDOW (0,0)-(255,MAXHT%)           'define window in viewport
490    FOR I%=0 TO 255                     'display histogram
500        IF HIST%(I%) > 0 THEN LINE (I%,0)-(I%,HIST%(I%))
510    NEXT I%
520    VIEW (80,172)-(600,180)             'define viewport for axis
530    DELTA = 255/5                       '5 tick marks
540    FOR I%=0 TO 5                       'x axis tick marks and labels
550        X2%=I%*DELTA
560        LINE (X2%,0)-(X2%,MAXHT%)
570        LOCATE 24,I%*13+9
580        PRINT X2%;
```

Program 12.3 (Part 1 of 2). Histogram of an Image.

the histogram between these two values to fill the entire value range. For example, if most of the histogram density is between 120 and 200 in a possible range of 0 to 255, we can increase the contrast by mapping the values of 120 to 200 into a new histogram from 0 to 255. Any values of less than 120 are then mapped as zero, and

```
590    NEXT I%
600    LINE (0,MAXHT%)-(255,MAXHT%)          'x axis
610    LOCATE 25,40
620    PRINT "BRIGHTNESS";                   'label x-axis
630    LOCATE 1,1
640 RETURN
```

Program 12.3 (Part 2 of 2).   Histogram of an Image.

values greater than 200 are mapped as 255. Program 12.4 illustrates how this is done, and Figure 12.4 illustrates the effect of such an operation.

Notice that Program 12.4 essentially contains Program 12.3. However, in addition, in lines 150 to 200, we locate the minimum and maximum bins in the histogram that contain any counts. The user is then given the option of changing the minimum and maximum (lines 210 to 250). The new minimum and maximum are then used to calculate a factor (line 260) that is used in a subroutine (lines 880 to 980) that expands the histogram to fill the new range. The user can expand the histogram until it is satisfactory. At that point, we wish to create a new image. The most efficient manner of doing this is to create a lookup table that allows one to map values in the original image to their values in the new image; this avoids performing a floating-point calculation on each pixel in the image. The program shows how to perform the mapping in lines 1010 to 1080. A new data file is then created using this table (lines 1110 to 1230).

12.4.3.2 Other Histogram Enhancement

Many other operations can also be used to enhance the histogram of an image. For example, we may choose to stretch or shrink the entire histogram. In more sophisticated operations, we may choose to modify only certain regions of the histogram; for example, we may wish to stretch the dark areas and shrink the light areas to emphasize details in the dark areas of the image. In any case, these operations are designed to improve the appearance of the image, and hence the decision about which histogram operations to use is best left to the end user rather than to the programmer.

12.4.4 Filtering

There are many occasions when we may wish to filter an image. Filtering of images is analogous to filtering one-dimensional data, as described in Chapter 4. However, because of the huge volume of data involved, we often limit ourselves to computationally simple operations in order to be able to filter the data in a reasonable time period. Usually, this means using some sort of **convolution** operation.

In essence, convolution of images is the two-dimensional analog of the moving average and related filters discussed in Chapter 4. As in the one-dimensional case,

Number of Pixels

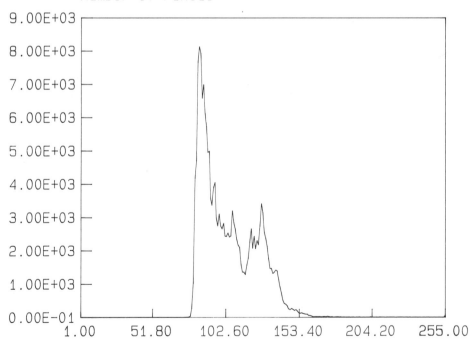

Figure 12.3. **Histogram of an Image.** The image on the top has been used to calculate the histogram on the bottom.

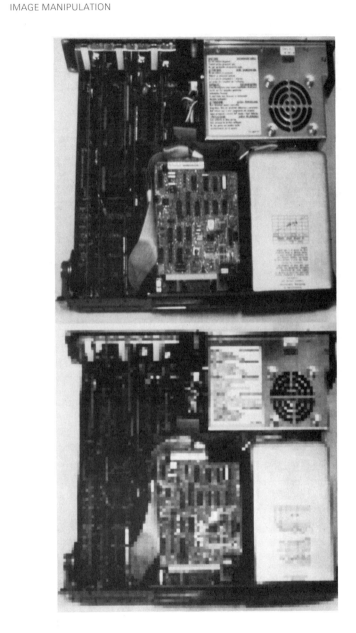

Figure 12.4.   Contrast Enhancement.   Top. A low contrast image, as recorded by the frame grabber. **Bottom**. A contrast-enhanced version of the same image.

```
10 REM Program Contrast_Enhancement
20 REM LABORATORY AUTOMATION USING THE IBM PC
30 REM Program to increase the contrast of an image
40 REM Must be run on graphics monitor!
50 REM *******************************************************
60 DIM IMAGE%(1000),HIST%(255),HIST2%(255)
70 SCREEN 0:CLS
80 REM                Main program
90    GOSUB 360                                'get user input
100   FACTOR!=1!                               'histogram expansion factor
110   GOSUB 430                                'calculate histogram
120   GOSUB 630                                'plot the histogram
130   Y$="n"
140   WHILE Y$ <> "y" AND Y$ <> "Y"
150      MAX%=0:MIN%=255                       'calculate min, max of histogram
160      FOR I%=0 TO 255
170         IF HIST%(I%) > 0 AND (I% < MIN%) THEN MIN%=I%
180         IF HIST%(I%) > 0 AND (I% > MAX%) THEN MAX%=I%
190      NEXT I%
200      PRINT "Current histogram minimum = ";MIN%," maximum = ";MAX%
210      MIN%=-1
220      WHILE ((MIN% < 0) OR (MAX% > 255) OR (MIN% >= MAX%))
230         INPUT "Enter the new histogram minimum (0-255) ",MIN%
240         INPUT "Enter the new histogram maximum (0-255) ",MAX%
250      WEND
260      FACTOR!=255/(MAX%-MIN%)
270      GOSUB 870                             'calculate new histogram
280      GOSUB 630                             'display new histogram
290      INPUT "Is this new mapping satisfactory  <Y or N > ",Y$
300   WEND
310   GOSUB 1010                               'create new mapping
320   GOSUB 1110                               'create new data file
330 END
340 REM ********************
350 REM Get user input
360   INPUT "Enter the name of the data file ",FILN$
370   OPEN FILN$ FOR INPUT AS #1
380   INPUT "Enter the number of pixels per row in your image ",PIXELS%
390   INPUT "Enter the number of rows in your image ",ROWS%
400 RETURN
410 REM ********************
420 REM Calculate the histogram
430   FOR I=0 TO 255                           'zero HIST% array
440      HIST%(I)=0
450   NEXT I
460   REM Get the input line and calculate its histogram
470   FOR I% = 1 TO ROWS%
480      GOSUB 570                             'get a row of data
490      FOR J% = 1 TO PIXELS%
500         K%= IMAGE%(J%)                     'calculate bin number
510         HIST%(K%)=HIST%(K%)+1              'increment count in that bin
520      NEXT J%
530   NEXT I%
540 RETURN
550 REM ********************
560 REM Read in a line of data
570   FOR K% = 1 TO PIXELS%
580      INPUT #1, IMAGE%(K%)
590   NEXT K%
600 RETURN
610 REM ********************
620 REM Plot histogram on high res screen
630   SCREEN 2:CLS:KEY OFF                     'high resolution graphics
640   MAXHT%=HIST%(1)
650   FOR I%=1 TO 254                          'get maximum histogram height
660      IF HIST%(I%) > MAXHT% THEN MAXHT%=HIST%(I%)
670   NEXT I%
```

Program 12.4 (Part 1 of 2).   Contrast Enhancement.

```
680    VIEW (80,12)-(600,172)                  'define viewport
690    WINDOW (0,0)-(255,MAXHT%)               'define window in viewport
700    FOR I%=0 TO 255                         'display histogram
710        IF HIST%(I%) > 0 THEN LINE (I%,0)-(I%,HIST%(I%))
720    NEXT I%
730    VIEW (80,172)-(600,180)                 'define viewport for axis
740    DELTA = 255/5                           '5 tick marks
750    FOR I%=0 TO 5                           'tick marks with labels
760        X2%=I%*DELTA
770        LINE (X2%,0)-(X2%,MAXHT%)
780        LOCATE 24,I%*13+9
790        PRINT X2%;
800    NEXT I%
810    LINE (0,MAXHT%)-(255,MAXHT%)            'x axis
820    LOCATE 25,40
830    PRINT "BRIGHTNESS";                     'label x-axis
840    LOCATE 1,1:VIEW
850 RETURN
860 REM ********************
870 REM Recalculate histogram
880    FOR I%=0 TO 255
890        HIST2%(I%)=HIST%(I%)
900        HIST%(I%)=0
910    NEXT I%
920    FOR I%=0 TO 255
930        K%=(I%-MIN%)*FACTOR!
940        IF K% < 0 THEN K%=0
950        IF K% > 255 THEN K%=255
960        HIST%(K%)=HIST%(K%)+HIST2%(I%)
970    NEXT I%
980 RETURN
990 REM ********************
1000 REM Map old data into new
1010    CLS
1020    FOR I%=0 TO 255
1030        K%=(I%-MIN%)*FACTOR!
1040        IF K% < 0 THEN K%=0
1050        IF K% > 255 THEN K%=255
1060        HIST2%(I%)=K%
1070    NEXT I%
1080 RETURN
1090 REM ********************
1100 REM Create new data file
1110    CLOSE #1                               'rewind input file
1120    OPEN FILN$ FOR INPUT AS #1
1130    INPUT "Enter the name of the desired output file ",FILN2$
1140    OPEN FILN2$ FOR OUTPUT AS #2
1150    FOR I%=1 TO ROWS%
1160        GOSUB 550
1170        FOR J%=1 TO PIXELS%
1180            IMAGE%(J%)=HIST2%(IMAGE%(J%))
1190            PRINT #2,IMAGE%(J%);
1200        NEXT J%
1210        PRINT #2,""
1220    NEXT I%
1230 RETURN
```

Program 12.4 (Part 2 of 2).   Contrast Enhancement.

we make use of the fact that adjoining pixels in the image have some sort of relationship to one another. For example, we may design a low-pass filter to take advantage of the fact that the change in intensity from pixel to pixel in a typical image is relatively slow; hence, very rapid changes may be due to high-frequency noise and can be filtered out with a low-pass filter.

We can show a simple convolution operation using a **convolution mask**, or set of convolution coefficients. We start by constructing the 3 x 3 mask. For example, to implement a low-pass filter, we might use a mask of:

$$1 \ 1 \ 1$$
$$1 \ 1 \ 1$$
$$1 \ 1 \ 1$$

We now pass the mask across the image. We can visualize the mask centered above a single pixel. To apply our filter, we simply multiply each pixel by the coefficient in the mask directly above the pixel. The sum of the nine products of these operations is then divided by the sum of the coefficients to yield a single number, which is the new value for the central pixel. To convolve the entire image, we repeat this process above each pixel in the image, using its 8 nearest neighboring pixels plus the pixel of interest.

In this case, with all of the coefficients being 1, we are simply averaging 9 pixels together and using the averaged value for the central pixel. However, it is important to note that the new averaged value is not used in computing any other pixels, but is simply placed in an output image. Furthermore, we must make special provisions for the four edges of the image; usually, we simply duplicate each of the edges before convolving them.

All of this can be rather confusing, so as usual we illustrate the point with a computer program, namely Program 12.5. You may notice in this program that we only have three lines (rows) of data in memory at one time. Furthermore, to reduce computation time, we use pointers to the rows so that we do not need to move them around in memory after each line of pixels has been convolved.

Hence, the main loop controlling the convolution (lines 490-590) assumes that 3 rows of data are in memory and calls a routine to convolve the data (lines 830 to 980). The points to the 3 rows are reset (lines 510 to 550) and a new line of data is read from the disk (lines 750-800). This process is repeated for each set of three lines in the image. To be sure that the edges are convolved, the first and last pixel of each row (lines 780 to 790) and the first and last row (lines 380 to 400 and 680 to 700) are each duplicated.

We can devise a large number of different types of masks for different purposes. For example, in addition to the low-pass filter described above, we can describe a mask for high-pass filters:

$$-1 \ -1 \ -1$$
$$-1 \ \ 9 \ -1$$
$$-1 \ -1 \ -1$$

or

```
10 REM Program Convolve
20 REM LABORATORY AUTOMATION USING THE IBM PC
30 REM Program to convolve image with 3 x 3 convolution kernel
40 REM ****************************************************************
50 DIM IMAGE%(1000,3),POINTER%(3),NEWIMAGE%(1000),COEFF%(3,3)
60 SCREEN 0:CLS
70 REM                    Main program
80     GOSUB 150                            'get user input
90     GOSUB 360                            'set up for first row of image
100    GOSUB 490                            'convolve image
110    GOSUB 620                            'convolve last row of image
120 END
130 REM **********************
140 REM get user input
150    INPUT "Enter the name of the data file ",FILN$
160    OPEN FILN$ FOR INPUT AS #1
170    FILN2$=FILN$                         'avoid using input file as output
180    WHILE FILN$ = FILN2$
190        INPUT "Enter the name of the output data file ",FILN2$
200    WEND
210    OPEN FILN2$ FOR OUTPUT AS #2
220    INPUT "Enter the number of pixels per row in your image ",PIXELS%
230    INPUT "Enter the number of rows in your image ",ROWS%
240    PRINT "Enter the set of coefficients for the convolution kernel"
250    SUM%=0
260    FOR I%=1 TO 3
270        FOR J%=1 TO 3
280            LOCATE 5+I%,J%*6
290            INPUT "",COEFF%(I%,J%)
300            SUM%=SUM%+COEFF%(I%,J%)
310        NEXT J%
320    NEXT I%
330 RETURN
340 REM **********************
350 REM First row is special case; use rows 1,1,2 as 3 rows
360    ROW%=1
370    GOSUB 750                            'read in first row of data
380    FOR I%=0 TO PIXELS% + 1              'duplicate first row
390        IMAGE%(I%,2)=IMAGE%(I%,1)
400    NEXT I%
410    ROW%=3
420    GOSUB 750                            'read in second row
430    POINTER%(1)=1                        'set up pointers to current 3 rows
440    POINTER%(2)=2
450    POINTER%(3)=3
460 RETURN
470 REM **********************
480 REM Main convolving loop
490    FOR I%=1 TO ROWS%-2
500        GOSUB 830                        'convolve image
510        FOR J%=1 TO 3                    'reset pointers to 3 rows
520            POINTER%(J%)=POINTER%(J%)+1
530            IF POINTER%(J%) > 3 THEN POINTER%(J%)=1
540        NEXT J%
550        ROW%=POINTER%(3)
560        GOSUB 750                        'get next row of data
570    NEXT I%
580    GOSUB 830                            'convolve next-to-last row
590 RETURN
600 REM **********************
610 REM Last row is special case; use rows n-1, n, n as 3 rows of data
620    FOR J%=1 TO 3
630        POINTER%(J%)=POINTER%(J%)+1
640        IF POINTER%(J%) > 3 THEN POINTER%(J%)=1
650    NEXT J%
660    K%=POINTER%(3)
670    L%=POINTER%(2)
```

Program 12.5 (Part 1 of 2). Convolution.

```
680    FOR I%=1 TO PIXELS%                       'duplicate last row
690        IMAGE%(I%,K%)=IMAGE%(I%,L%)
700    NEXT I%
710    GOSUB 830                                 'convolve last row
720 RETURN
730 REM ********************
740 REM Subroutine to read in a line of data
750    FOR K% = 1 TO PIXELS%
760        INPUT #1,IMAGE%(K%,ROW%)
770    NEXT K%
780    IMAGE%(0,ROW%)=IMAGE%(1,ROW%)             'duplicate first pixel into zeroth
790    IMAGE%(PIXELS%+1,ROW%)=IMAGE%(PIXELS%,ROW%) 'duplicate last pixel also
800 RETURN
810 REM ******************
820 REM subroutine to convolve data
830    FOR Q%=1 TO PIXELS%
840        NEWIMAGE%(Q%)=0                        'initialize sum
850        FOR P%=1 TO 3                          'convolve
860            FOR R%=1 TO 3
870                NEWIMAGE%(Q%)=NEWIMAGE%(Q%)+IMAGE%(Q%+P%-2,POINTER%(R%))*
                   COEFF%(P%,R%)
880            NEXT R%
890        NEXT P%
900        IF SUM% <> 0 THEN NEWIMAGE%(Q%)=NEWIMAGE%(Q%)/SUM%
910        IF NEWIMAGE%(Q%) < 0 THEN NEWIMAGE%(Q%)=0
920        IF NEWIMAGE%(Q%) > 255 THEN NEWIMAGE%(Q%)=255
930    NEXT Q%
940    FOR Q%=1 TO PIXELS%                       'print results of convolving
950        PRINT #2,NEWIMAGE%(Q%);
960    NEXT Q%
970    PRINT #2,""
980 RETURN
```

Program 12.5 (Part 2 of 2). Convolution.

$$
\begin{array}{rrr}
0 & -1 & 0 \\
-1 & +5 & -1 \\
0 & -1 & 0
\end{array}
$$

With very little modification of the mask, we can produce an omnidirectional edge enhancement filter referred to as a Laplacian filter:

$$
\begin{array}{rrr}
-1 & -1 & -1 \\
-1 & 8 & -1 \\
-1 & -1 & -1
\end{array}
$$

or

$$
\begin{array}{rrr}
0 & 1 & 0 \\
1 & -4 & 1 \\
0 & 1 & 0
\end{array}
$$

The effects of these filters are shown in Figure 12.5.

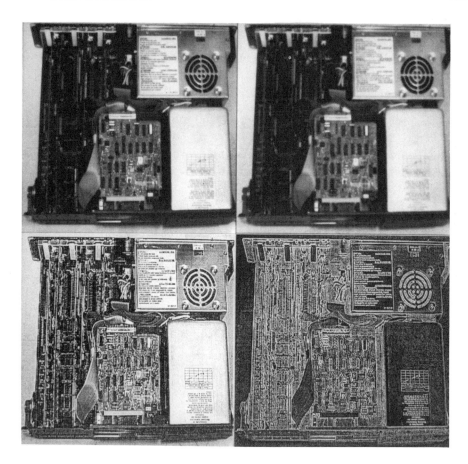

Figure 12.5.   Convolution Filters.   Top Left. Original image. **Top right**. The same image
          after using a low-pass convolution filter. **Bottom left**. The same image after
          using a high-pass convolution filter. **Bottom right**. The same image after
          using a Laplacian convolution filter and histogram equalization.

So far, all of the filters we have shown have used 3 x 3 masks. It is indeed possible to devise larger masks, in much the same manner as described in Chapter 4 for
one-dimensional data. However, 3 x 3 masks are often used because of their speed in
implementation compared to larger filters.

12.4.5 False Color

Most image acquisition systems currently available are designed to obtain and store images that are black-and-white, rather than color. However, it is often possible and desirable to display the image with colors replacing one or more of the shades of gray. For example, we may use shades of blue, rather than shades of gray, for the entire image. However, to emphasize certain features, we often add multiple colors, or add colors only to certain intensities in the image. For example, to an infrared satellite image of land masses, we might add a range of colors to aid the human eye in distinguishing rivers from mountains from cultivated fields. This type of **false color** can be very useful.

For many systems, addition of false color can be quite easy. This is because the image display hardware uses hardware **lookup tables** that map the values in the original image into specific colors in the final display. For example, many systems allow 8 bits of color information to be sent to each of three different sources in the monitor; the colors are added together to form a single image. Hence, up to 2^{24} or 16.7 million colors can be displayed; the color lookup tables select 256 of these to be used at any one time to correspond to the 2^8 values in the original digitized image.

12.5 IMAGE ANALYSIS

The final step in the processing of digital images is **image analysis**. This includes all of the processing that we do to obtain the desired scientific information from an image. This often includes such tasks as measurement of features in the image, comparison of features among images, comparison of images to model data, three-dimensional rendering of data, and so forth. This step is obviously highly dependent upon the type of image being analyzed, and is usually the most sophisticated step to perform. It is therefore beyond the scope of this text, but there are numerous texts and journals for those interested in pursuing this field.

12.6 IMPROVING PERFORMANCE

Before leaving this topic, we should point out that image processing, because it deals with very large data sets, often consumes large amounts of CPU time. Hence, in our own laboratory we often use one of three approaches for handling this amount of data.

The first technique is to use special-purpose array processors or special-purpose imaging systems to process the data at the workstation.

The second technique is to use assembly language for key parts of the program. Often, converting 5 percent or less of the code from BASIC to assembly language will greatly improve the performance. You may therefore wish to read Chapter 14 to better understand how to interface assembly language routines to BASIC.

The third technique is to use a large computer to perform the numerically intensive computing involved in image processing and let the microcomputer do the user interaction and image displays. This involves using tightly-coupled cooperative processing, as described in Chapter 13.

EXERCISES

1. Modify the programs in this chapter to input data from your own image data file and to output the results to the display or graphics printer you are using.
2. Create a menu-driven program to input an image and perform any or all of the operations described in this chapter in response to user commands.
3. Modify Program 12.1 to interpolate the data to fill the screen in the vertical (Y) direction. Note that the program already interpolates the data in the X direction.
4. Write a program to perform histogram equalization. This process attempts to improve the contrast of an image by shifting pixel values so that the histogram becomes essentially flat; that is, so that there is approximately the same number of pixels of each intensity in the image.

PART THREE

ADVANCED TECHNIQUES

13

LOCAL AREA NETWORKS IN THE LABORATORY

If you work in an environment with several computers, you undoubtedly routinely encounter situations where you are using one computer, but wish to use programs, printers, plotters, disk storage, or other features of another computer system in your environment. If so, you may be ready to consider a **local area network (LAN)**. LANs are one of the current "hot topics" in laboratory automation, with LAN sessions at scientific meetings attracting large crowds.

Part of the reason for the interest in this topic is that so few people understand it, and in particular, the special role that it can have in the scientific or engineering laboratory. What we will try to do is to help you understand the basics of LANs and how they might be applied in your own environment. As usual, our emphasis is upon the things that you will need to know to use LANs, rather than upon the intricate details of LAN hardware or software. In addition, we shall emphasize the unique uses of LANs in the laboratory; i.e., applications that allow us to improve the acquisition, analysis and distribution of scientific data.

What, then, is a local area network? Simply put, it is a set of hardware and software connections that provide digital communications over relatively short distances. In practice, it is usually essentially a high-speed serial or parallel interface among several computer systems (and their associated peripheral hardware). It is usually distinguished from long-distance communications systems (such as telephone systems) and from high-speed networks that tightly couple processors together; these types of networks are outside of the scope of this text.

13.1 LAN HARDWARE

Local Area Network systems can be more easily understood if we examine the hardware and software separately. Let's discuss hardware first.

In principle, there is an infinite number of possible types and arrangements of hardware in LANs. In practice, fortunately, only a few of these arrangements have been implemented. We can conveniently examine the components of LAN hardware by examining the physical cabling and associated transmission rates, network topology, control hardware and control protocols, and interconnect hardware. We shall only consider hardware arrangements from systems that are widely commercially available.

13.1.1 LAN Cabling

Three cabling alternatives are generally available: **twisted pair wire**, **coaxial cable**, and **optical fiber**. Twisted pair wire is widely used in networks where the maximum data rate is less than 1 Mbps (million bits per second). Its major advantage is that it is cheap, and hence it is often used in small LANs. As we discussed in another context in Chapter 4, however, twisted wire is relatively susceptible to noise problems, so it must be used over short distances and at slower speeds than coaxial cable. It has thus far been the most popular transmission medium for LANs.

Coaxial cable, because of its lower susceptibility to noise, is able to sustain much higher rates than twisted pair wires, with maximum rates in the tens of Mbps. Coaxial cable can also be used over longer distances than twisted pair wires, and can support more devices. However, coaxial cable is more expensive than twisted pair.

Optical fiber is the most expensive type of cable, but has maximum transmission rates of hundreds of Mbps to a few Gbps, depending upon the cable used. It also is smaller and generates almost no emissions that might interfere with signals on other cables. Its primary disadvantage is its higher cost.

In general, for small or low-speed networks, you will probably want to use twisted pair. For larger networks, such as those found in large universities and industrial settings, coax is probably currently the medium of choice. However, for larger networks, you should consider not only the cost of the cable, but the cost of installing it; the installation cost is often high enough that you may wish to install fiber optic cable in addition to coax. In other words, you may save money in the long run by installing cable that will provide sufficient bandwidth for future, as well as present, networking needs. Currently fiber is often used for the "backbone" or central cabling, of the network, even if it is not used for the cabling to individual users.

There are two generally-used signaling techniques, and these depend upon the type of cable selected. **Baseband** techniques, which can be used with any of the three types of wire, use constant-voltage pulses; i.e., they use the entire frequency spectrum. Baseband systems suffer from blurring of the pulses when the pulses are sent over long distances (e.g., more than a kilometer), particularly at high frequencies.

Broadband systems, on the other hand, can use frequency-division multiplexing (FDM), which divides the frequencies into channels; this allows transmission of TV, radio, and digital data signals on the same physical cable. In fact, the cable customarily used is 75-ohm coax cable, which is also used for standard community antenna television (CATV) systems.

Generally speaking, baseband is simpler to install but more limited in the number of signals and the length of the cables allowed. Broadband, by contrast, is often used in systems designed to cover large physical areas with large numbers of users.

13.1.2 LAN Topologies

One of the major features distinguishing LANs is their **topology**, or physical shape. Five common topologies are star, bus, tree, ring, and star-ring. These are illustrated in Figure 13.1.

In the **star** topology, all of the locations, or **nodes** on the network, are connected to a central location. The central location contains a switching system which, upon request from a node, will connect the requesting node to the desired destination node.

The **bus** topology requires the use of a multiple-access broadcast medium. Transmissions from one node are in the form of **packets**, which contain not only the data but also source and destination addresses. Each node then monitors the network for packets addressed to that node, and copies them from the network.

The **tree** topology is really a more general case of the bus topology; often the two are discussed together as the bus/tree topology.

The **ring** topology consists of a single ring or loop. Each node is attached to the ring via hardware **repeaters** that process each packet of data arriving at that node. The repeater hardware serves to copy data from the ring addressed to that node and to send packets along to the next node in the ring.

In the **star-ring** topology, all of the wiring is passed through a central location called a **ring wiring concentrator**. In this topology, all of the repeaters are located in the same physical location.

Not surprisingly, some types of wiring are more appropriate with some topologies than others. For example, baseband systems (using twisted pair or coax) are not suitable for tree topologies. Broadband coax is generally not useful in ring systems. Most star systems use twisted pair wiring.

Before leaving this topic, we should mention that there is no one topology that is most suitable for all applications. Generally, each vendor supports one or two of these topologies and hence will encourage users to consider that particular topology.

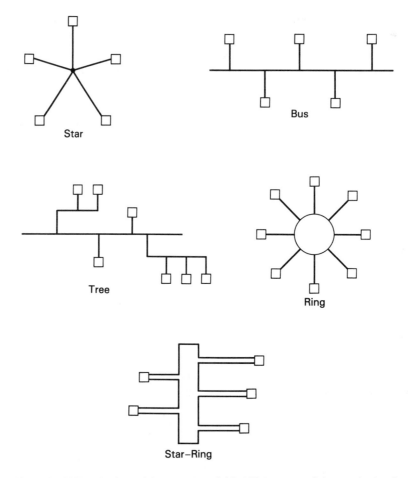

Figure 13.1. LAN Topologies. Most commercial LANs use one of the topologies shown.

13.1.3 LAN Access

In all LANs there is the possibility of contention for the network; i.e., it may happen that two or more users attempt to send a message at the same time. In these cases, there must be a method for determining who will send data first. There are many possible methods; two commonly-used medium access control protocols are CSMA/CD and token. **CSMA/CD** stands for carrier sense multiple access with collision detection. In this technique, a station wishing to transmit information first listens to the network to ascertain if another station is already transmitting. If the network is idle, then the station transmits its message. It then waits for an acknowledgment from the receiving station.

The possibility exists, however, that when the message is transmitted, another station also thinks the network is idle, and starts transmitting a message at the same time; i.e., a collision of data packets occurs. To minimize the effects of such collisions in CSMA/CD, stations monitor the network while they are sending. If a station discovers that another station is sending at the same time, then it ceases transmission immediately and transmits a jamming signal to notify other stations that a collision has occurred. The station then waits a random amount of time and tests the network to see if it can begin another transmission.

In the **token** system, one or more special control packets, or tokens, circulate on the bus or ring. When a station receives the token, it is given permission to send one or more packets of data (including source and destination addresses). When it is finished, the token is passed to another station desiring access to the network.

Each of these control protocols can be quite complicated in order to allow for fault monitoring and correction, network initialization, and addition or deletion of stations from the network. The decision about which protocol to utilize is again one that is largely related to your preference of vendors; in general, however, CSMA/CD is considered to perform better at low network loads, and token to perform better at high network loads. This is because CSMA/CD involves less overhead and hence is initially more efficient. Under high load conditions for CSMA/CD, however, there are so many collisions that much time is wasted recovering from the collisions. Collisions do not occur in the token system, and so performance is better than for CSMA/CD under high-load conditions.

13.1.4 LAN Interconnection

For a variety of reasons, it is often desirable or necessary to have several LANs, often ones that use different protocols. Two pieces of hardware designed to make this possible are bridges and gateways.

Bridges provide connections between two networks of the same type. The function of the bridge is not only to connect the networks electrically, but also to provide control functions as well. Hence, for example, the bridge normally allows messages to pass from one network to another only if addressed to a station within the second network. Thus, messages from one station to another on the same physical LAN never leave that LAN.

A **gateway**, on the other hand, permits messages from one network to be sent to stations on a physically quite different network; e.g., from a ring network that uses tokens to a bus network using CSMA/CD. It does this by performing whatever protocol conversions are necessary and by keeping track of the addresses of users on a given network. Gateways are becoming increasingly important in the multi-vendor environments often found in engineering and scientific laboratories.[8]

---

8   Some vendors distinguish between gateways and routers. Both serve the function of connecting dissimilar systems. Often one term refers to connection to a mainframe and the other to connection to another LAN.

13.2 LAN PROTOCOL STANDARDS

Because LANs are a relatively new and very rapidly growing field, one of the difficulties to be faced by users is the variety of protocols; there are almost more protocols than LAN vendors. Hence, several organizations have tried to develop standard protocols that can be used on multiple networks. These standards are particularly important if you are trying to connect together several different networks.

One standard that was developed to ensure intercommunications among the U.S. Department of Defense networks is the Transmission Control Protocol/Internet Protocol (TCP/IP). This standard has been implemented by a large number of vendors, and is particularly common in U.S. universities. The TCP/IP standard is concerned with such things as standard user functions and a standard network addressing scheme.

A second standard has been developed by the International Institute of Electrical Engineers, namely the IEEE 802 standard (which actually includes several competing protocols). The IEEE standard is concerned primarily with the details of the hardware (e.g., CSMA/CD and token passing) and relatively little about the user applications.

A third standard, Open Systems Interconnection (OSI) is being developed by the International Standards Organization (ISO). Although the OSI standard has been only slowly moving toward finalization, it is expected eventually to supplant the earlier standards. This standard attempts to define all of the components of the network system, as shown in Figure 13.2. It provides a simplified view of the features needed in a LAN system, including both hardware and software.

The OSI standard essentially defines seven layers, each of which is built upon the layer immediately below it. These seven layers include:

1. **Physical**. This defines the signal characteristics of the particular LAN, and includes encoding and transmission of the data across physical links.

2. **Link**. The link layer provides point-to-point transfer of data over the physical links in the LAN. It establishes, maintains and releases the link connections.

3. **Network.** This layer defines inter- and intra-network communications, including all possible routes that a message can take to its destination.

4. **Transport.** The transport layer provides a logical (not a physical) connection from a process on one machine to another. This includes error detection and correction, message sequencing and end-to-end flow control.

5. **Session**. This layer organizes and synchronizes the dialog between two network nodes, enables access to procedures on other nodes, and validates the identity and authority for the communicating systems.

6. **Presentation**. This layer ensures a common syntax among nodes, so that all devices can understand one another at the application layer level.

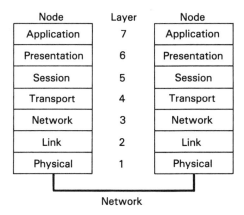

Figure 13.2.   OSI Standard.   The OSI Standard for LANs includes seven layers, each with a specific function.   By comparison, TCP/IP basically is concerned with layers 3 through 7, while the IEEE 802 standard is concerned primarily with layers 1 through 3.

7.   **Application**.   The application layer provides standard user functions and network interfaces.   This is the layer that the user sees; all other layers are seen only by the LAN programmer.

13.3 APPLICATIONS OF LANS

Now that we have discussed LANs enough so that you have seen most of the commonly-used terminology, let's see how to use a LAN in the laboratory.

13.3.1 PC to PC Communications

One common way to use a LAN in the laboratory is to provide communications among several PCs.   In many cases, these may be PCs from different manufacturers; all must understand the same protocols (e.g., TCP/IP) and have LAN adapter boards to perform the necessary connections to the network.   We might see a system such as that shown in Figure 13.3.

In this system, we can use the network to perform a number of functions:

■   **Sharing of expensive resources**, such as laser printers or plotters.   This is done relatively straightforwardly using standard LAN software.   The computer to

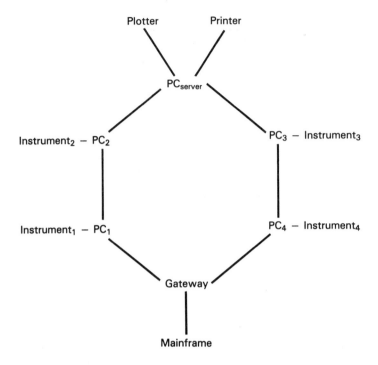

Figure 13.3. Typical Small Laboratory Network. A typical single-laboratory network might be used to provide PC to PC communications and common access to central resources.

which the printer or plotter is attached is often referred to as a **server**, and the system requesting the service as the **client**.

- **Sharing of files or programs**. Some vendors market software that allows multiple users on a network to share files or even programs. These programs are usually designed to provide some measure of security against unauthorized access to files.

- **Mail**. In larger laboratories, the network is often used to send messages from one person to another. Often, this is done in conjunction with a central mainframe computer system, as described below, which can be used to provide the centralized storage, retrieval, and handling that is usually desired for mail systems.

- **Updating software.** Systems are now beginning to become available that automate the distribution and updating of software. This can be particularly important in large laboratories, because it can remove the burden of software maintenance from the individual user.

- **Remote access to functions**. Another growing area is the use of the LAN as a means of tying together the computational power of several machines. In this mode of operation, a program running on one machine can request that a certain task or program be run on another machine.

- **Remote access to experiments**. A special case of the preceding that has special usefulness in the laboratory is using the LAN to provide remote access to data being collected in an experiment. and even to In this mode of operation, for example, we might use the LAN to provide access to an experiment running in an environmentally-controlled laboratory. The scientist can then view data from the experiment as it is being collected without entering the controlled environment. Remote control of the experiment is also possible, although it should be approached cautiously because of potential safety and security problems.

All of these functions together can be used to provide what is often termed **distributed processing**, the performing of computing tasks by a number of loosely-connected processors.

13.3.2 PC to Mainframe Communications

When a central mainframe computer or powerful workstation is connected to the network, the processing power of the network can be greatly expanded. Essentially all of the functions described for the PC-only LAN are expanded by including a system with much larger memory, computational power, and storage capacity. For example, corporate or campus-wide mail systems usually require large storage capacity for as-yet-undelivered mail, and to allow archiving of messages on an automatic basis. In addition, the LAN-connected PC can be used as a terminal to the mainframe system, assuming that a suitable terminal-emulator program is available.

In our experience, one of the most attractive ways of using networks with powerful servers is to allow the PC to treat the server *as though it is a plug-in high-performance processor for the PC*. Its presence should be just as transparent to the user as the math coprocessor chip or an array processor card; the server gives better performance to numerically intensive tasks, but the user sees only the improved performance and not the details of how that performance is achieved. These applications are sometimes referred to as **tightly-coupled cooperative processing**, although the more generic term distributed processing is probably preferable. In this mode, we use the high-speed connection of the LAN to move data rapidly between the PC and the server, but use the best features of each type of system. Specifically, we often use the PC for its friendly user interface and high-resolution displays, and the server for its large memory and high-speed computational ability. For example, we may collect data at the workstation, move it to a supercomputer for sophisticated analysis, and then return the results to the PC for display; all of this is controlled from a single program on the PC. Hence, to the user it appears as though all processing is done at the PC, even though part of the processing was done on a mainframe system.

13.4 LAN SUPPORT

We would be remiss if we did not mention that adding a LAN to your laboratory also adds complexity. LANs, just like any other hardware, can break down. Furthermore, the installation and routine maintenance of even small networks can be time-consuming. Hence, many sites have an individual or group responsible for LAN maintenance. These LAN managers are typically responsible for adding nodes to the network as needed, locating troubles on the network, starting up any central hardware after shutdowns, etc. Presumably, as networks become more common, this process will become more and more automated. However, if you are planning a LAN, be sure to plan how it will be supported within your organization.

13.5 LANS IN THE LABORATORY

How then does a laboratory LAN differ from a LAN in a typical business environment? First, the laboratory computing hardware environment is typically much more heterogeneous than a business environment. The typical laboratory may have computer hardware from a dozen or more vendors. Hence, it is especially important to select LAN systems and protocols that will support a wide range of vendors.

A second difference is that the software environment is also more heterogeneous in a laboratory environment. Many business offices use a very small number of programs (e.g., editor, spreadsheet and database program), while the laboratory scientist may use programs much more tailored to a particular application. Hence, many business LAN systems use diskless workstations, so that the individual user is required to get both programs and data only from the network server. Clearly the diskless workstation is inappropriate in the vast majority of scientific laboratories. One possible exception to this is in industries such as the pharmaceutical industry, where there is an emphasis upon being able to prove what software was in use at a particular time during the data collection and analysis process. In these situations, it becomes important to have some form of centralized distribution of software to workstations.

A third difference is that scientific LANs may need to be able to support much higher data transmission rates, particularly under heavily loaded conditions, than business LANs. One of the trends we see in the scientific environment is that networks will increasingly be used for the processing of large data sets, such as data bases and images. Given scientists' almost insatiable demand for more and more sophisticated data processing, it is likely that LANs will be used to help make the additional resources available to the scientist. For this reason, as we mentioned above, many larger laboratories are currently considering fiber optic cables for their future LANs.

13.6 LAN PROGRAMMING

Unlike our practice in previous chapters, we shall not show you any specific examples of LAN programs, because there is so much variation in LANs from lab to lab. However, we can give you some hints that may help you in developing your own applications.

First, we would suggest that you attempt to find a standard application environment; i.e., use one of the standard protocols, a standard method of cooperative processing, a standard windowing environment, etc. The advantage of this is you will not have to worry about as many of the details of your specific LAN, and moving applications from one LAN to another is much easier.

Second, we again recommend that you use commercial software whenever possible. Programming in a LAN environment is not easy, even with many of the newer programmer toolkits available.

Third, because of the nature of LANs, it is difficult to ensure that a message will be transmitted at exactly the desired time on a LAN. Hence, the LAN should in general not be used for real-time control of processes that are very rapid (above a few kHz). Real-time control and data acquisition are best performed by a dedicated processor. Data from such a process can be made available to the network, but should be made available in a fashion that can be asynchronous with the data collection and control process.

EXERCISES

1. Use your own LAN system to send a file to another computer for printing. Explain what happens to the file during the transfer process, using the OSI model.

2. Write a program that will process a data file, send it to another processor for further analysis, and then return it to the original program for display.

3. Describe how a simple network might be built with just serial interfaces and twisted pair wire. What would be the advantages and disadvantages of such a network?

4. One of the advantages of the star network is that if one arm of the network fails, it does not affect the other arms. Examine each of the other topologies, and describe what might happen if a portion of the network fails.

5. For your own LAN, write a program that will invoke a procedure on another computer connected to the LAN. (Depending upon the software in your system, this may be either very simple or very complex.) What are the advantages and limitations to this type of operation?

14

USING ASSEMBLY LANGUAGE FOR HIGH SPEED

14.1 INTRODUCTION

After you have made your programs function properly, one of the inevitable questions you will ask yourself is, "How can I make this program run faster?" There are many ways to do this. For example, you will certainly want to convert your program to run in compiled BASIC if you have been using only interpreter BASIC. Another easy-to-implement option is to buy a faster computer (e.g., the PC/AT instead of the PC/XT); this appears to be expensive, but the cost is probably far less than the cost of your time in rewriting your programs. A third option is to buy one of the many "accelerator" chips or boards currently available that use a faster CPU, a faster math coprocessor, higher clock speeds, no-wait-state memory, or some combination of these; these may be quite satisfactory for speeding execution of programs you have written but may cause problems with some commercial programs or boards that depend upon a fixed processor speed for timing. We generally do not recommend this last approach.

After you have done all of these things, you may still find that the program does not run fast enough. This is a problem that we all face because as scientists and engineers we continually push our equipment for peak performance. It is at this point that someone usually suggests switching to a language other than BASIC: C, Pascal, Forth, Modula-2, Fortran, or some other.

The approach we generally take in our own laboratories, however, is to leave the majority of the program in BASIC (or whatever higher level language it was originally written) and try instead to identify the portion of the program that is time-critical. We then rewrite just this small section of the program in assembly language.

To give you an idea about how a program's execution speed may improve, let's look at a short example program segment:

```
50 FOR X=1 TO 10000 STEP 0.5
60    Y=MX + B
70 NEXT X
```

This segment takes approximately 74 seconds to execute when run with BASICA on a PC/XT. Using the BASIC Compiler improves the speed to 20 seconds. However, coding in assembly language reduces the execution time to 2 seconds! A significant part of this improvement is because the assembler version of the program uses the 8087 math coprocessor.

Before you get too excited about learning assembly language, we should make one additional suggestion to improve the speed of your programs: for programs that perform extensive floating-point computations, you should consider a language that directly supports use of the math coprocessor. For example, the above program segment written in either compiled QuickBASIC or IBM Professional Fortran requires approximately 4 seconds to execute; this is almost as fast as the assembly-language version. Our own preference is to use some dialect of BASIC (e.g., QuickBASIC) for most programs, and C for very large programs or those on which multiple programmers will work. However, choosing computer languages is much like buying a car; the choice of models is a subject of considerable debate but ends up being largely a matter of what you personally like and feel comfortable using.

In any case, the fastest possible execution is always achievable with **assembly language**. What is assembly language? It is a computer language that is very close to the machine language the computer uses. However, instead of specifying the instructions in the binary form that the computer understands, the user writes instructions as mnemonics that the "assembler" program converts to binary machine-language instructions. Unlike BASIC and other higher level languages, assembly language requires that you understand in considerable detail the hardware of the computer CPU; assembly language is specific to the type of CPU you are using.

In short, assembly-language programming is difficult to do! It is much harder to develop programs and much harder to detect problems ("bugs") when programming in assembly language than when using a high-level language such as BASIC. However, assembly-language programs have one extremely important advantage: they are fast. This is because assembly language programming allows you to have complete control over precisely how the CPU and other hardware achieve their assigned tasks. Such control can allow you to produce programs that are optimized for the specific task you wish performed. This advantage is particularly useful in achieving very high rates of speed for data collection and instrument control.

A second advantage to users of BASICA and IBM Compiled BASIC is that assembly-language programming allows use of the 8087 or 80287 math coprocessors, whereas these implementations of BASIC do not. For programs that require extensive floating-point computations, the latter advantage can be very important. (However, recall that we recommend using dialects of BASIC which provide this support.)

It is not our intention to teach you assembly-language programming. Several excellent books are available on the subject. However, these books generally show only general-purpose examples. We would like to show examples that will help you in the laboratory.

14.2 INTERFACING TO HIGH LEVEL LANGUAGES

As we have mentioned, we do not recommend writing programs entirely in assembly language; instead, we suggest you write only the critical portion in assembly language and leave the remainder in BASIC. This clearly implies that there must be some way to connect assembly language routines to BASIC programs. One such connection, and the one we prefer to use, is the CALL statement in BASIC. This is illustrated in Program 14.1 which is designed as a skeleton program that you can adapt for your own purposes.

This is of course a very simple program, but it illustrates several important features. First, *all variables must be defined before being used in the CALL statement.* Second, we place all variables that are to be communicated to or from the assembly-language routine within the parentheses of the CALL statement. And third, arrays are included by referencing the first element of the array; you should not use the array name by itself.

Now we are ready to illustrate the assembly language routine called from BASIC. Program 14.2 shows the skeleton of an assembly language program called from Program 14.1.

Program 14.2 is designed to perform a very simple task. It reads one variable of each of the types permitted in BASICA and copies the values to another variable of the same type. (It may seem like this is a do-nothing program, but we find that very often the major problems encountered in using assembly language with BASIC are problems with communicating between the two languages.)

The first thing you may notice about the assembly language subroutine is that there is a FRAME structure listing each of the variables passed by Program 14.1. The BASIC call statement works by pushing each variable onto the stack in the order they are listed in the CALL statement, so that the last variable listed in the CALL statement is at the top of the stack. Then six bytes of information about the calling program are pushed onto the stack. Thus, the FRAME sets aside 6 bytes and then

```
10 REM PROGRAM DEMOCALL
20 REM LABORATORY AUTOMATION USING THE IBM PC
30 REM Illustrates call to assembly-language subroutine
40 REM ********************************************************
50 DIM D%(100),H%(100)
60    A%=10                          'MUST define all variables in CALL statement
70    B!=20!
80    C$="hello"
90    D%(0)=100
100   D%(1)=1
110   E%=0                           'even those returned by CALL
120   F!=0
130   G$=SPACE$(5)
140   N%=2
150   PRINT "Before calling the subroutine, the variables are";
      A%;B!;C$;" "; D%(0);D%(1)
160   CALL BAS2ASSM(A%,B!,C$,D%(0),E%,F!,G$,H%(5),N%)
170   PRINT "After calling the subroutine, the variables are";
      E%;F!;G$;" "; H%(5);H%(6)
180 END
```

Program 14.1.   Calling an Assembly Language Routine from BASIC.

lists each of the variables *in the reverse order they were given in the CALL statement*. The address of each parameter in the CALL statement is assumed to occupy two bytes of space. (Most other languages use four bytes for each parameter so that they can pass both a segment and an offset for each parameter address, but BASIC passes only an offset. The segment is assumed to be the Data Segment (DS) established by the calling BASIC program.)

At this stage, we should also emphasize one very important point. Program 14.1 and Program 14.2 *are both designed for Compiled BASIC only* (since we assume you are trying to achieve higher speeds). All subsequent examples in this chapter are also for the IBM BASIC Compiler, Version 2.0 or later. If you are using a different BASIC compiler, please consult the appropriate user's manual. We have also included sample routines on the Scientific Routines Diskette (see Appendix B) to illustrate calling assembly language subroutines from several high level languages other than BASIC.

14.3 AN ASSEMBLY-LANGUAGE ROUTINE FOR DATA COLLECTION

To illustrate a practical use of assembly language, let's examine a program to collect data using the A/D of the DT2801 card. You may wish to examine Program 3.2 and Program 9.2 again to refresh your memory about how the A/D on the DT2801 card is programmed.

Program 14.3 is a simple BASIC program to collect data and display them. *Note that the program must be run from a graphics display*, such as the IBM Color Graphics Display.

```
;PROGRAM BAS2ASSM
;LABORATORY AUTOMATION USING THE IBM PC
;Skeleton subroutine to illustrate calls from IBM BASIC Compiler
;*************************************************************************
;
;         assumes call of form CALL BAS2ASSM(A%,B!,C$,D%(d),E%,F!,G$,H%(h),N)
;                  copies A% to E%, B! to F!, C$ to G$, array D% to H%
;                  N is the number of values to be transferred from D% to H%
;                  and d and h are the subscripts of the first element of interest
;                      (d and h are usually either 0 or 1)
;
;.8087                               ;put this in routine if using 8087 instructions
;
;set up structure for easy reference to parameters used in CALL
;
FRAME           STRUC
OLDBP           DW      ?                   ;old BP
OLDIP           DW      ?                   ;IP for return
OLDCS           DW      ?                   ;CS for return
;now set aside two bytes for each parameter, in reverse order listed in CALL
NPTR            DW      ?                   ;pointer to N
HPTR            DW      ?                   ;pointer to H array
GPTR            DW      ?                   ;pointer to G descriptor block
FPTR            DW      ?                   ;pointer to F
EPTR            DW      ?                   ;pointer to E
DPTR            DW      ?                   ;pointer to D array
CPTR            DW      ?                   ;pointer to C descriptor block
BPTR            DW      ?                   ;pointer to B
APTR            DW      ?                   ;pointer to A
FRAME           ENDS
PARAMSIZE       EQU     OFFSET APTR - OFFSET OLDCS    ;bytes in parameters
;
CSEG    SEGMENT  WORD PUBLIC 'CODE'
        ASSUME CS:CSEG, DS:NOTHING,  ES:NOTHING
;
;define a data area for any variables or constants not passed in CALL
AA      DW      ?
BB      DW      ?
        DW      ?
CLEN    DW      ?
CA      DW      ?
GLEN    DW      ?
GA      DW      ?
ARRAYLOC DW     ?
;
;define routine as a public procedure
        PUBLIC  BAS2ASSM
BAS2ASSM PROC   FAR
;       INT     03H                 ;remove comment if you wish to DEBUG
        PUSH    BP                  ;save BP
        MOV     BP,SP               ;set base of parameter list
;first, get any parameters you need from BASIC
        MOV     DI,[BP].APTR        ;get A% from BASIC
        MOV     AX,WORD PTR [DI]
        MOV     AA,AX               ;and store it in AA
        MOV     DI,[BP].BPTR        ;get B! from BASIC
        MOV     AX,WORD PTR [DI]    ;   (remember, it is 4 bytes long)
        MOV     BB,AX               ;and store both words of it in BB
        MOV     AX,WORD PTR [DI]+2
        MOV     BB+2,AX
        MOV     DI,[BP].CPTR        ;get string descriptor block
        MOV     AX,WORD PTR [DI]    ;get string length
        MOV     CLEN,AX
        MOV     AX,WORD PTR [DI]+2  ;get string address
        MOV     CA,AX
        MOV     DI,[BP].DPTR        ;get array address
```

Program 14.2 (Part 1 of 2). Assembly Language Program Called from BASIC.

```
          MOV        ARRAYLOC,DI          ;store address (not value)
          ;
          ;  put your code here
          ;
;when you are finished, return values to BASIC
          MOV        DI,[BP].EPTR         ;move number in AA into E%
          MOV        AX,AA
          MOV        WORD PTR [DI],AX
          MOV        DI,[BP].FPTR         ;move number in BB into F!
          MOV        AX,BB
          MOV        WORD PTR [DI],AX     ;including both words
          MOV        AX,BB+2
          MOV        WORD PTR [DI]+2,AX
          MOV        DI,[BP].GPTR         ;get string descriptor block for G$
          MOV        AX,WORD PTR [DI]
          MOV        GLEN,AX              ;save its length
          MOV        AX,WORD PTR [DI]+2   ;put its initial address into DI
          MOV        DI,AX
          MOV        CX,CLEN              ;put length of C$ into CX
          CMP        CX,GLEN              ;Basic Compiler does not allow
          JLE        C1                   ;   you to change the length
          MOV        CX,GLEN              ;   of a string in assembly routine,
C1:       CMP        CX,0                 ;   so check to make sure string
          JE         C2                   ;   can be copied
          MOV        SI,CA                ;put address of C$ into SI
          MOV        AX,DS                ;make ES=DS
          MOV        ES,AX                ;move from DS:[SI] to ES:[DI]
          REP        MOVSB                ;   (i.e., from C$ to G$)
C2:       MOV        DI,[BP].NPTR         ;get count of items in array to move
          MOV        CX,[DI]              ;put in CX for use with LOOP
          MOV        DI,[BP].HPTR         ;get address of H array into DI
          MOV        SI,ARRAYLOC          ;and address of D array into SI
L1:       MOV        AX,WORD PTR [SI]
          MOV        WORD PTR[DI],AX
          ADD        SI,2                 ;go to next word of each (2 bytes)
          ADD        DI,2
          LOOP       L1
          POP        BP
          RET        PARAMSIZE            ;number of bytes used in parameters
BAS2ASSM ENDP
CSEG     ENDS
          END
```

Program 14.2 (Part 2 of 2).   Assembly Language Program Called from BASIC.

Notice that Program 14.3 performs all of the operations that are not time-critical, such as user input, calculating the rates of the two timers, and post-run plotting. These tasks could of course all be done in assembly language as well, but programming them in BASIC is much simpler.

A simple part of Program 14.3 that may require some explanation is the use of the P% variable. P% is used to enable or disable real-time plotting. The real-time plotting routine in the assembly language program described below can not always plot as rapidly as the A/D is collecting data. Hence, for the maximum data acquisition rates, you will want to disable the real-time plotting by setting P% to zero. In this program we have arbitrarily decided to limit real-time plots to acquisition rates below 1000 Hz.

Program 14.4 is the routine that performs the time-critical tasks of data collection and real-time plotting. It begins with the FRAME structure described for

```
10 REM PROGRAM A2D
20 REM LABORATORY AUTOMATION USING THE IBM PC
30 REM Program to demonstrate data collection under program control
40 REM        Compile using  BASCOM PROG143;
50 REM        Link using       LINK PROG143+PROG144,,,BASRUN20
60 REM Note:  This program must be run on a graphics display!
70 REM Uses the DT2801
80 REM ***************************************************************
90 DIM A%(5000)
100 REM Define all registers
110 BASEADD%=&H2EC                        'base address of board
120 COMMAND%=BASEADD%+1                   'command register
130 STATUS% =BASEADD%+1                   'status register
140 DATUM%  =BASEADD%                     'data register
150 REM ***********************************
160 REM                Main Program
170    GOSUB 310                          'get user input
180    INPUT "Type the <RETURN> key to start data collection.",Y$
190    CLS:SCREEN 2                       'Set up for real-time plot routine
200    GOSUB 690                          'initialize board
210    GOSUB 780                          'set up A/D, clock
220    GOSUB 1120                         'wait for previous command to finish
230    OUT COMMAND%,&HE                   'give READ A/D command
240    CALL A2DASM (A%(1),P%,N%)          'collect data (and real-time plot)
250    GOSUB 500                          'do any post-run processing
260    ERRORCHECK%=INP(STATUS%)           'check for error
270    IF (ERRORCHECK% AND &H80) THEN PRINT "Error reading A/D":GOSUB 1170
280    OUT COMMAND,&HF                    'stop the board
290 END
300 REM ***********************************
310 REM Get user input
320    WHILE N% < 3 OR N% > 5000
330       INPUT "Enter the number of data points to collect (3 to 5000) ",N%
340    WEND
350    INPUT "Enter the starting A/D channel desired ",C%
360    INPUT "Enter the ending A/D channel desired ",E%
370    WHILE RATE! < 6 OR RATE! > 13700
380       INPUT "Enter the rate in points/second (7-13700) ",RATE!
390    WEND
400    TICK!=400000!/RATE!                'Calculate number of ticks at 400 kHz
410    IF TICK! < 32768! THEN TICK%=TICK! ELSE TICK%=TICK!-65536!
420    IF RATE! < 1000 THEN P%=1 ELSE P%=0   'Enable real-time plot if < 1 kHz
430    WHILE GAIN% < 1 OR GAIN% > 8
440       INPUT "Enter the desired A/D gain (1, 2, 4 or 8) ",GAIN%
450    WEND
460    GAINCODE%=LOG(GAIN%)/LOG(2)        'convert to gain code (0,1,2,3)
470 RETURN
480 REM ***********************************
490 REM Display normalized data.
500    SCREEN 1:COLOR 9,0
510    VIEW:CLS:VIEW (50,10)-(270,180),2,B
520    YMIN%=A%(1)                        'find min, max in data
530    YMAX%=A%(1)
540    FOR I%=2 TO N%
550       IF A%(I%)<YMIN% THEN YMIN%=A%(I%)
560       IF A%(I%)>YMAX% THEN YMAX%=A%(I%)
570    NEXT I%
580    IF YMIN%=YMAX% THEN PRINT "All data are the same":RETURN
590    WINDOW (1,YMIN%)-(N%,YMAX%)        'display values normalized to min, max
600    PSET (1,A%(1))
610    FOR I%=2 TO N%
620       LINE -(I%,A%(I%)),3
630    NEXT I%
640    LOCATE 24,1:PRINT "Minimum= ";YMIN%; "     Maximum= ";YMAX%;
650    LOCATE 1,1:PRINT "Gain = ";GAIN%
660    Y$=INKEY$:IF Y$=""THEN 660
```

Program 14.3 (Part 1 of 2).   BASIC Program for Data Collection Using Assembly Language.

```
 670 RETURN
 680 REM ************************************
 690 REM set up board
 700    OUT COMMAND%,&HF                    'stop board
 710    TEMP%=INP(DATUM%)                   'clear data register
 720    REM wait for command to finish, then clear errors
 730    GOSUB 1120
 740    OUT COMMAND%,&H1                    'clear the command register
 750 RETURN
 760 REM ************************************
 770 REM Set up clock and A/D
 780    GOSUB 1120
 790    OUT COMMAND%,&H3                 'write SET CLOCK PERIOD command
 800    CLOCKHIGH%=(TICK% AND &HFF00)/256
 810    CLOCKLOW%=TICK% AND &HFF
 820    IF CLOCKHIGH% < 0 THEN CLOCKHIGH%=CLOCKHIGH% +256
 830    GOSUB 1090
 840    OUT DATUM%,CLOCKLOW%                'output low byte
 850    GOSUB 1090
 860    OUT DATUM%,CLOCKHIGH%               'output high byte
 870    GOSUB 1120
 880    ERRORCHECK%=INP(STATUS%)        'check for error
 890    IF (ERRORCHECK% AND &H80) THEN PRINT "Error setting clock":GOSUB 1170
 900    OUT COMMAND%,&HD                     'write SET A/D PARAMETERS command
 910    GOSUB 1090
 920    OUT DATUM%,GAINCODE%                'set gain
 930    GOSUB 1090
 940    OUT DATUM%,C%                       'set starting a/d channel
 950    GOSUB 1090
 960    OUT DATUM%,E%                       'set ending a/d channel
 970    HIGHPOINTS%=(N% AND &HFF00)/256 'get low, high bytes of number of points
 980    LOWPOINTS%=N% AND &HFF
 990    GOSUB 1090
1000    OUT DATUM%,LOWPOINTS%
1010    GOSUB 1090
1020    OUT DATUM%,HIGHPOINTS%
1030    GOSUB 1120
1040    ERRORCHECK%=INP(STATUS%)        'check for error
1050    IF (ERRORCHECK% AND &H80) THEN PRINT "Error setting A/D":GOSUB 1170
1060 RETURN
1070 REM ************************************
1080 REM check to see if ready to set a register
1090    WAIT STATUS%,&H2,&H2                'wait for bit 1 to be reset
1100 RETURN
1110 REM check to see if ready to send another command
1120    WAIT STATUS%,&H2,&H2                'wait for bit 1 to be reset
1130    WAIT STATUS%,&H4                    'wait for bit 2 to be set
1140 RETURN
1150 REM ************************************
1160 REM Determine source of error
1170    OUT COMMAND%,&HF                    'stop the board
1180    TEMP%=INP(DATUM%)
1190    GOSUB 1120
1200    OUT COMMAND%,&H2                    'send READ ERROR REGISTER command
1210    GOSUB 1090
1220    VALUE%=INP(DATUM%)
1230    GOSUB 1090
1240    PRINT "Value of error register low byte is: ";VALUE%
1250    VALUE2%=INP(DATUM%)
1260    PRINT "Value of error register high byte is: ";VALUE2%
1270    IF VALUE2%=4 THEN PRINT "You have tried to collect data at too fast"
1280 STOP
```

Program 14.3 (Part 2 of 2). BASIC Program for Data Collection Using Assembly Language.

Program 14.2. It then defines a series of constants, namely the registers on the DT2801 board. You may wish to change the value of BASE if you have more than one DT2801 card in your computer.

The main code segment of Program 14.4 begins by setting aside memory locations for temporary variables used by the routine. It then saves the BP register (so it can be used in retrieving parameters from the stack) and calls two subroutines sequentially. Notice that the main routine is declared PUBLIC and FAR so that it can be called from the BASIC program, while the subroutines are all local, NEAR routines.

The data collection routine is very similar to Program 9.2. Again the use of the timer allows data to be taken at precisely-defined intervals. In addition, the data collection routine will, at the user's option, call a real-time plotting routine. This routine uses a call to the BIOS interrupt 10 routine, which plots a single datum (see the IBM Technical Reference Manual for a listing of the BIOS routines and their functions). Note that the data are scaled so that they never go off-scale. However, the data will usually look more informative during the post-run plotting; this is when they are scaled to fill the plotting area.

14.4 AN ASSEMBLY-LANGUAGE ROUTINE FOR SIGNAL AVERAGING

Another major application for assembly-language routines is real-time floating-point computations. As you may recall, BASICA and the IBM BASIC compiler do not support use of the math coprocessor. Hence, if you wish to perform rapid floating-point calculations using these languages, you will need to write assembly-language routines to do the calculations on the math coprocessor (8087 or 80287). In addition, you may find it possible to achieve faster floating point computations by hand coding time-critical floating point routines rather than depending upon the routines produced by languages that support the use of the math coprocessor. Again, however, we caution you that routines using the math coprocessor are often hard to write and even harder to debug.

A simple but useful example of an application of the math coprocessor is shown in Program 14.5 and Program 14.6, which together perform real-time averaging of data collected by the DT2801 A/D. These are modifications of Program 14.3 and Program 14.4, respectively.

The only major modifications contained in Program 14.5 are the addition of a request for the number of points to average (in line 390) and the additional parameter in the CALL statement (line 250). However, the assembly language routine in Program 14.6 has been extensively modified from that in Program 14.4. All of these modifications have to do with the averaging of the data points as they are collected.

First of all, the assembler is notified of the presence of the 8087 code with the directive ".8087". The other changes are in the COLLECT subroutine. As the first operation in this routine, the 8087 is initialized to ensure that it is in a known state. The 8087 is loaded with the number of points to be averaged; this number will be

```
                    TITLE   A2DASM
;LABORATORY AUTOMATION USING THE IBM PC
;       subroutine to do timed data collection from the DT2801
;       call from BASIC with call of form:
;               CALL A2DASM (A%(1),P%,N%)
;       where   A% is array where data are to be stored
;               P% is 0 to omit real-time plot, other to plot
;               N% is number of points to be collected
;       ********************************************************************
FRAME           STRUC
OLDBP           DW      ?       ;old BP
OLDIP           DW      ?       ;IP for return
OLDCS           DW      ?       ;CS for return
;now set aside two bytes for each parameter, in reverse order listed in CALL
NPTR            DW      ?
PPTR            DW      ?
APTR            DW      ?
FRAME           ENDS
PARAMSIZE       EQU     OFFSET APTR - OFFSET OLDCS   ;bytes in parameters
;       ********************************************************************
;definitions:
BASE            =02ECH          ;base of DT2801 board (assuming Adapter 0)
COMMAND         =BASE+1         ;a/d control low byte
STATUS          =BASE+1         ;a/d status
DATUM           =BASE           ;a/d datum
;       ********************************************************************
CSEG    SEGMENT  WORD PUBLIC 'CODE'
        ASSUME CS:CSEG,  DS:NOTHING,   ES:NOTHING
;data storage
PLOT    DW      ?               ;plot flag
TEMPSI  DW      ?               ;temp storage for si register
;       ********************************************************************
;start of executable code
        PUBLIC  A2DASM
A2DASM  PROC    FAR
        PUSH    BP              ;save BP
        MOV     BP,SP           ;set base parameter list
        INT     03H             ;use during debugging with DEBUG
        CLI                     ;turn off system interrupts
        CALL    NEAR PTR SETUP  ;set up for data collection
        CALL    NEAR PTR COLLECT;collect data
        STI                     ;turn interrupts back on
        POP     BP              ;restore BP
        RET     PARAMSIZE       ;return to basic program
A2DASM  ENDP
;       ********************************************************************
;                       SETUP ROUTINE
SETUP   PROC    NEAR
        MOV     SI,[BP].NPTR    ;get number of points to collect
        MOV     CX,[SI]
        MOV     DI,[BP].PPTR    ;get plot flag
        MOV     AX,[DI]
        MOV     PLOT,AX         ;save in PLOT
        MOV     TEMPSI,0        ;start plotting at x=0 on screen
        MOV     DI,[BP].APTR    ;get address of data array into DI
X1:     RET
SETUP   ENDP
;       ********************************************************************
;                       DATA COLLECTION ROUTINE
;note: CX already has count of data points, DI points to data array
COLLECT PROC    NEAR
BEGIN:
        MOV     DX,STATUS       ;test bits 0 and 2 to see if data ready
B1:     IN      AL,DX           ;loop until ready
        AND     AL,5H
        JE      B1
        MOV     DX,DATUM        ;get low byte
```

Program 14.4 (Part 1 of 2).   Assembly Language Program for Data Collection.

```
            IN      AL,DX
            MOV     BL,AL           ;store datum
            MOV     DX,STATUS
B2:         IN      AL,DX
            AND     AL,5H
            JE      B2
            MOV     DX,DATUM        ;get high byte
            IN      AL,DX
            MOV     BH,AL
            MOV     AX,BX
;store datum in data array
            MOV     WORD PTR [DI],AX ;store data in A% array
            ADD     DI,2            ;increment pointer by 2 (2 bytes /datum)
            CMP     PLOT,0          ;don't plot if plot flag=0
            JZ      NOPLOT
            CALL    NEAR PTR PLOTPOINT  ;call plotting routine
NOPLOT:
            LOOP    BEGIN
            RET
COLLECT   ENDP
;****************************************************************
;                       PLOTTING ROUTINE
;Uses BIOS interrupt 10H to plot a single point on the IBM CGA screen.
;Note: Assumes that BASIC has already invoked the plotting mode (SCREEN 2)
;            INPUT:     AX=DATUM TO PLOT
;                       TEMPSI contains x value of previous point
PLOTPOINT PROC NEAR
            PUSH    SI
            PUSH    DX
            PUSH    CX
            PUSH    BX
;calculate y-value to plot = 199-(datum/21)  so a/d values of 0-4095 fit screen
            CWD                     ;make AX into a double word
            MOV     BX,21           ;divide by 21--quotient in AX
            DIV     BX
            MOV     DX,AX           ;result into DX
            NEG     DX              ;negate and add to 199
            ADD     DX,199
            CMP     DX,199          ;check if values within limits of screen (0-199)
            JL      OKHIGH
            MOV     DX,199
            JMP     OKLOW
OKHIGH:CMP     DX,0
            JG      OKLOW
            MOV     DX,0
OKLOW: MOV     SI,TEMPSI       ;get x-value of last point on screen
            INC     SI              ;go to next location on screen
            CMP     SI,640          ;test if at right edge of 640 x 200
            JL      M1              ;   screen
            MOV     SI,0            ;if so, go to left edge to plot
M1:     MOV     CX,SI           ;get x-value into cx
            MOV     TEMPSI,SI       ;save x value
            MOV     AX,3073         ;AH=12,AL=1 to write dot to screen
            INT     10H             ;plot point
            POP     BX
            POP     CX
            POP     DX
            POP     SI
            RET
PLOTPOINT       ENDP
CSEG    ENDS
            END
```

Program 14.4 (Part 2 of 2). Assembly Language Program for Data Collection.

```
10 REM  PROGRAM AVERAGE
20 REM  LABORATORY AUTOMATION USING THE IBM PC
30 REM  Program to demonstrate data collection with averaging
40 REM          Compile using  BASCOM PROG145;
50 REM          Link using     LINK PROG145+PROG146,,,BASRUN20
60 REM  Note:  This program must be run on a graphics display!
70 REM  Runs on a DT2801
80 REM  *******************************************************
90 DIM A%(5000)
100 REM Define all registers
110 BASEADD%=&H2EC                   'base address of board
120 COMMAND%=BASEADD%+1              'command register
130 STATUS% =BASEADD%+1              'status register
140 DATUM%  =BASEADD%                'data register
150 REM *********************************
160 REM          Main Program
170    GOSUB 330                     'get user input
180    INPUT "Type the <RETURN> key to start data collection.",Y$
190    M%=N%*NAVG%                   'calculate total data to be collected
200    CLS:SCREEN 2                  'Set up for real-time routine
210    GOSUB 750                     'initialize board
220    GOSUB 830                     'set up A/D, clock
230    GOSUB 1170                    'check if ready for next command
240    OUT COMMAND%,&HE              'send READ A/D command
250    CALL A2DAVG (A%(1),P%,N%,NAVG%) 'call assembly routine
260    GOSUB 550                     'Do any post-run processing
270    ERRORCHECK%=INP(STATUS%)      'check for error
280    IF (ERRORCHECK% AND &H80) THEN PRINT "Error reading A/D":GOSUB 1220
290    OUT COMMAND,&HF               'stop the board
300 END
310 REM *********************************
320 REM Get user input
330    WHILE N% < 1 OR N% > 5000
340       INPUT "Enter the number of data points to collect (1 to 5000) ",N%
350    WEND
360    INPUT "Enter the A/D channel desired ",C%
370    E%=C%                         'use only one channel
380    WHILE NAVG% < 1
390      INPUT "How many data do you wish to average for each one stored? ",NAVG%
400    WEND
410    WHILE RATE! < 6 OR RATE! > 13700
420       INPUT "Enter the rate in points/second (7-13700) ",RATE!
430    WEND
440    TICK!=400000!/RATE!           'Calculate number of ticks at 400 kHz
450    IF TICK! < 32768! THEN TICK%=TICK! ELSE TICK%=TICK!-65536!
460    IF RATE! < 1000 THEN P%=1 ELSE P%=0  'Enable real-time plot if < 1 kHz
470    WHILE GAIN% < 1 OR GAIN% > 8
480       INPUT "Enter the desired A/D gain (1, 2, 4 or 8) ",GAIN%
490    WEND
500    GAINCODE%=LOG(GAIN%)/LOG(2)   'convert to gain code (0,1,2,3)
510    IF RATE!/NAVG% < 1000 THEN P%=1 ELSE P%=0   'Enable real-time plot if slow
520 RETURN
530 REM *********************************
540 REM Routine to display normalized data.
550    SCREEN 1:COLOR 9,0
560    VIEW:CLS:VIEW (50,10)-(270,180),2,B
570    YMIN%=A%(1)
580    YMAX%=A%(1)
590    FOR I%=2 TO N%
600       IF A%(I%)<YMIN% THEN YMIN%=A%(I%)
610       IF A%(I%)>YMAX% THEN YMAX%=A%(I%)
620    NEXT I%
630    IF YMIN%=YMAX% THEN PRINT "All data are the same":GOTO 690
640    WINDOW (1,YMIN%)-(N%,YMAX%)
650    PSET (1,A%(1))
660    FOR I%=2 TO N%
670       LINE -(I%,A%(I%)),3
```

Program 14.5 (Part 1 of 3). Real-Time Signal Averaging Program.

```
680    NEXT I%
690    LOCATE 24,1:PRINT "Minimum= ";YMIN%; "      Maximum= ";YMAX%;
700    LOCATE 1,1:PRINT "Gain = ";GAIN%
710    Y$=INKEY$:IF Y$=""THEN GOTO 710
720  RETURN
730  REM **********************************
740  REM set up board
750    OUT COMMAND%,&HF                      'stop board
760    TEMP%=INP(DATUM%)                     'clear data register
770    REM wait for command to finish, then clear errors
780    GOSUB 1170
790    OUT COMMAND%,&H1                      'clear the command register
800  RETURN
810  REM **********************************
820  REM set up clock and A/D
830    GOSUB 1170
840    OUT COMMAND%,&H3                      'write SET CLOCK PERIOD command
850    CLOCKHIGH%=(TICK% AND &HFF00)/256
860    CLOCKLOW%=TICK% AND &HFF
870    IF CLOCKHIGH% < 0 THEN CLOCKHIGH%=CLOCKHIGH% +256
880    GOSUB 1140
890    OUT DATUM%,CLOCKLOW%                  'output low byte
900    GOSUB 1140
910    OUT DATUM%,CLOCKHIGH%                 'output high byte
920    GOSUB 1170
930    ERRORCHECK%=INP(STATUS%)         'check for error
940    IF (ERRORCHECK% AND &H80) THEN PRINT "Error setting clock":GOSUB 1220
950    OUT COMMAND%,&HD                      'write SET A/D PARAMETERS command
960    GOSUB 1140
970    OUT DATUM%,GAINCODE%                  'set gain
980    GOSUB 1140
990    OUT DATUM%,C%                         'set starting a/d channel
1000   GOSUB 1140
1010   OUT DATUM%,E%                         'set ending a/d channel
1020   HIGHPOINTS%=(M% AND &HFF00)/256  'get low, high bytes of number of points
1030   LOWPOINTS%=M% AND &HFF
1040   GOSUB 1140
1050   OUT DATUM%,LOWPOINTS%
1060   GOSUB 1140
1070   OUT DATUM%,HIGHPOINTS%
1080   GOSUB 1170
1090   ERRORCHECK%=INP(STATUS%)         'check for error
1100   IF (ERRORCHECK% AND &H80) THEN PRINT "Error setting A/D":GOSUB 1220
1110 RETURN
1120 REM **********************************
1130 REM check to see if ready to send next command or to set a register
1140   WAIT STATUS%,&H2,&H2              'wait for bit 1 to be reset
1150 RETURN
1160 REM check to see if ready to send another command
1170   WAIT STATUS%,&H2,&H2              'wait for bit 1 to be reset
1180   WAIT STATUS%,&H4                  'wait for bit 2 to be set
1190 RETURN
```

Program 14.5 (Part 2 of 3). Real-Time Signal Averaging Program.

used as the denominator in calculating the mean value of the data. For each set of points to be averaged a zero is loaded onto the 8087's internal register. Each of the points is then collected and added to the sum of previously collected points on the 8087's internal set of registers. Note that direct transfers from the CPU registers to the 8087 registers are not permitted; hence, the data must first be stored in memory and then loaded from memory into the 8087 register. After all of the data have been summed, they are then divided by the number of points and the average is then stored in the data array and plotted if desired.

```
1200 REM **************************************
1210 REM Determine source of error
1220   OUT COMMAND%,&HF              'stop the board
1230   TEMP%=INP(DATUM%)
1240   GOSUB 1170
1250   OUT COMMAND%,&H2              'send READ ERROR REGISTER command
1260   GOSUB 1140
1270   VALUE%=INP(DATUM%)
1280   GOSUB 1140
1290   PRINT "Value of error register low byte is: ";VALUE%
1300   VALUE2%=INP(DATUM%)
1310   PRINT "Value of error register high byte is: ";VALUE2%
1320   IF VALUE2%=4 THEN PRINT "You have tried to collect data at too fast"
1330 STOP
```

Program 14.5 (Part 3 of 3).   Real-Time Signal Averaging Program.

Two cautions about using the 8087 should be observed. First, because the 8087 operates asynchronously from the CPU, you need to place FWAIT statements immediately before any statements that use data placed in memory by the 8087. Second, you should be careful to remove any remaining items from the 8087 registers when you exit the routine. You may also find it very helpful to document in your code exactly what is in the 8087 registers at all times, as shown in this example.

A subtle point about this program needs to be mentioned. Notice that the timer is still used, even for the points to be averaged. You may think that it would be better to just take points as fast as possible and average them together. However, recall from our discussion of noise reduction in Chapter 4 that if possible you will want to collect data at some frequency that is a submultiple of the noise frequency (e.g., at 10 Hz if your major source of noise is 50 or 60 Hz line frequency). Hence, data should be taken at a controlled rate rather than simply as fast as the A/D can go.

EXERCISES

1.   Modify Program 14.4 to wait for a digital I/O line on the DT2801 card to be toggled before beginning data collection.

2.   Modify Program 14.4 to rescale the data as it is being plotted. Use the multiple gain feature of the A/D and switch to another gain, but always use the highest gain that does not result in off-scale data.

3.   Modify the programs in this chapter to use the A/D card of your choice.

```
                    TITLE   AVGASM
.8087                                        ;tells assembler using 8087 codes
;LABORATORY AUTOMATION USING THE IBM PC
;      Subroutine to do averaged, timed data collection from the DT2801
;      call from BASIC with call of form:
;            CALL A2DAVG (A%(1),P%,N%,V%)
;      where   A% is array where data are to be stored
;              P% is 0 to omit real-time plot, other to plot
;              N% is number of points to be collected
;              V% is number of data to average per datum stored
;This routine averages together several data and stores as a single point
; *******************************************************************************
;
FRAME           STRUC
OLDBP           DW      ?                    ;old BP
OLDIP           DW      ?                    ;IP for return
OLDCS           DW      ?                    ;CS for return
;now set aside two bytes for each parameter, in reverse order listed in CALL
VPTR            DW      ?
NPTR            DW      ?
PPTR            DW      ?
APTR            DW      ?
FRAME           ENDS
PARAMSIZE       EQU     OFFSET APTR - OFFSET OLDCS    ;bytes in parameters
; *******************************************************************************
;definitions:
BASE            =02ECH                       ;base of DT2801 board (assuming Adapter 0)
COMMAND         =BASE+1                      ;a/d control low byte
STATUS          =BASE+1                      ;a/d status
DATUM           =BASE                        ;a/d datum
; *******************************************************************************
CSEG    SEGMENT  WORD PUBLIC 'CODE'
        ASSUME CS:CSEG,  DS:NOTHING,   ES:NOTHING
;data storage
PLOT    DW      ?                            ;plot flag
TEMPSI  DW      ?                            ;temp storage for SI register
AVERG   DW      ?                            ;number of times to average data
TEMP    DW      ?                            ;temporary storage of datum
;  *******************************************************************************
;start of executable code
        PUBLIC  A2DAVG
A2DAVG PROC   FAR
        PUSH    BP                           ;save BP
        MOV     BP,SP                        ;set base parameter list
        CALL    NEAR PTR SETUP               ;set up for data collection
        CALL    NEAR PTR COLLECT             ;collect data
        STI                                  ;turn interrupts on again
        POP     BP                           ;restore BP
        RET     PARAMSIZE                    ;return to basic program
A2DAVG ENDP
;  *******************************************************************************
;                          SETUP ROUTINE
SETUP  PROC    NEAR
;  get information to set up a/d
        FINIT                                ;initialize 8087
        MOV     DI,[BP].VPTR                 ;get number of data to average
        MOV     AX,[DI]                      ;
        MOV     AVERG,AX                     ;save in AVERG
        MOV     DI,[BP].PPTR                 ;get plot flag
        MOV     AX,[DI]
        MOV     PLOT,AX                      ;save in PLOT
        MOV     DI,[BP].NPTR                 ;get number of points to collect
        MOV     CX,[DI]                      ;    into CX register
        MOV     DI,[BP].APTR                 ;get address of data array into DI
        MOV     TEMPSI,0                     ;start plotting at x=0 on screen
        CLI                                  ;turn off system interrupts
```

Program 14.6 (Part 1 of 3). Assembly Language Program for Signal Averaging.

```
          RET
SETUP     ENDP
;    *****************************************************************
;                        DATA COLLECTION ROUTINE
;
;         On entry:          DS:DI = address of array where data are to be stored
;                            CX    = number of points to collect
;                            PLOT  = plot flag
;                            AVERG = number of points to average
;
COLLECT PROC    NEAR
          FILD    AVERG               ;load 8087 with no. times to average (divisor)
BEGIN:
          MOV     SI,AVERG            ;get number of times to average
          FLDZ                        ;zero top of 8087 stack
AVG:      MOV     DX,STATUS           ;test bits 0 and 2 to see if data ready
B1:       IN      AL,DX               ;loop until ready
          AND     AL,5H
          JE      B1
          MOV     DX,DATUM            ;get low byte
          IN      AL,DX
          MOV     BL,AL               ;store datum
          MOV     DX,STATUS
B2:       IN      AL,DX
          AND     AL,5H
          JE      B2
          MOV     DX,DATUM            ;get high byte
          IN      AL,DX
          MOV     BH,AL
          MOV     TEMP,BX             ;store data in location TEMP
          FIADD   TEMP                ;8087 stack:  ST(0)=averaged data
                                      ;             ST(1)=divisor (AVERG)
          DEC     SI                  ;check if have averaged sufficient data
          JNE     AVG
          FDIV    ST,ST(1)            ;divide to get average
                                      ;8087 stack:  ST(0)=averaged datum
                                      ;             ST(1)=divisor
          FISTP   WORD PTR [DI]       ;store data in A% array
                                      ;now 8087 stack: ST(0)=divisor
          CMP     PLOT,0              ;don't plot if plot flag=0
          JZ      NOPLOT
          FWAIT                       ;wait for FISTP to finish
          MOV     AX,WORD PTR [DI]    ;get averaged datum
          CALL    NEAR PTR PLOTPOINT  ;call plotting routine
NOPLOT:
          FWAIT
          ADD     DI,2                ;increment pointer by 2 (2 bytes /datum)
          LOOP    BEGIN
          FISTP   TEMP                ;free 8087 stack
          FWAIT
          RET
COLLECT   ENDP
;*******************************************************************
;                        PLOTTING ROUTINE
;
;Uses BIOS interrupt 10H to plot a single point on the IBM CGA screen.
;Note:  Assumes that BASIC has already invoked the plotting mode (SCREEN 2)
;          INPUT:    AX=DATUM TO PLOT
;                    TEMPSI contains x value of previous point
;
PLOTPOINT       PROC            NEAR
          PUSH    SI
          PUSH    DX
          PUSH    CX
          PUSH    BX
;calculate y-value to plot = 199-(datum/21)   so a/d values of 0-4095 fit screen
```

Program 14.6 (Part 2 of 3).   Assembly Language Program for Signal Averaging.

```
           CWD                        ;make AX into a double word
           MOV      BX,21             ;divide by 21--quotient in AX
           DIV      BX
           MOV      DX,AX             ;result into DX
           NEG      DX                ;negate and add to 199
           ADD      DX,199
           CMP      DX,199            ;check if values within limits of screen (0-199)
           JL       OKHIGH
           MOV      DX,199
           JMP      OKLOW
OKHIGH:CMP          DX,0
           JG       OKLOW
           MOV      DX,0
OKLOW: MOV          SI,TEMPSI         ;get x-value of last point on screen
           INC      SI                ;go to next location on screen
           CMP      SI,640            ;test if at right edge of 640 x 200
           JL       M1                ;  screen
           MOV      SI,0              ;if so, go to left edge to plot
M1:        MOV      CX,SI             ;get x-value into cx
           MOV      TEMPSI,SI         ;save x value
           MOV      AX,3073           ;AH=12,AL=1 to write dot to screen
           INT      10H               ;plot point
           POP      BX
           POP      CX
           POP      DX
           POP      SI
           RET
PLOTPOINT           ENDP
CSEG   ENDS

           END
```

Program 14.6 (Part 3 of 3). Assembly Language Program for Signal Averaging.

15

INTERRUPTS AND DIRECT MEMORY ACCESS

As we briefly mentioned when discussing methods of operation of the A/D (see page 36), two of the possible modes of operation of many A/Ds and other laboratory boards are **interrupts** and **direct memory access** (DMA). These two modes both allow the computer to accomplish two or more tasks "simultaneously."

We shall shortly be discussing the theory of each mode in considerable detail. However, from a practical, laboratory automation point of view, the uses of each of these modes are as follows:

- **DMA**. DMA is most useful in the laboratory for "smart," very high speed boards (e.g., the IBM GPIB board or the Data Translation DT2700 series of data acquisition boards). These boards need to transfer data to or from the IBM PC at very high rates and have on-board microprocessors that manage the data transfer process. Using this mode data transfer rates in excess of 100 kHz are possible. However, if you are interested only in high speed transfers of data, then you may wish to use programmed I/O, which generally can go as fast as DMA on the PC but is often simpler to program.

- **Interrupts.** Interrupts are most often used for relatively slow-speed data acquisition (rates under 1 kHz). This mode requires less "intelligence" on the laboratory board but can provide the means of managing and prioritizing several different tasks. Because this is done by the CPU of the IBM PC, it is inherently slower than DMA, where the task management is done by the micro-processor on the laboratory board. It can allow for more interaction among tasks and more real-time decision-making about priorities of various devices than the DMA mode, however.

One note of caution is needed. We have found occasions where one or the other of these two modes is required. However, programming either of these two modes is almost always very difficult because of the use of assembly language, the difficulty in testing routines that run asynchronously or on another microprocessor, and the lack of clear documentation on programming in these modes. We think you will find our examples useful in developing such routines, but we suggest that you undertake developing programs to support these two modes of operation only if you are an experienced programmer and have plenty of time to spend. *These are clearly not techniques for the beginner.*

15.1 INTERRUPTS

Before going any further, we should try to explain what interrupts are. Available on almost all computers, an **interrupt** is a combination of hardware and software by which an important process can tell the computer to halt whatever it is doing and do something else instead. Interrupts thus allow more than one process to occur asynchronously.

A simple example on the IBM PC is the clock interrupt; whenever the clock "ticks" (once every 18.2 seconds on the PC), the CPU is told to stop what it is doing and update the time it has stored in memory. It also checks to see if the date needs to be changed and checks whether the diskette drive needs to be turned off. Once these tasks have been completed, the processor resumes whatever it had been doing before the interrupt occurred.

15.1.1 Interrupts on the IBM PC

Although there are several ways to classify the interrupts on the IBM PC, the most useful is probably to separate interrupts into "software" versus "hardware" interrupts.

Software interrupts are handled directly by the CPU. Most of the interrupts on the IBM PC are software interrupts. For example, writing a number or letter on the monitor screen can be accomplished with a software interrupt. For those who are familiar with programming in higher level languages, software interrupts are similar but not identical in concept to subroutines.

Hardware interrupts, on the other hand, are those that are directed to a special interrupt controller chip, the Intel 8259, which in turn communicates with the CPU (Figure 15.1). There are only eight interrupt lines leading to the 8259, so the number of sources of hardware interrupts is limited to a maximum of eight at any one time (unless there are multiple 8259s in the system, as there are in the PC/AT). The assignments for these are shown in Table 15.1.

Table 15.1. IBM PC Hardware Interrupt Assignments

8259 Line	Interrupt Number	Address (Hex)	Interrupt Name
0	8	20-23	Time of Day[a]
1	9	24-27	Keyboard[a]
2	A	28-2B	Reserved
3	B	2C-2F	Communications[b]
4	C	30-33	Communications[c]
5	D	34-37	Disk
6	E	38-3B	Diskette
7	F	3C-3F	Printer

[a] These interrupt lines are not available except to the system board. All other lines are available on the "I/O" channel that is connected to each of the "slots" in the IBM PC.

[b] Assigned for use as secondary asynchronous communications or synchronous data link control (SDLC) communications or secondary binary synchronous communications (BSC), which are three different communications protocols frequently used on computer systems.

[c] Assigned for use as primary asynchronous communications, SDLC communications, or primary BSC communications. Notice that one SDLC adapter uses both interrupts B and C.

Essentially, each of the eight hardware interrupt lines is connected to the 8259. When any of the hardware devices generates an interrupt signal, *it is the function of the 8259 to decide whether the interrupt should be serviced immediately, later, or not at all*. If the interrupt is to be serviced immediately, the 8259 generates a signal to the CPU that an interrupt is pending and provides the CPU with information about which of the eight interrupt lines generated the interrupt.

Interrupt processing can thus involve interrupts generated either by the devices attached to the system (hardware interrupts) or by the programs already running in the system (software interrupts). It is not uncommon to use both types in combination. However, we will make the most use of hardware interrupts, because *they allow an external device (e.g., an A/D) to request immediate service.*

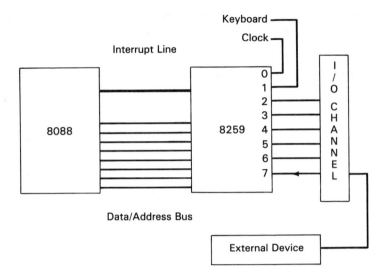

Figure 15.1. Interrupt hardware on the IBM PC. Interrupt requests on the IBM PC are directed to the 8259 Interrupt Controller that determines the priority of each interrupt. Two of the lines to the 8259 are permanently connected to the keyboard and the system clock. The 8259 signals the CPU that an interrupt is ready using a single interrupt line. If the interrupt request is accepted by the CPU, the information about which device is requesting the interrupt is passed to the CPU from the 8259 via the data/address bus. In more advanced systems such at the IBM PC/AT, there may be multiple 8259s available.

15.1.2 Should the Interrupt be Serviced?

Unfortunately for the programmer, it isn't that simple to use interrupts. First of all, not all eight hardware interrupts will always be needed, or even desired. For example, the user may wish to disable the clock or keyboard while very high speed interrupt processing is occurring; we'll see an example of disabling the clock interrupts in our next program. Each of the eight interrupts can be selectively enabled or disabled by accessing the I/O (input/output) port 33 (21 in hexadecimal). The following process can be used to select which port is to be used:

1.   Read port 33 (one byte of data) to find out which interrupts are enabled. Each bit of the byte tells whether the corresponding interrupt is on or off. For example, a value of 11000001 (binary) would mean that interrupts 0, 6 and 7 are turned off (disabled), while interrupts 1 through 5 are enabled.

2.   Turn on or off additional interrupts as desired. A way to do this would be to perform a logical AND between the current status byte and a byte that has

zeros at each bit corresponding to an interrupt that is to be turned on, and then a logical inclusive OR with a byte that has ones in it at each location that corresponds to an interrupt to be turned off. The resulting byte must be written back to port 33.

An example of this would be the following, where all numbers are shown in binary:

	10101100	Original byte in port 33
AND	11110011	To turn on interrupts 2 and 3
	10100000	Intermediate result
OR	00000001	To turn off interrupt 0
	10100001	Result: interrupts 0, 5, and 7 are disabled, all others are enabled.

Note that bits are numbered from 7 to 0 (left to right), and that the final result must be written to port 33 in order to be effective.

The second factor used in deciding whether an interrupt is to be allowed is its priority. Associated with each of the eight possible hardware interrupts is a priority: the lower the interrupt number, the higher its priority. Thus, the clock interrupt (interrupt 0) will normally have priority over all other interrupts. This means that no other process can interrupt in the middle of an interrupt by the clock, but that the clock can interrupt any other hardware interrupt. This makes sense because it is more important that the clock "tick" be recorded than, say, that a character be received from the keyboard -- otherwise, the clock might "tick" again before the keyboard interrupt had been completed and the information about the clock "tick" thus lost.

The third factor that determines whether an interrupt will be serviced is the status of the interrupt flag bit in the processor flag register word. Normally, an STI (set interrupt flag) instruction in assembly language is used to set the flag bit, which enables all interrupts. The CLI (clear interrupt flag) instruction disables all interrupts. Generating any interrupt normally disables further interrupts by clearing the interrupt flag until an STI instruction is executed or until the flag register is restored as part of the return from interrupt (IRET) instruction.

15.1.3 Save Information on the Current Program

Once an interrupt has been generated and the CPU notified that the interrupt should be serviced, the CPU must store sufficient information about the program currently executing so that control can be returned to the program once the interrupt has been serviced. This is done transparently to the user by the interrupt hardware. Essentially, however, the current flag register and the information about the instruction in the main program that is to be executed next are stored on the stack.

15.1.4 Interrupt Service Routines

Now control must be switched to an **interrupt service routine.** This is the routine that actually processes the information generated by the interrupting device.

What makes the interrupt process unique is the method by that control is passed to the interrupt service routine. This is done by having a unique "vector" associated with each interrupt (see Table 15.1). In order for an interrupt to function properly, your program must store the address of the beginning of the interrupt service routine at the appropriate vector location.

Because there are other interrupts than just the hardware interrupts, the eight hardware interrupts use vectors 8 through 15 in the IBM PC. It takes four bytes to store the address of the interrupt service routine; hence, each vector is located in memory beginning at an address four times its vector number. Thus, for example, vector 8, which is associated with hardware interrupt 0 by the 8259, is located in memory beginning at address 32 (20 hex).

Notice that four bytes of memory are required for the vector. This is because the complete address is specified by using a new value for the code segment (CS) register and a 16-bit offset; this allows the positioning of the interrupt service routine anywhere in the 1 megabyte address space of the IBM PC (although of course much of that space would be inappropriate).

Once an interrupt occurs and the information about the current status of the processor is stored, then the processor reads the appropriate vector, jumps to that location in memory, and begins executing the code there.

What must a hardware interrupt service routine do? The following steps must be performed, although there is some flexibility in the order in which they may be performed:

1. Restore interrupts. Because a jump to an interrupt service routine resets the interrupt flag and thus disables further interrupts, the routine should as soon as possible set the interrupt flag to allow other (higher priority) interrupts to occur.

2. Save registers. Any registers that will be used in the interrupt routine should be saved (usually on the stack). This is because those same registers are probably being used in the portion of the program that was running when the interrupt occurred, and they have to be the same when the interrupt starts and finishes. If segment registers (usually ES and DS) are to be modified, these must be saved as well.

3. Establish new registers. If any registers (other than CS) are to be used during the interrupt service routines, they must be either read from memory or popped off of a stack that is local to the interrupt routine. The system stack should not be used for this purpose because there is no way of predicting where the stack pointer will be pointing at the time the interrupt occurs.

4. Service the interrupting device. This code will be unique to the device causing the interrupt. For example, if it is a laboratory data collecting device, then the

interrupt service routine might input data from the device and store it somewhere in memory.

5.   Acknowledge receipt of interrupt. Before the 8259 can process another hard-
     ware interrupt at the same or lower priority, the processor must acknowledge
     receipt of the previous interrupt. This is done in the IBM PC by writing a 32
     (20 hex) to port 32 (20 hex).

6.   Restore registers. Any registers saved must be restored to their original values.
     This is usually done by popping off of the stack any registers saved in step 2.

15.1.5 Return of Control

The last step on the IBM PC is actually also part of the interrupt service routine. It
must always be an IRET instruction in assembly language; this returns control to
whatever instruction the processor was ready to execute when the interrupt occurred.

15.1.6 Preparation for Interrupts

Careful reading of the above implies that before the first interrupt can be executed,
three types of preparation must occur:

1.   The interrupt vectors must be prepared. The addresses of the interrupt service
     routines must be stored at the appropriate vector location. However, it is pos-
     sible that more than one type of device will use the same hardware interrupt
     line; it is therefore good programming practice to save the vectors already at
     the vector location before creating the desired vector. The vectors should be
     restored upon finishing the program or upon any kind of abnormal exit.

2.   The 8259 must be programmed. As we have described, the 8259 interrupt con-
     troller chip must be told which interrupts to allow.

3.   The interrupting device must be electrically connected to the appropriate inter-
     rupt line. Of the eight hardware interrupt lines, lines 0 and 1 are dedicated to
     the clock and the keyboard, respectively. Hence, only interrupt lines 2 through
     7 are available to the user on the "I/O Channel," which is the set of lines avail-
     able to any boards installed on the IBM PC. As shown in Table 15.1, many of
     these interrupt lines are already used by standard IBM devices such as the disk
     controller or serial (communications) interface.

If the device you are connecting gives a choice of interrupts, you should choose
one not used by the devices already in your system. However, this is not always possible, so you should at least choose an interrupt corresponding to a device that will
not be operating during the time while your new device is operating. For example,
the interrupts reserved for communications often will not be used during other proc-

essing. It is especially important to remember to save and restore vectors if you are connecting two devices to the same interrupt line.

Table 15.2.   Registers on the IBM DACA Used for Interrupts

Register[a]	Name	Address[b]	Meaning of Bits
0	A/D Control	02E2	bit 0=start conversion
			bit 2=enable interrupt on end-of-conversion
			bits 8-15=channel number
0	A/D Status	02E2	bit 0=A/D is busy still
2	A/D Datum	22E2	bits 0-11=datum
			bits 12-15 are 0s
8	Timer 0	82E2	datum counter 0
9	Timer 1	92E2	datum counter 1
10	Timer 2	A2E2	datum counter 2
11	Clock Control	B2E2	bit 0=0 for binary, 1 for BCD
			bits 3-1=mode
			(mode 0=interrupt on terminal count)
			(mode 2=rate generator)
			(mode 3=square wave generator)
			(mode 4=software triggered strobe)
			bits 5-4=0 to latch counter,
			2 to read/load 2 bytes, low byte first
			bits 7-6=counter selected (0-2)
12	Device Number	C2E2	bits 0-7=device number
			Set to 9 for A/D
13	Interrupt	D2E2	bit 0=enable timer-generated interrupt
			bit 1=enable counter-generated interrupt
			bit 2=enable end-of-conversion interrupt
			bits 3-6 =0000
			bit 7=reset interrupt circuits of DACA

a Only those registers and bit meanings are shown that are used in programming the A/D in the interrupt mode.

b The addresses shown are for Adapter 0. If you have more than one DACA card in your IBM PC, then each DACA must have a different adapter number. Because the addresses shown are in hex, you should add 0400 (hex) to the address for each higher numbered adapter. For example, the A/D control register of Adapter 1 is referenced at address 06E2 (hex).

15.1.7 Example of a Laboratory Application

If you are like we are, you will find it difficult to do all of these steps correctly without an example. Hence, we will describe data collection using interrupts and the A/D on the IBM Data Acquisition and Control (DACA) board; the DT2801 does not use interrupts. The registers on the DACA used in this program are described in Table 15.2.

Basically, the goal of this program is to allow data collection to occur independently of data processing. This is achieved by using three interrupts: interrupt upon data ready for storage; interrupt upon the user typing a CTRL-C or CTRL-BREAK on the keyboard to abort the run; and interrupt when finished with all data collection.

The three interrupts are handled by only two interrupt service routines: the "data ready" interrupt by the data service routine and the other two by the exit service routine.

15.1.8 Program Logic

Program 15.1 is the BASIC program TESTINT that processes data from the A/D. Program 15.2 is the assembly language routine DACAINT that sets up and performs the actual data collection using the A/D of the DACA. TESTINT is written to be used with the IBM PC BASIC Compiler; hence, it must be compiled and linked with the assembled form of DACAINT in order to work properly. This can be done by typing the following (assuming that the BASIC compiler and the linker are in the current subdirectory):

> BASCOM TESTINT;
> ASM DACAINT;
> LINK TESTINT+DACAINT;

The flow of logic in the program is fairly straightforward, as illustrated in Program 15.2.   TESTINT defines the necessary variables and then calls DACAINT. DACAINT establishes the interrupt vectors (i.e., tells the system where the interrupt service routines are located in memory), sets up the timer in the DACA adapter that is used to trigger data collection at specific intervals, and then begins data collection. Control is returned to TESTINT, which is then interrupted every time a datum has been collected and is ready for storage. TESTINT plots the data on the screen using standard IBM BASIC graphics calls (assuming a color/graphics monitor board is present and a color monitor is used). This continues until all data have been plotted, at which point the data are scaled to fill the plot area and replotted.

```
10 REM PROGRAM TESTINT
20 REM LABORATORY AUTOMATION USING THE IBM PC
30 REM Program to demonstrate using interrupt-driven data collection
40 REM      Compile using  BASCOM PROG151;
50 REM      Link using      LINK PROG151+PROG152,,,BASRUN20
60 REM Note:  This program must be run on a graphics display!
70 REM **************************************************************
80 DIM A%(5000)
90 P%=0                                'count of points collected so far
100 F%=0                               'error flag
110 REM                  Main Program
120    GOSUB 210                       'get user input
130    INPUT "Type the <RETURN> key to start data collection.",Y$
140    GOSUB 370                       'set up for real-time routine
150    CALL DAACINT (A%(1),TIMER1%,F%,P%,N%,C%,TIMER0%)
160    GOSUB 520                       'go to real-time routine
170    GOSUB 680                       'do any post-run processing
180 END
190 REM ***************************
200 REM Get user input
210    WHILE N% < 1 OR N% > 5000
220       INPUT "Enter the number of data points to collect (1 to 5000)",N%
230    WEND
240    INPUT "Enter the A/D channel desired",C%
250    INPUT "Enter the rate in points/second",RATE!
260    REM Calculate times to send to DAAC timers, assuming that
270    REM      the timers "tick" at 1.023 MHz.
280    TICK!=1023000!/RATE!            'calculate number of ticks
290    TIMER1!=INT(TICK!/65536!)+2     'each timer counts down from max to 0
300    TIMER0!=TICK!/TIMER1!           '    and two timers are chained together
310    REM Integers above 32767 must be stored as negative numbers
320    IF TIMER0! > 32767 THEN TIMER0%=TIMER0!-65536! ELSE TIMER0%=TIMER0!
330    TIMER1%=TIMER1!
340 RETURN
350 REM ***************************
360 REM Set up screen for Real-time Routine
370    CLS:SCREEN 1
380    LOCATE 1,1:PRINT "Points collected so far=";
390    WINDOW (1,0)-(N%,4096)
400    VIEW (100,50)-(200,150),2,B
410    COLOR 9,0
420 RETURN
430 REM ***************************
440 REM Real-time plot
450 REM This subroutine displays a plot of the points as they are
460 REM      collected.  However, it could do any other useful task
470 REM      that you choose.  It will be interrupted at intervals
480 REM      by the data collection process, but you will not see
490 REM      this occur.
500 REM First "while" loop checks whether all points have been collected
510    J%=1
520    WHILE P% < N%
530       REM second "while" loop plots any points collected so far
540       WHILE J% <= P%
550          PSET (J%,A%(J%)),1
560          J%=J%+1
570       WEND
580       REM print number of points collected so far
590       LOCATE 1,35:PRINT P%;
600       IF F% <>0 AND P%< N% THEN PRINT "Run aborted by user after ";
          P%;" points":GOTO 630
610    WEND
620    PRINT "Run successfully completed."
630    INPUT "Type the <RETURN> key to continue",Y$
640 RETURN
650 REM ***************************
660 REM Post-run processing; displays the data after normalizing it.
```

Program 15.1 (Part 1 of 2). Using an Interrupt-Driven Subroutine.

```
670    IF P% < 2 THEN RETURN
680    VIEW:CLS:VIEW (100,50)-(200,150),2,B
690    YMIN%=A%(1)
700    YMAX%=A%(1)
710    FOR I%=2 TO P%
720        IF A%(I%)<YMIN% THEN YMIN%=A%(I%)
730        IF A%(I%)>YMAX% THEN YMAX%=A%(I%)
740    NEXT I%
750    WINDOW (1,YMIN%)-(P%,YMAX%)
760    PSET (1,A%(1))
770    FOR I%=2 TO P%
780        LINE -(I%,A%(I%)),3
790    NEXT I%
800    LOCATE 23,1:PRINT "Minimum= ";YMIN%; "     Maximum= ";YMAX%;
810    Y$=INKEY$:IF Y$=""THEN 810
820 RETURN
```

Program 15.1 (Part 2 of 2).   Using an Interrupt-Driven Subroutine.

15.1.9 Communications

One of the major problems in an interrupt-driven process is communications between the interrupting process and the main program. Because the interrupts occur asynchronously with execution of the main program, it is usually important for the main program to be able to find out what has been accomplished by the interrupt service routine(s) at any given time. An obvious way in which to do this is to have one or more variables accessible to each routine for the purpose of inter-routine communications.

Communication between the TESTINT and DACAINT routines is handled by two variables. One variable, P%, is used to keep track of the number of data points collected so far; that is, it is incremented each time a datum is collected. The other variable, F%, is used to tell whether DACAINT has stopped collecting data for any reason. It is initially set to zero and remains that value unless the data collection process has been finished or has been aborted by a CTRL-BREAK, at which point F% is set to the number of data points actually collected. If the program is aborted before any points have been collected, then F% is set to a value of -1.

TESTINT periodically checks both P% and F% to determine the current status of the data collection process. This is done so that the plotting routine in TESTINT never tries to plot a datum before it has been collected by DACAINT. It also does not wait for another datum to be collected when data collection has been prematurely terminated by the user typing a "control-break."

15.1.10 Assembly Language Routine

Each of the interrupts requires that an interrupt vector for it be established. Because IBM has defined that the vectors are located in memory locations 0 through 7F

```
                        TITLE   DACAINT
;LABORATORY AUTOMATION USING THE IBM PC
;Subroutine to use the ibm data acquisition and control board to
;    acquire data (using A/D) and store in memory
;Runs using interrupts
;**********
;NOTE! FOR PROPER FUNCTIONING, YOU MUST CONNECT "DELAY OUT" TO
;       "A/D CE" ON DISTRIBUTION PANEL (USES TIMER TO TRIGGER A/D)
;**********
;
;        CALL FROM COMPILED BASIC WITH CALL OF FORM:
;               CALL DACAINT (A%(1),I%,F%,P%,N%,C%,S%)
;        NOTE:   ALL PARAMETERS MUST BE INTEGER*2   *******************
;        WHERE   A%   IS ARRAY WHERE DATA ARE TO BE STORED
;                I%   IS TIMER0 VALUE
;                F%   IS OVERRUN FLAG..SET TO ZERO UPON NORMAL EXIT
;                     OTHERWISE SET TO NUMBER OF POINTS ACTUALLY
;                     COLLECTED
;                P%   IS COUNTER OF NO. OF POINTS COLLECTED SO FAR..
;                     USED TO COMMUNICATE BACK TO MAIN PROGRAM
;                N%   IS NUMBER OF POINTS TO BE COLLECTED
;                C%   IS CHANNEL NUMBER OF A/D
;                S%   IS TIMER1 VALUE
;
;in general, this routine uses interrupt-driven data collection.
;    When the clock counts to zero it generates an interrupt, which vectors
;    to location "DONE."  After the datum has been stored, control is
;    returned to the main (BASIC) program unless all of the data points have
;    been collected.  If all of the points have been collected, or the
;    overrun flag is set because the a/d was going too fast, then an
;    interrupt is generated, and routine "OVERRUN" is executed.  This
;    turns off data collection and signals the main program using F%
;    that data collection has been terminated.  Typing control-break
;    (control-C) during data collection will also terminate the data
;    collection in a similar manner.  All interrupt vectors are
;    restored upon termination of data collection.  Note also that the
;    system clock is turned off during data collection.
;
; *********************************************************************
FRAME              STRUC
OLDBP        DW        ?                  ;old BP
OLDIP        DW        ?                  ;IP for return
OLDCS        DW        ?                  ;CS for return
SPTR         DW        ?
CPTR         DW        ?
NPTR         DW        ?
PPTR         DW        ?
FPTR         DW        ?
IPTR         DW        ?
APTR         DW        ?            ;POINTERS TO VARIABLES PASSED
FRAME              ENDS
PARAMSIZE    EQU       OFFSET APTR-OFFSET OLDCS   ;bytes in parameters
; *********************************************************************
;DEFINITIONS:  (ASSUMES ADAPTER 0)
BASE .           =02E2H                   ;base of daac board
DEVICE           =0C2E2H                  ;device select register
ADCCONTROL       =02E2H                   ;a/d status & control low byte
ADCCONTRO2       =02E3H                   ;a/d  control high byte (channel no.)
ADCDATA          =22E2H                   ;a/d datum
CLOCK0           =82E2H                   ;clock 0
CLOCK1           =92E2H                   ;clock 1
CLOCKCONTROL     =0B2E2H                  ;clock control port
INTERRUPT        =0D2E2H                  ;interrupt register
INT0             =20H                     ;8259 interrupt chip ports
INT1             =21H
START            =5                       ;start, stop codes for a/d (w/interrupts)
NOGO             =0
```

Program 15.2 (Part 1 of 5).   Assembly-Language Interrupt Routine.

```
; **********************************************************************
ABS0    SEGMENT AT 0
;INTERRUPT VECTORS (OCCUPY 4 BYTES EACH)
STGLOCO LABEL BYTE                          ;for overrun, timer interrupts
        ORG 0AH*4                           ;  as vectors 0A, 0F hex
ADOVERRUN LABEL DWORD                       ;note that overrun should be higher
        ORG 0FH*4                           ;  priority than a/d done
ADDONE    LABEL DWORD
        ORG 1BH*4                           ;control-break vector
CTRLBREAK  LABEL DWORD
ABS0    ENDS
; **********************************************************************
CSEG    SEGMENT  WORD PUBLIC 'CODE'
        ASSUME CS:CSEG,  DS:NOTHING,  ES:NOTHING
; MEMORY LOCATIONS USED:
TEMPVECT DW   20   DUP  (?)                 ;space to save vectors
DSTEMP   DW   ?                             ;stores DS
CHANNEL  DW   ?                             ;channel number
ARRAYPTR DW   ?                             ;pointer to current position in data (A%)
COUNTPTR DW   ?                             ;pointer to counter of pts. stored (P%)
FLAGPTR  DW   ?                             ;pointer to overrun flag (F% in params)
NPTS     DW   ?                             ;number of points desired (N% in params)
         EVEN                               ;align on word boundary
; **********************************************************************
;START OF EXECUTABLE CODE  .. MAIN ROUTINE
        PUBLIC  DACAINT
DACAINT PROC    FAR
;       INT     03H                         ;remove comment if you wish to DEBUG
        PUSH    BP                          ;save BP
        MOV     BP,SP                       ;set base of parameter list
        MOV     AX,DS                       ;save DS register
        MOV     DSTEMP,AX
        CALL    NEAR PTR DACAINTX           ;set up interrupt vectors
        CALL    NEAR PTR DACAINTY           ;set up DACA
        POP     BP
        RET     PARAMSIZE                   ;number of bytes used in parameters
DACAINT ENDP
; **********************************************************************
;SAVE CURRENT INTERRUPT VECTORS, THEN SET UP NEW VECTORS
DACAINTX PROC    NEAR
        PUSH    DS                          ;save DS
        PUSH    ES                          ;save ES
        PUSH    CS                          ;set ES=CS
        POP     ES
        SUB     AX,AX                       ;set DS=0
        MOV     DS,AX
        MOV     CX,2                        ;save a/d vectors
        MOV     SI,OFFSET ADOVERRUN
        MOV     DI,OFFSET TEMPVECT
        CLD                                 ;do MOVSW in forward direction
        REP     MOVSW
        MOV     CX,2
        MOV     SI,OFFSET ADDONE
        REP     MOVSW
        MOV     CX,2                        ;save control-break vector
        MOV     SI,OFFSET CTRLBREAK
        REP     MOVSW
        PUSH    CS                          ;set DS=CS
        POP     DS
        MOV     ES,AX                       ;set ES=0
        MOV     CX,2                        ;establish new a/d vectors
        MOV     SI,OFFSET VECTORTABLE
        MOV     DI,OFFSET ADOVERRUN
        CLD
        REP     MOVSW
        MOV     CX,2
        MOV     DI,OFFSET ADDONE
```

Program 15.2 (Part 2 of 5).   Assembly-Language Interrupt Routine.

```
        REP      MOVSW
        MOV      CX,2                        ;establish new control-break vector
        MOV      DI,OFFSET CTRLBREAK
        REP      MOVSW
        POP      ES
        POP      DS
        RET
DACAINTX ENDP
;    ***********************************************************************
;  GET INFORMATION TO SET UP A/D
DACAINTY PROC    NEAR
        MOV      DX,DEVICE                   ;select a/d (device=9)
        MOV      AL,9
        OUT      DX,AL
        MOV      DX,ADCCONTROL               ;inhibit convert start
        MOV      AL,NOGO
        OUT      DX,AL
        MOV      SI,[BP].CPTR                ;get channel number
        MOV      CX,[SI]                     ;
        MOV      CHANNEL,CX                  ;save channel number
        MOV      DX,ADCCONTRO2               ;and send channel no.
        MOV      AL,CL
        OUT      DX,AL                       ; (use only lower byte)
;
        MOV      SI,[BP].FPTR                ;get overrun flag address
        MOV      FLAGPTR,SI
        MOV      WORD PTR [SI],0             ;zero the flag
        MOV      SI,[BP].NPTR                ;get number of points desired
        MOV      AX,WORD PTR [SI]
        MOV      NPTS,AX
;
        MOV      SI,[BP].PPTR                ;get point counter address
        MOV      COUNTPTR,SI
;
;SET UP TIMER
NOINC:  MOV      DX,CLOCKCONTROL             ;set to timer1, mode 2, binary counter
        MOV      AL,74H
        OUT      DX,AL
        MOV      SI,[BP].SPTR                ;get timer1 value
        MOV      CX,[SI]
        MOV      AL,CL                       ;get low byte & send
        MOV      DX,CLOCK1
        OUT      DX,AL
        MOV      DX,CLOCK1+1                 ;now high byte
        MOV      AL,CH
        OUT      DX,AL
;NOW SET UP A/D FOR INTERRUPTS
        MOV      DX,ADCCONTROL
        MOV      AL,START                    ;set bits 0,2
        OUT      DX,AL
        MOV      DX,ADCCONTRO2
        MOV      AX,CHANNEL
        OUT      DX,AL
; ALMOST READY TO GO....
        MOV      AX,[BP].APTR                ;save address where first datum stored
        MOV      ARRAYPTR,AX
        STI                                  ;enable interrupts on 8088
        IN       AL,INT1                     ;enable interrupts 2+7 on 8259 chip
        AND      AL,7BH
        OR       AL,001H                     ;turn off system clock interrupts
        OUT      INT1,AL
        MOV      DX,CLOCKCONTROL             ;set to timer 0,mode 2, binary counter
        MOV      AL,34H
        OUT      DX,AL
        MOV      SI,[BP].IPTR                ;get timer0 value
        MOV      CX,[SI]
        MOV      DX,CLOCK0                   ;output low, high bytes
```

Program 15.2 (Part 3 of 5).   Assembly-Language Interrupt Routine.

```
        MOV     AL,CL
        OUT     DX,AL
        MOV     DX,CLOCK0+1             ;outputting timer value starts clock
        MOV     AL,CH
        OUT     DX,AL
        MOV     DX,INTERRUPT           ;set interrupt register for a/d generated
        MOV     AL,84H                 ; interrupt & clear interrupts (bits 2,7)
        OUT     DX,AL
        MOV     AL,04H
        OUT     DX,AL
        RET                            ;finished..go back to basic program
DACAINTY ENDP
;       ****************************************************************
;INTERRUPT SERVICE ROUTINE; COLLECT UPON EXTERNAL INTERRUPT FROM A/D
DONE:
        PUSH    AX                     ;save registers used in this routine
        PUSH    DX
        PUSH    DI
        PUSH    DS
        MOV     AX,DSTEMP              ;DS may not be same as originally,
        MOV     DS,AX                 ;    so restore
        MOV     DX,ADCCONTROL         ;turn off a/d
        MOV     AL,NOGO
        OUT     DX,AL
        MOV     DX,ADCCONTRO2
        MOV     AX,CHANNEL
        OUT     DX,AL
        MOV     DX,ADCDATA            ;get both bytes of datum into memory
        IN      AX,DX
        MOV     DI,ARRAYPTR           ;get data array pointer
        MOV     WORD PTR [DI],AX      ;and store datum
        ADD     ARRAYPTR,2            ;increment pointer by 2
        MOV     DI,COUNTPTR           ;update count of no. pts collected so far
        INC     WORD PTR [DI]
        MOV     AX,DS:[DI]            ;compare to total pts. desired
        MOV     DI,NPTS
        CMP     AX,DI
        JL      DONE2                 ;finished collecting?
        INT     0AH                   ;yes, close up shop
DONE2:  MOV     AL,START              ;set a/d ready to trigger again
        MOV     DX,ADCCONTROL
        OUT     DX,AL
        MOV     DX,ADCCONTRO2
        MOV     AX,CHANNEL
        OUT     DX,AL
        MOV     DX,INTERRUPT          ;reset interrupt on daac board
        MOV     AL,84H                ;(bit 7)
        OUT     DX,AL                 ;This must follow resetting a/d trigger!!
        MOV     AL,04H
        OUT     DX,AL
        MOV     AL,20H                ;acknowledge to 8259 interrupt rec'd
        OUT     INT0,AL
        POP     DS
        POP     DI                    ;restore all registers
        POP     DX
        POP     AX
        STI                           ;re-enable interrupts
        IRET
;       ****************************************************************
;INTERRUPT SERVICE UPON FINISH OR ABORT
OVERRUN:
        PUSH    ES                    ;save registers used in this routine
        PUSH    DS
        PUSH    AX
        PUSH    CX
        PUSH    DX
        PUSH    SI
```

Program 15.2 (Part 4 of 5).   Assembly-Language Interrupt Routine.

```
            PUSH    DI
            MOV     AX,DSTEMP               ;restore DS to original value
            MOV     DS,AX
            MOV     DX,INTERRUPT            ;turn off timer interrupts
            MOV     AL,0
            OUT     DX,AL
            IN      AL,INT1                 ;get which interrupts enabled
            OR      AL,84H                  ;disable interrupts 2,7 on 8259
            AND     AL,0FEH                 ;turn on system clock again
            OUT     INT1,AL
            INT     03H
            MOV     DI,COUNTPTR             ;get points collected so far
            MOV     AX,DS:[DI]
            CMP     AX,NPTS                 ;if points collected = NPTS then
            JGE     T2                      ;    is normal exit, else
            CMP     AX,0                    ;if no points taken yet, store -1
            JG      T2
            MOV     AX,-1
    T2:     MOV     SI,FLAGPTR              ;set overrun flag if abnormal exit
            MOV     WORD PTR DS:[SI],AX     ;    or to NPTS if normal exit
            SUB     AX,AX                   ;now reverse process to restore vectors
            MOV     ES,AX                   ;ES=0
            PUSH    CS                      ;set DS=CS
            POP     DS
            MOV     CX,2
            MOV     SI,OFFSET TEMPVECT
            MOV     DI,OFFSET ADOVERRUN
            CLD
            REP     MOVSW
            MOV     CX,2
            MOV     DI,OFFSET ADDONE
            REP     MOVSW
            MOV     CX,2
            MOV     DI,OFFSET CTRLBREAK
            REP     MOVSW
            MOV     AH,14                   ;ring the bell to indicate finished
            MOV     AL,7
            MOV     BH,0
            INT     10H
            MOV     AL,20H                  ;signal interrupts all received
            OUT     INT0,AL
            POP     DI                      ;restore registers
            POP     SI
            POP     DX
            POP     CX
            POP     AX
            POP     DS
            POP     ES
            STI
            IRET
    ;     ************************************************************************
    VECTORTABLE LABEL   WORD        ;TABLE FOR VECTOR ADDRESSES
            DW      OFFSET OVERRUN
            DW      CSEG
            DW      OFFSET DONE
            DW      CSEG
            DW      OFFSET OVERRUN
            DW      CSEG
    CSEG    ENDS
            END
```

Program 15.2 (Part 5 of 5). Assembly-Language Interrupt Routine.

(hex), the first step is to be able to address these low memory locations. This is done by defining a segment beginning at location 0, i.e.,

```
ABS0        SEGMENT        AT 0
```

All of the vectors are then loaded into this segment at the appropriate locations. However, the current vectors at those locations must be saved before new ones are loaded, so that the system can be restored to its original state when the program is finished. This is accomplished using the following type of routine:

```
SUB     AX,AX            ;set AX=0
MOV     DS,AX            ;set data segment register to 0
MOV     CX,2             ;use CX to count no. words to transfer
MOV     SI, OFFSET V1    ;transfer vector at V1
MOV     DI, OFFSET T1    ;to location T1 for storage
CLD                      ;sets direction of transfer
REP     MOVSW            ;moves words from V1 to T1
```

Once the current vectors are saved, new ones are moved into place:

```
MOV     ES,AX            ;set extra segment register to 0
PUSH    CS               ;set data segment to current
POP     DS               ;   code segment
MOV     CX,2             ;CX counts no. of words to transfer
MOV     SI, OFFSET R1    ;address of interrupt service routine
MOV     DI, OFFSET V1    ;address of vector
REP     MOVSW            ;execute transfer
```

It is useful to remember that the MOVSW instruction moves words from location DS:[SI] to ES:[DI]; that is, from a location in the data segment to a location in the extra segment. Hence, the DS and ES registers must be carefully set to the appropriate values. In the case of the DACAINT routine, the vector is located in a segment starting at location zero; the old vector is stored in a data segment created by the program; and the address of the interrupt service routine is in the code segment. In the example shown above, the location in memory of some variables is not specified. Also notice that each vector is four bytes (two words) in length and includes both an offset and a segment value.

15.1.11 Setting Segment Registers

How then are segment registers set? Assuming that you have defined each segment to start on an address divisible by 16 (using the PARA align-type), you can use the following approach:

```
MOV     AX, SEG T1       ;get segment location of T1
MOV     DS, AX           ;and save it in data segment register
```

Alternatively, if you have named the segment S1, for example, you can simply use:

```
MOV        AX, S1
MOV        DS, AX
```

You need to remember that because interrupts occur asynchronously from the main program, there is no way of telling what the values of any of the segment registers will be. Hence, the general-purpose and segment registers (other than CS) must each be specifically saved and set to the desired values each time an interrupt occurs. In fact, it is this saving and resetting of registers that is the primary reason that interrupt service routines are relatively slow.

Notice also that if a MOVS or MOVSW statement is to be used, the direction flag should be set to the appropriate direction during each interrupt, because the main program may have set it in the opposite direction.

One final reminder: don't forget that each time a hardware interrupt occurs, you must both clear the interrupt flag (with an STI instruction or with the IRET instruction) and acknowledge to the interrupt controller (the 8259 chip) that you have received the interrupt.

15.1.12 Program performance

It is important for us to note that the interrupt service routine for data collection in DACAINT is not as fast as the corresponding routine using program control of the A/D (see Program 14.3). This is because of the additional overhead involved in acknowledging interrupts and saving registers. Thus, this program is most useful at rates below 1 kHz, where 75% or more of the processor time can be used for processes other than data collection.

15.2 DIRECT MEMORY ACCESS

Now that you have suffered through interrupts, we will show you how to use a much more modern technique, namely **direct memory access**, or DMA. In DMA, a processor on the data acquisition board sends data to, or receives data from, the PC's memory **without intervention by the PC's CPU.** In other words, the laboratory interface board has direct access to the PC's memory.

This process is mediated by a special chip, the **DMA controller**. Recall that in both software polling and interrupts, the entire process is under control of the CPU, whereas for DMA, data is transmitted without involving the CPU (except to initiate the process). Hence, the primary advantage of DMA is that it **uses fewer CPU resources than other methods of data acquisition and control**. DMA is often used with either A/Ds and D/As, or both.

DMA as a technique is almost always preferable to interrupts because it is easier to program and allows much faster data acquisition or control rates. However, software polling or similar direct control of data acquisition is even simpler than (and often as rapid as) DMA.

In a typical program involving DMA, we will see the following:

- Setup of the DMA controller.
- Setup of the data acquisition or control device.
- Initiation of DMA transfers.
- Transfer of data **with simultaneous running of other parts of of your program**.
- Termination of DMA transfers.

15.2.1 DMA on the IBM PC

From a practical point of view, DMA, like interrupts, allows simultaneous execution of two processes; one is controlled by the CPU from your program, while the other is controlled by the processor of the laboratory device. What makes DMA possible is a special controller chip: the 8237-2 on the PC, the 8237A-5 on the PC/XT, and two 8237A-5s on the PC/AT. The 8237A-5 controller, which we shall consider in most of our examples, allows up to 4 channels, or input/output streams, of DMA. The uses of the DMA channels are shown in Table 15.3. Note that Channel 0 is used for memory refresh by the PC and PC/XT; the PC/AT has a separate refresh controller. Most data acquisition boards built for the PC/XT allow use of Channels 1 and 3 for your DMA device, but if your PC has a hard disk, you normally should only use Channel 1 for data acquisition. Data acquisition boards for the PC/AT usually only allow use of DMA channels 5, 6, or 7.

Table 15.3.   IBM PC DMA Hardware Channels

Channel[a]	Function
0	Dynamic memory refresh (PC, XT), Spare (AT)
1	Spare (or Communications)
2	Diskette
3	Disk
4	Cascade for Channels 0-3
5	Spare
6	Spare
7	Spare

[a]Channels 4 to 7 are available on the PC/AT but not on the PC or PC/XT.

Many other devices (such as communications adapters, GPIB, etc.) use DMA. Hence, **you should check to be sure that there is no conflict in DMA channel usage**.

Instructions can be given to the DMA controller using a series of ports that access registers in the DMA controller, much as we have controlled the DT2801 board using port I/O. There are six registers of interest to us:

- **Page Select Register**. Together with the Base Address Register, this determines the starting address in memory for the DMA data transfers. A **page** is a 64 Kbyte area of memory (64 Kword for the second controller of the PC/AT) within which all of the DMA transfers will take place. The **base address** is the offset within this segment where the first data transfer will take place. For example, page 1, base address 0 refers to absolute address 65,536 on the XT, but 131,072 on the second controller of the AT (recall that 64K means 64 x 1024, and that addresses begin at page 0, base address 0).

- **Base Address Register**. Together with the Page Select Register, this determines the starting address in memory for the DMA data transfers. The base address is a byte address for the PC and PC/XT, but a word address for the second controller of the PC/AT.

- **Byte/Word Count Register**. This is the count of the number of bytes (PC and PC/XT) or words (PC/AT) of data to be transferred. Normally, your program should check to make sure that the transfers do not cross any page boundaries.

- **Mask Register**. The mask register is used to select the DMA channel. There is one mask register associated with each DMA controller of the PC/AT.

- **Mode Register**. The mode register specifies the type of transfer (read or write), channel, and whether to autoinitialize. Autoinitialization refers to setting the the DMA so that when it has transferred the desired amount of data, the controller resets to allow data transfers to/from the same area of memory. There is one mode register associated with each DMA controller.

- **Byte Pointer Flip/Flop Register**. This register is used to signal the controller that a two-byte address is being sent, low byte first. It should be used to ensure that two-byte transfers are properly interpreted by the controller.

The ports for the DMA controller addresses are shown in Table 15.4. We have shown only the addresses for the commonly-used channels.

Table 15.4. DMA Controller Access

Address(Hex)	Function[a]
02	Base Address, Channel 1
06	Base Address, Channel 3
C4	Base Address, Channel 5
C8	Base Address, Channel 6
CC	Base Address, Channel 7
03	Byte count, Channel 1
07	Byte count, Channel 3
C6	Word count, Channel 5
CA	Word count, Channel 6
CE	Word count, Channel 7
83	DMA page select, Channel 1
82	DMA page select, Channel 3
8B	DMA page select, Channel 5
89	DMA page select, Channel 6
8A	DMA page select, Channel 7
0A	Mask register, Channels 0-3
D4	Mask register, Channels 4-7
0B	Mode register, Channels 0-3
D6	Mode register, Channels 4-7
0C	Byte pointer flip/flop, Channels 0-3
D8	Byte pointer flip/flop, Channels 4-7

[a]Channels 4 to 7 are available on the PC/AT but not on the PC or PC/XT.

The mode registers can be programmed with values that determine the functions of the DMA controller. Bits 0 and 1 give the channel number, bit 2 is set when writing to memory, bit 3 is set when reading from memory, bit 4 is set to enable autoinitialization, bit 5 is set to cause decrementing of the address or is cleared to cause incrementing the address during DMA transfers, and bits 6 and 7 determine the mode of operation (only one mode, called single mode, is normally used). The common combinations of these bits are shown in Table 15.5.

Table 15.5.   DMA Mode Register

Value[a]	Function	Channel[b]	Autoinitialize
45	Write	1 or 5	No
49	Read	1 or 5	No
55	Write	1 or 5	Yes
59	Read	1 or 5	Yes
46	Write	2 or 6	No
4A	Read	2 or 6	No
56	Write	2 or 6	Yes
5A	Read	2 or 6	Yes
47	Write	3 or 7	No
4B	Read	3 or 7	No
57	Write	3 or 7	Yes
5B	Read	3 or 7	Yes

[a]Values shown are in hexadecimal.

[b]Note that the same commands are written to each of the two DMA controllers in the PC/AT. All commands to mode register 0B affect channels 0 to 3, while commands to mode register D6 affect channels 4 to 7.

15.2.1.1 Setup of DMA Controller

To use the DMA controller, we typically perform the following steps:

1. Set the appropriate mode register to indicate the function (read or write to memory), channel and whether to autoinitialize.

2. Write any value to the appropriate flip/flop register to signal that the next bytes written to the base address register will be in the order low byte then high byte.

3. Write the low, then high byte of the base address. This is the offset of the address from the page specified to the page register; it is a byte address for channels 0 to 3 and a word address for channels 4 to 7 (i.e., a word address is a byte address shifted one bit to the right).

4. Write the low, then high byte of the byte count. This is a count of the number of bytes to be transferred, which is calculated as one less than twice the number of integer data to be transferred. If you are using channels 4 to 7, then it is the count of the number of words to be transferred rather than the number of bytes.

5.   Write the page number. Don't forget that pages start at zero, like most numbering on the PC.

6.   Set the DMA channel.

15.2.1.2 Other DMA Steps

Once the DMA controller has been programmed, the remaining steps are quite similar to those used with programmed transfers of data (Chapter 9). We first must set up the board for DMA; the details of this clearly depend on the laboratory board being used. Often, it involves only setting a hardware flag. Next, we initiate the DMA transfers; again, this is dependent upon the particular board being used. Once the process is initiated, however, no further control by the PC's CPU is required until the DMA transfers have been completed. Hence, the CPU can perform other, perhaps unrelated, tasks during this time. Finally, once the transfers have been completed, the CPU again resumes complete control of the processing in the system.

Thus, most of the unusual programming for DMA involves programming the DMA controller chip. In fact, other modes of operation of the DMA controller chip are possible (although seldom used); the interested reader is referred to the technical manuals published by any of several chip makers for additional details on other modes of operation of the 8237 chip.

15.2.2 Application to Data Collection Problems

As we have throughout this textbook, we shall now illustrate this process with specific programming examples. In order to do this, we must first understand how to program the data acquisition board for DMA, then discuss the differences between interpreter and compiled BASIC, and then finally show our programs.

We will use the DT2801 board as our example. Operating the DT2801 in DMA mode is not much different than operating it in programmed I/O mode. Hence, you may wish to review Chapter 9, and in particular re-examine Program 9.2. The only real difference in programming the A/D for DMA is that the DMA bit of the Command Register must be set; this is bit 4, and should be set to a one. Also, because the number of bytes transferred is determined by the DMA controller chip, rather than the A/D, the count sent to the A/D is not used by the A/D.

Using the DMA mode with BASICA or similar interpreter BASICs is somewhat difficult. This is because BASICA allocates space for arrays after all of the non-dimensioned variables; thus, the physical location of an array can change as new variables are added to the program. For example, consider the following very simple program:

```
10 DIM A%(100)
20 PRINT VARPTR(A%(1))
30 B!=1
```

40 PRINT VARPTR(A%(1))

The VARPTR function in BASICA returns the address of a variable within BASICA's data space. If this program is run, it prints out two different addresses for the location of A%(1)!

Hence, it is very difficult to tell the DMA controller where an array is located into which it should store the data from the A/D. An alternative solution is to set aside a space in memory above the BASIC program, and then use PEEK statement of BASIC to retrieve the data once it has all been stored there. That is the approach we shall use here. As we shall see, a somewhat different approach is used with compiled BASIC and other languages.

15.2.3 DMA with Interpreter BASIC

Using DMA with Interpreter BASICA is illustrated by Program 15.3, which has been adapted from Program 9.2. We shall describe only differences between the two programs.

The program begins, in line 80, by CLEARing a space for the data to be stored. This limits the amount of space used by BASICA for data to 20,000 bytes; you may wish to change the 20,000 to some other value, depending upon the amount of memory your PC has, the version of DOS, other memory-resident programs, and the size of your BASIC program (in fact, if you have large amounts of memory, you may not need the CLEAR statement at all). To test this, you can use the following program to find out the beginning of BASICA's data space:

```
10 DEF SEG=0
20 J%=PEEK(&H510)
30 K%=PEEK(&H511)
40 ADD!=256!*K%+J%
50 PRINT "Start of BASIC Data Segment = ";ADD!
```

Knowing then the amount of memory you have in your system, you can determine whether it is permissible for BASIC to allocate its normal 64K for its data space, or whether you need to use CLEAR to reserve some of that 64K for your own use.

To prepare for using DMA, we first (in line 360) calculate the number of bytes of data to be transferred. We will actually use a number one less than the byte count, because the DMA controller counts down from that number to -1 as it transfers data.

In the subroutine beginning at line 640, we set up the DT2801. However, we send a dummy count of the data, as described above; in this case, we send a count consisting of 1s in both bytes.

The subroutine beginning at line 1620 is the major new code added to this program. We have set up the program to use channel 1 of the DMA; the other addresses are then as described in Table 15.4 and Table 15.5 for channel 1.

```
10 REM PROGRAM DMA
20 REM LABORATORY AUTOMATION USING THE IBM PC
30 REM Program to demonstrate the use of A/D in DMA mode
40 REM Stores data in memory, then retrieves with PEEK commands
50 REM Uses the DT2801          REQUIRES GRAPHICS MONITOR!
60 REM Run using Interpeter BASIC
70 REM ***********************************************************
80 CLEAR,20000                        'set aside space for data storage
90 DIM VALUE%(1000)
100 REM Define all registers, etc.
110 BASEADD%=&H2EC                     'base address of board
120 COMMAND%=BASEADD%+1                'command register
130 STATUS%=BASEADD%+1                 'status register
140 DATUM%=BASEADD%                    'data register
150 REM ***********************************
160 REM          Main Program
170   SCREEN 0:CLS:NPOINTS%=0
180   GOSUB 270                        'get user input
190   GOSUB 400                        'set up board
200   GOSUB 480                        'set up clock
210   GOSUB 640                        'set up A/D
220   GOSUB 1620                       'set up DMA
230   GOSUB 820                        'display data
240 END
250 REM ***********************************
260 REM Get user input
270   INPUT "From what channel of the A/D do you wish to start? ",SCHANA2D%
280   INPUT "Upon which channel of the A/D do you wish to end? ",FCHANA2D%
290   INPUT "What gain of the A/D do you wish (1,2,4 or 8)? ",ADGAIN%
300   GAINCODE%=LOG(ADGAIN%)/LOG(2)    'convert gains 1,2,4,8 to 0,1,2,3
310   INPUT "At what rate do you wish to collect data, in Hz? ",RATE!
320   WHILE (NPOINTS% < 2) OR (NPOINTS% > 10000)
330       INPUT "For how many seconds do you wish to collect data? ",NSECS!
340       NPOINTS%=NSECS! * RATE!
350   WEND
360   BYTECOUNT%=NPOINTS%*2-1          'byte count for DMA
370 RETURN
380 REM ***********************************
390 REM Set up board
400   OUT COMMAND%,&HF                 'stop board
410   TEMP%=INP(DATUM%)                'clear data register
420   REM wait for command to finish, then clear errors
430   GOSUB 1360
440   OUT COMMAND%,&H1                 'clear the command register
450 RETURN
460 REM ***********************************
470 REM Set up clock
480   GOSUB 1360
490   OUT COMMAND%,&H3                 'write SET CLOCK PERIOD command
500   TICKS!=400000!/RATE!             'clock is 400 kHz
510   IF TICKS! < 3 OR TICKS!>65535! THEN PRINT "Illegal clock rate":GOSUB 1190
520   CLOCKHIGH%=INT(TICKS!/256)       'calculate high, low bytes of clock value
530   CLOCKLOW%=TICKS!-CLOCKHIGH%*256! 
540   GOSUB 1330
550   OUT DATUM%,CLOCKLOW%             'output low byte
560   GOSUB 1330
570   OUT DATUM%,CLOCKHIGH%            'output high byte
580   GOSUB 1360
590   ERRORCHECK%=INP(STATUS%)         'check for error
600   IF (ERRORCHECK% AND &H80) THEN PRINT "Error setting clock":GOSUB 1190
610 RETURN
620 REM ***********************************
630 REM Set up A/D for DMA mode
640   GOSUB 1360                       'wait until ready for command
650   OUT COMMAND%,&HD                 'write SET A/D PARAMETERS command
660   GOSUB 1330
670   OUT DATUM%,GAINCODE%             'set gain
```

Program 15.3 (Part 1 of 3).   DMA from Interpreted BASIC.

```
680    GOSUB 1330
690    OUT DATUM%,SCHANA2D%              'set starting a/d channel
700    GOSUB 1330
710    OUT DATUM%,FCHANA2D%             'set ending a/d channel
720    GOSUB 1330
730    OUT DATUM%,1                     'do dummy writes of low, high bytes of count
740    GOSUB 1330
750    OUT DATUM%,1
760    GOSUB 1360
770    ERRORCHECK%=INP(STATUS%)         'check for error
780    IF (ERRORCHECK% AND &H80) THEN PRINT "Error setting A/D":GOSUB 1190
790 RETURN
800 REM ***********************************
810 REM Get data from A/D
820    SCREEN 2:CLS:KEY OFF:PRINT "Beginning data collection...please wait"
830    WINDOW (0,0)-(NPOINTS%,4095)     'set up for real-time plot
840    DEF SEG =&H4000                  'data are stored in this segment by DMA
850    COUNT%=NPOINTS%*2:NEXTPOINT%=1:P%=0
860    GOSUB 1140                       'start A/D in DMA mode
870    WHILE COUNT%>0                   'wait for data to be ready
880       OUT DMACLEAR%,0               'make sure get low byte, then high byte
890       J%=INP(DMACOUNT1%)            'get current byte count from DMA controller
900       K%=INP(DMACOUNT1%)
910       COUNT!=K%*256!+J%             'convert to signed integer value
920       IF COUNT! > 32767 THEN COUNT%=COUNT!-65536! ELSE COUNT%=COUNT!
930       COUNT%=(COUNT%+2)\2           'make sure on an even byte
940       L%=NPOINTS%-COUNT%            'controller counts down, so convert
950       FOR X%=NEXTPOINT% TO L%       'plot all points collected so far
960          MEM%=(X%-1)*2
970          M%=PEEK(MEM%)              'get point from memory
980          N%=PEEK(MEM%+1)
990          Y!=256!*N%+M%              'convert to signed integer
1000            IF Y! > 32767 THEN VALUE%(X%)=Y!-65536! ELSE VALUE%(X%)=Y!
1010            PSET (X%,VALUE%(X%))    'plot the point
1020       NEXT X%
1030       IF L%>=NEXTPOINT% THEN NEXTPOINT%=L%+1 'keep track of where finished
1040    WEND
1050    PRINT "Finished acquiring data."
1060    DEF SEG
1070    GOSUB 1360
1080    CHECK%=INP(STATUS%)
1090    IF (CHECK% AND &H80) THEN PRINT "Error acquiring data": GOSUB 1190
1100    GOSUB 1410                      'plot data
1110 RETURN
1120 REM ***********************************
1130 REM start A/D taking data
1140    GOSUB 1360
1150    OUT COMMAND%,(&HE+&H10)         'write READ A/D command (with DMA mode bit)
1160 RETURN
1170 REM ***********************************
1180 REM Determine source of error
1190    OUT COMMAND%,&HF                'stop the board
1200    TEMP%=INP(DATUM%)
1210    GOSUB 1360
1220    OUT COMMAND%,&H2                'send READ ERROR REGISTER command
1230    GOSUB 1330
1240    VALUE%=INP(DATUM%)
1250    GOSUB 1330
1260    PRINT "Value of error register low byte is: ";VALUE%
1270    VALUE2%=INP(DATUM%)
1280    PRINT "Value of error register high byte is: ";VALUE2%
1290    IF VALUE2%=4 THEN PRINT "You have tried to collect data at too fast"
1300 STOP
1310 REM ***********************************
1320 REM Check to see if ready to set a register
1330    WAIT STATUS%,&H2,&H2            'wait for bit 1 to be reset
```

Program 15.3 (Part 2 of 3). DMA from Interpreted BASIC.

```
1340 RETURN
1350 REM Check to see if ready to send another command
1360    WAIT STATUS%,&H2,&H2          'wait for bit 1 to be reset
1370    WAIT STATUS%,&H4              'wait for bit 2 to be set
1380 RETURN
1390 REM *********************************
1400 REM Display normalized data
1410    SCREEN 1:COLOR 9,0
1420    VIEW:CLS:VIEW (50,10)-(270,180),2,B
1430    YMIN%=VALUE%(1)
1440    YMAX%=VALUE%(1)
1450    FOR I%=2 TO NPOINTS%
1460        IF VALUE%(I%)<YMIN% THEN YMIN%=VALUE%(I%)
1470        IF VALUE%(I%)>YMAX% THEN YMAX%=VALUE%(I%)
1480    NEXT I%
1490    WINDOW (1,YMIN%)-(NPOINTS%,YMAX%)
1500    NUMCHAN%=FCHANA2D%-SCHANA2D%+1  'calculate number of channels
1510    FOR J%=1 TO NUMCHAN%
1520        PSET (1,VALUE%(J%))
1530        FOR I%=J% TO NPOINTS% STEP NUMCHAN%
1540            LINE -(I%-J%+1,VALUE%(I%)),3
1550        NEXT I%
1560    NEXT J%
1570    LOCATE 24,1:PRINT "Minimum= ";YMIN%; "    Maximum= ";YMAX%;
1580    LOCATE 1,1
1590 RETURN
1600 REM *********************************
1610 REM Set up DMA channel 1 for use with DT2801
1620    DMABASE1%   =    2            'address of DMA base address reg., ch. 1
1630    DMACOUNT1%  =    3            'address of DMA byte count reg., ch. 1
1640    DMAMASK%    =    10           'address of DMA mask register
1650    DMAMODE%    =    11           'address of DMA mode register
1660    DMACLEAR%   =    12           'address to clear DMA pointer flip/flop
1670    DMAPAGE%    =    131          'address of DMA page select, channel 1
1680    DMACHAN%    =    1            'use DMA channel number 1
1690    OUT DMAMODE%,&H45             'set dma mode=read a/d,ch.1, no autoinit
1700    OUT DMACLEAR%,0              'clear flip-flop
1710    OUT DMABASE1%,0             'select offset address &H0000
1720    OUT DMABASE1%,&H0
1730    OUT DMAPAGE%,4               'set dma page count = 0 for page 4
1740    LOWCOUNT%=BYTECOUNT% AND &HFF
1750    HIGHCOUNT%=(BYTECOUNT% AND &HFF00)/256
1760    OUT DMACOUNT1%,LOWCOUNT%
1770    OUT DMACOUNT1%,HIGHCOUNT%     'send count to dma count register
1780    OUT DMAMASK%,DMACHAN%         'set dma channel
1790 RETURN
```

Program 15.3 (Part 3 of 3).   DMA from Interpreted BASIC.

We begin by setting the mode register to specify that we are setting channel 1 to read data from the DT2801 into the PC's memory in line 1690. We then need to write a two-byte address to the base address register. Before doing this, we write any value to clear the flip-flop, in line 1700. This is done to ensure that the DMA controller interprets the two bytes as being in the order low-byte, then high byte. In this specific program, we wish to have the data stored beginning at location 40000 hex (262,144 decimal). Hence, in lines 1710 to 1730 we send the offset of 0000 to the base address register, and the page 4 value to the page register. We have picked this location arbitrarily, high enough to avoid the space used by BASIC. (If you have other memory-resident programs, or if you have less than 640K of system memory, then you should use the short program above to find out where BASIC's data space begins, CLEAR a space above that, and use the space that has been CLEARed for storing the data). Note that if we were using DMA channels 4 to 7, we would use word addresses rather than the byte addresses used here for channels 0 to 3.

Next, we send the byte count we have previously calculated. In lines 1740 to 1770, we first take the 2-byte integer count and split the low and high bytes of that count. We then send it to the appropriate DMA register.

The last operation is to set the channel to be used by sending it to the mask register of the DMA controller (line 1780).

Finally, we are ready to collect data. To start the data acquisition board collecting and transferring data, in line 1150 we set the DMA bit in the command register of the DT2801, and send the READ A/D command. Once this command is sent, the DMA transfer begins and occurs as fast as the data are collected, which is determined by the on-board clock of the DT2801.

After data collection has been initiated, we have two choices. First, we can simply test to see when the last datum has been transferred. To do this, we could use a program segment such as the following

```
FLAG%=1
WHILE FLAG%=1              'wait for data to be ready
   FLAG%=INP(STATUS%)
   FLAG%=FLAG% AND &H2         'check bit 1
WEND
```

Alternatively, we can do something useful while the data are being transferred. This is normally preferable, unless we are doing very high speed data transfers; it is the approach we have used in Program 15.3. In this case, we have chosen to perform plotting of the data while they are being received.

In order to plot the data, we must know when they are received in memory. In lines 890 to 930, we get the count of bytes of data remaining to be transferred by the DMA controller. Because the data are transferred one byte at a time, but our data are 2-byte integers, we essentially wish to get a count of the number of complete 2-byte words transferred; this is done in line 930. We then convert this count of words yet to be transferred to a count of the number of words transferred so far, as shown in line 940. Lines 950 to 1020 then get all of the data that has been transferred but not yet plotted, and read it into a BASIC array using the PEEK statement. Notice that line 840 had set the BASIC data segment register to &H4000; this means that PEEK statements get data from addresses beginning at &H40000, which is where we had told the DMA controller to store the data. (If you have not used segment registers, you need to know that you must multiply the value of the segment register by 16 (10 hex) to get the segment address; hence, a register of &H4000 refers to address &H40000.) Notice also that when we read a two-byte address that is a *signed* integer, we must make sure we handle negative numbers properly.

Once the data have been collected, we can then process them further; in the case shown here, we just replot them in color.

15.2.4 DMA with Compiled BASIC

Before leaving the topic of DMA, we need to show you how to use DMA with compiled BASIC. In essence, this process is very similar to that with the BASICA interpreter; in fact, you can compile Program 15.3 and run it with no changes.

However, compiled BASIC generally uses arrays that are fixed in their memory position at compile time (unlike interpreter BASICA, which uses variable positions for arrays, as we have described). Hence, we can take advantage of this and have the DMA controller put the data directly into a BASIC array, instead of having to PEEK the data and transfer it to the array. In order to do this, we need to know the address of the first element of the array. We can use the BASIC **VARPTR** function to get the address of the first element of the array. However, this address is the address within the BASIC data space; we also need to know the starting address of the data space to calculate the entire address. Some dialects of BASIC provide a special command for this purpose; however, we have included an assembly routine for this purpose. In fact, the beginning location of the BASIC data space is nothing more than the address pointed to by the **DS** register of the 8088 family of CPUs. Hence, in Program 15.4, line 1620, we call the assembly routine, which is shown in Program 15.5, to get the value of the DS register. As noted above, to convert this to the address of the data segment, we multiply the segment register value by 16 in line 1650, and then add to that value the address of the array within the data space. The result is a 20-bit address. In lines 1660 to 1690, we then test to make sure that the data we will be storing will not cross from one 64K segment to another, because the DMA controller only can write data to memory in one 64K segment at a time. Much as in Program 15.3, we then output both the page and offset addresses to the DMA controller, initiate data collection, and plot the data as it is being collected. Note that now, in line 950, we can get the data just by looking at the data array where we told the DMA controller to store the data.

EXERCISES

1. Modify Program 15.1 to perform some other task of interest to you other than plotting the data.

2. Modify Program 15.1 and Program 15.2 to collect data from more than one channel. Display data from each channel in a different color in the real-time plot.

3. Modify Program 15.2 to increase its speed by removing real-time plotting and any other features you think are unnecessary.

4. Program 15.2 contains features that slow the program below the maximum rate obtainable on the IBM DACA. Consult the IBM Technical Reference

```
10 REM PROGRAM DMA_DIRECT
20 REM LABORATORY AUTOMATION USING THE IBM PC
30 REM Program to demonstrate the use of A/D in DMA mode with
40 REM    direct storage of data in a BASIC data array
50 REM Compile using:        BASCOM PROG154;
60 REM Link using:           LINK PROG154+PROG155,,,BASRUN20
70 REM Uses the DT2801                      REQUIRES GRAPHICS MONITOR!
80 REM ****************************************************************
90 DIM VALUE%(1000)
100 REM Define all registers, etc.
110 BASEADD%=&H2EC                    'base address of board
120 COMMAND%=BASEADD%+1               'command register
130 STATUS%=BASEADD%+1                'status register
140 DATUM%=BASEADD%                   'data register
150 REM ***********************************
160 REM            Main Program
170   SCREEN 0:CLS:NPOINTS%=0
180   GOSUB 270                       'get user input
190   GOSUB 400                       'set up board
200   GOSUB 480                       'set up clock
210   GOSUB 640                       'set up A/D
220   GOSUB 1550                      'set up DMA
230   GOSUB 820                       'display data
240 END
250 REM ***********************************
260 REM Get user input
270   INPUT "From what channel of the A/D do you wish to start? ",SCHANA2D%
280   INPUT "Upon which channel of the A/D do you wish to end? ",FCHANA2D%
290   INPUT "What gain of the A/D do you wish (1,2,4 or 8)? ",ADGAIN%
300   GAINCODE%=LOG(ADGAIN%)/LOG(2)   'convert gains 1,2,4,8 to 0,1,2,3
310   INPUT "At what rate do you wish to collect data, in Hz? ",RATE!
320   WHILE (NPOINTS% < 2) OR (NPOINTS% > 1000)
330       INPUT "For how many seconds do you wish to collect data? ",NSECS!
340       NPOINTS%=NSECS! * RATE!
350   WEND
360   BYTECOUNT%=NPOINTS%*2-1         'byte count for DMA
370 RETURN
380 REM ***********************************
390 REM Set up board
400   OUT COMMAND%,&HF                'stop board
410   TEMP%=INP(DATUM%)               'clear data register
420   REM wait for command to finish, then clear errors
430   GOSUB 1290
440   OUT COMMAND%,&H1                'clear the command register
450 RETURN
460 REM ***********************************
470 REM Set up clock
480   GOSUB 1290
490   OUT COMMAND%,&H3                'write SET CLOCK PERIOD command
500   TICKS!=400000!/RATE!            'clock is 400 kHz
510   IF TICKS! < 3 OR TICKS! > 65535! THEN PRINT "Illegal rate.":GOSUB 1120
520   CLOCKHIGH%=INT(TICKS!/256)      'calculate high, low bytes of clock value
530   CLOCKLOW%=TICKS!-CLOCKHIGH%*256!
540   GOSUB 1260
550   OUT DATUM%,CLOCKLOW%            'output low byte
560   GOSUB 1260
570   OUT DATUM%,CLOCKHIGH%           'output high byte
580   GOSUB 1290
590   ERRORCHECK%=INP(STATUS%)        'check for error
600   IF (ERRORCHECK% AND &H80) THEN PRINT "Error setting clock":GOSUB 1120
610 RETURN
620 REM ***********************************
630 REM Set up A/D for DMA mode
640   GOSUB 1290                      'wait until ready for command
650   OUT COMMAND%,&HD                'write SET A/D PARAMETERS command
660   GOSUB 1260
670   OUT DATUM%,GAINCODE%            'set gain
```

Program 15.4 (Part 1 of 3).   DMA from Compiled BASIC.

```
680   GOSUB 1260
690   OUT DATUM%,SCHANA2D%              'set starting a/d channel
700   GOSUB 1260
710   OUT DATUM%,FCHANA2D%             'set ending a/d channel
720   GOSUB 1260
730   OUT DATUM%,1                     'do dummy writes of low, hi bytes of count
740   GOSUB 1260
750   OUT DATUM%,1
760   GOSUB 1290
770   ERRORCHECK%=INP(STATUS%)         'check for error
780   IF (ERRORCHECK% AND &H80) THEN PRINT "Error setting A/D":GOSUB 1120
790   RETURN
800   REM **********************************
810   REM Get data from A/D
820   SCREEN 2:CLS:PRINT "Beginning data collection...please wait"
830   WINDOW (0,0)-(NPOINTS%,4095)  'set up for real-time plot
840   COUNT%=NPOINTS%:NEXTPOINT%=1:P%=0
850   GOSUB 1070                       'start A/D in DMA mode
860   WHILE COUNT%>0                   'wait for data to be ready
870      OUT DMACLEAR%,0               'make sure get low byte, then high byte
880      J%=INP(DMACOUNT1%)            'get current byte count from DMA controller
890      K%=INP(DMACOUNT1%)
900      COUNT!=K%*256!+J%             'convert to signed integer value
910      IF COUNT! > 32767 THEN COUNT%=COUNT!-65536! ELSE COUNT%=COUNT!
920      COUNT%=(COUNT%+2)\2           'change to word count (truncate remainder)
930      L%=NPOINTS%-COUNT%            'controller counts down, so convert
940      FOR X%=NEXTPOINT% TO L%       'plot all points collected so far
950            PSET (X%,VALUE%(X%))    'plot the point
960      NEXT X%
970      IF L% >= NEXTPOINT% THEN NEXTPOINT%=L%+1 'keep track of where finish
980   WEND
990   PRINT "Finished acquiring data."
1000  GOSUB 1290
1010  CHECK%=INP(STATUS%)
1020  IF (CHECK% AND &H80) THEN PRINT "Error acquiring data": GOSUB 1120
1030  GOSUB 1340                       'plot data
1040  RETURN
1050  REM **********************************
1060  REM Start A/D taking data
1070  GOSUB 1290
1080  OUT COMMAND%,(&HE+&H10)          'write READ A/D command (with DMA mode bit)
1090  RETURN
1100  REM **********************************
1110  REM Determine source of error
1120  OUT COMMAND%,&HF                 'stop the board
1130  TEMP%=INP(DATUM%)
1140  GOSUB 1290
1150  OUT COMMAND%,&H2                 'send READ ERROR REGISTER command
1160  GOSUB 1260
1170  VALUE%=INP(DATUM%)
1180  GOSUB 1260
1190  PRINT "Value of error register low byte is: ";VALUE%
1200  VALUE2%=INP(DATUM%)
1210  PRINT "Value of error register high byte is: ";VALUE2%
1220  IF VALUE2%=4 THEN PRINT "You have tried to collect data at too fast"
1230  STOP
1240  REM **********************************
1250  REM Check to see if ready to set a register
1260  WAIT STATUS%,&H2,&H2             'wait for bit 1 to be reset
1270  RETURN
1280  REM Check to see if ready to send another command
1290  WAIT STATUS%,&H2,&H2             'wait for bit 1 to be reset
1300  WAIT STATUS%,&H4                 'wait for bit 2 to be set
1310  RETURN
1320  REM **********************************
1330  REM Display normalized data.
```

Program 15.4 (Part 2 of 3).   DMA from Compiled BASIC.

```
1340    SCREEN 1:COLOR 9,0
1350    VIEW:CLS:VIEW (50,10)-(270,180),2,B
1360    YMIN%=VALUE%(1)
1370    YMAX%=VALUE%(1)
1380    FOR I%=2 TO NPOINTS%
1390        IF VALUE%(I%)<YMIN% THEN YMIN%=VALUE%(I%)
1400        IF VALUE%(I%)>YMAX% THEN YMAX%=VALUE%(I%)
1410    NEXT I%
1420    WINDOW (1,YMIN%)-(NPOINTS%,YMAX%)
1430    NUMCHAN%=FCHANA2D%-SCHANA2D%+1    'calculate number of channels
1440    FOR J%=1 TO NUMCHAN%
1450        PSET (1,VALUE%(J%))
1460        FOR I%=J% TO NPOINTS% STEP NUMCHAN%
1470            LINE -(I%-J%+1,VALUE%(I%)),3
1480        NEXT I%
1490    NEXT J%
1500    LOCATE 24,1:PRINT "Minimum= ";YMIN%; "     Maximum= ";YMAX%;
1510    Y$=INKEY$:IF Y$=""THEN 1510
1520 RETURN
1530 REM *********************************
1540 REM Set up DMA controller
1550    DMABASE1%    =    2            'address of DMA base address reg., ch. 1
1560    DMACOUNT1%   =    3            'address of DMA byte count reg., ch. 1
1570    DMAMASK%     =    10           'address of DMA mask register
1580    DMAMODE%     =    11           'address of DMA mode register
1590    DMACLEAR%    =    12           'address to clear DMA pointer flip/flop
1600    DMAPAGE%     =    131          'address of DMA page select, channel 1
1610    DMACHAN%     =    1            'use DMA channel number 1
1620    CALL GETSEG(DS%)              'get value of basic's ds register
1630    OUT DMAMODE%,&H45             'set DMA mode=read a/d,ch.1, no autoinit
1640    OUT DMACLEAR%,0               'clear flip-flop
1650    DS!=DS%*16!+VARPTR(VALUE%(1)) 'calculate 20-bit address of value% array
1660    TESTDS!=DS!+BYTECOUNT%
1670    TEST1%=INT(DS!/65536!)        'get segment of ds!
1680    TEST2%=INT(TESTDS!/65536!)    'get segment of testds!
1690    IF TEST2%<>TEST1% THEN PRINT "DATA WILL CROSS SEGMENT BOUNDARY. " :STOP
1700    LOWADD!=DS!-65536!*TEST1%     'get low word of address
1710    IF LOWADD! > 32767 THEN LOWADD%=LOWADD!-65536! ELSE LOWADD%=LOWADD!
1720    LOWBYTE%=LOWADD% AND &HFF
1730    HIGHBYTE%=(LOWADD% AND &HFF00)/256
1740    IF HIGHBYTE% < 0 THEN HIGHBYTE%=256+HIGHBYTE%
1750    OUT DMABASE1%,LOWBYTE%
1760    OUT DMABASE1%,HIGHBYTE%
1770    LOWCOUNT%=BYTECOUNT% AND &HFF
1780    HIGHCOUNT%=(BYTECOUNT% AND &HFF00)/256
1790    OUT DMACOUNT1%,LOWCOUNT%
1800    OUT DMACOUNT1%,HIGHCOUNT%     'send count to DMA count register
1810    OUT DMAPAGE%,TEST1%           'set DMA page count
1820    OUT DMAMASK%,DMACHAN%         'set DMA channel
1830 RETURN
```

Program 15.4 (Part 3 of 3).   DMA from Compiled BASIC.

Manual for detailed programming information and write a new routine designed for maximum speed.

5.   How easy would it be to write a program that operates two devices synchronously (e.g., outputting an I/O pulse after each A/D sampling) in either the DMA or interrupt modes, particularly if there must be a delay between the operation of the two devices?

```
;PROGRAM SEGMENT
;LABORATORY AUTOMATION USING THE IBM PC
;Returns the data segment used by BASIC
;**************************************************************************
;          Assumes call of form CALL GETSEG(A%)
;                   where A% is the data segment value returned by the routine
;
;set up structure for easy reference to parameters used in CALL
;
FRAME           STRUC
OLDBP           DW      ?                   ;old BP
OLDIP           DW      ?                   ;IP for return
OLDCS           DW      ?                   ;CS for return
;now set aside two bytes for parameter
APTR            DW      ?                   ;pointer to A%
FRAME           ENDS
PARAMSIZE       EQU     OFFSET APTR - OFFSET OLDCS   ;bytes in parameters
;
CSEG    SEGMENT  WORD PUBLIC 'CODE'
        ASSUME CS:CSEG,  DS:NOTHING,  ES:NOTHING
;
;define routine as a public procedure
        PUBLIC  GETSEG
GETSEG PROC     FAR
        INT     03H                         ;remove comment if you wish to DEBUG
        PUSH    BP                          ;save BP
        MOV     BP,SP                       ;set base of parameter list
        MOV     DI,[BP].APTR                ;get address of A% from BASIC
        MOV     AX,DS                       ;get value of DS
        MOV     WORD PTR [DI],AX            ;store the value
        POP     BP
        RET     PARAMSIZE                   ;number of bytes used in parameters
GETSEG ENDP
CSEG    ENDS
        END
```

Program 15.5.   Assembly Routine to Get BASIC's Data Segment Register

6.   Could you operate two devices, one using DMA and one using interrupts, simultaneously?

7.   Modify the programs in this chapter to support the A/D board of your choice.

16

CONCLUDING REMARKS

If you have waded through the last few chapters, you are probably ready for virtually any laboratory automation task that may befall you. We have a few final hints for you, however. First, think before doing (good advise in any experimental science). Careful hardware selection, clear planning of software design, and thoughtful debugging of software can greatly decrease the amount of time and effort required for automation. Watch someone who is excellent at automation; they rarely will start adjusting hardware or modifying software until they have tried to understand the problem first.

Similarly, if you are planning to automate a number of experiments, we suggest that you develop a strategy. Our strategy, evolved over a period of years, is to develop *tools* for automation. That is, where possible, we try to develop software that is highly modular and which can be rapidly adapted to a variety of experiments. Instead of approaching each experiment as a new experience in automation, we try to develop programs that are general-purpose and then modify them to suit a particular experimental problem.

Finally, we suggest that you try a variety of new ideas in your own work. One of the things that makes automation fun and exciting is the chance to try new approaches to problems. Certainly, the advent of the PC has markedly changed the way in which we do automation; as we move into an era of more powerful software, new workstations, distributed processing, local area networks, windowing, and high-resolution displays, automation will be changed even further. It's a fun time to be in laboratory automation!

PART FOUR

APPENDICES

A

BINARY AND
HEXADECIMAL ARITHMETIC

A.1 BINARY ARITHMETIC

Because computers "think" in binary, or base two, it is often useful to be able to do simple binary arithmetic. Binary arithmetic is used by computers because it is easy to build reliable electronic devices if they have only two states, low and high (or off and on). Often, the on and off states are represented as 1 and 0, respectively. The numbers 0 and 1 are the only values permitted in the binary system, but it is nonetheless possible to represent any size number in binary, including negative numbers and fractions. Let's see how.

Table A.1 shows a set of binary numbers and their decimal system equivalents. Notice that there is a pattern to the binary numbers. Just as in the decimal system, each column represents a different power of the base of the number system. In the decimal system, starting from the right, we have the ones column, the tens column, the hundreds column, and so forth. For example, 265 can be rewritten

$$(2 \times 100) + (6 \times 10) + (5 \times 1)$$

However, a more useful way of writing this for our purposes is

$$(2 \times 10^2) + (6 \times 10^1) + (5 \times 10^0)$$

Note the powers of ten, because the decimal system is a **base 10** system.

Table   A.1.   **Binary-Decimal-Hexadecimal Conversion**

Base 2	Base 10	Base 16
0000	0	0
0001	1	1
0010	2	2
0011	3	3
0100	4	4
0101	5	5
0110	6	6
0111	7	7
1000	8	8
1001	9	9
1010	10	A
1011	11	B
1100	12	C
1101	13	D
1110	14	E
1111	15	F

Similarly, binary is a **base 2** system, so each column represents a different power of 2. Thus, for example, the number 110 in the binary system (6 in the decimal system) can be represented as

$$(1 \times 2^2) + (1 \times 2^1) + (0 \times 2^0)$$

(Recall that 2^0 is equal to 1.) However, instead of calling the columns of a binary number "digits," the term **bits** is used. The bits of a binary number are often designated by the power of two that they represent; thus, for example, the 0 in 110 (binary) is called bit 0 or the least significant bit (LSB), while the left-most bit is bit 2, or the most significant bit (MSB).

Binary numbers can be added and subtracted in a manner similar to that used with decimal numbers. As in the decimal system, we must carry or borrow when necessary. For example,

1100	in binary is	12	in decimal.
+0101		+5
10001		17

Notice that $1 + 1 = 0$, carry 1, in binary, just as $9 + 1 = 0$, carry 1 in decimal.

Difficulties arise, however, when we try to write a negative number in binary, because the computer does not have an electronic state corresponding to negative (remember, only "on" and "off" are permitted). Hence, what is done in most computers is to represent negative numbers by designating the left-most (most significant) bit as the one that gives the sign. In the IBM PC (and other 16-bit computers), integers are normally represented by sixteen binary bits, the left-most of which gives the sign. If the left-most bit is a 1, then the number is negative, and if the left-most bit is a 0, then the number is positive.

That would be easy except that computers usually use **two's complement** arithmetic to represent negative numbers. Hence a number equal to the decimal value -1 is written in binary as 1111111111111111, not 1000000000000001 as you might expect.

The reason for this seeming oddity is that it makes addition and subtraction of numbers easier (for the computer, not for us). This is seen by listing a series of numbers near zero:

Decimal Value	Two's Complement Value
-3	1111 1111 1111 1101
-2	1111 1111 1111 1110
-1	1111 1111 1111 1111
0	0000 0000 0000 0000
1	0000 0000 0000 0001

Notice that each number in this list can be obtained from the one above it by adding 0000 0000 0000 0001 (the bits are written in groups of four for easier reading). The only apparent exception is that when 1 is added to -1, the result should be

$$
\begin{array}{r}
1111\ 1111\ 1111\ 1111 \\
+\ 0000\ 0000\ 0000\ 0001 \\
\hline
1\ 0000\ 0000\ 0000\ 0000
\end{array}
$$

However, the extra 1 (the "carry bit") is normally ignored when using signed numbers because it is outside the 16 bits used by the computer to store the result of mathematical operations.

The two's complement of a number can be easily calculated. To do so we simply change each bit in the number to the opposite value (i.e., find its complement) and add 1 to the result. For example, 4 in binary is 0000 0000 0000 0100. The complement of this number is 1111 1111 1111 1011. Adding one to the complement gives the result 1111 1111 1111 1100, which is how -4 (decimal) would be represented.

Thus, the advantage of two's-complement arithmetic is that the same hardware is used to add or subtract numbers. To subtract any number b from a, the two's com-

plement of *b* is added to *a*. This simplifies the computer hardware design significantly.

A.2 HEXADECIMAL ARITHMETIC

Although all digital computers use binary arithmetic, binary numbers are very cum-
bersome for humans. Hence, when using computers like the IBM PC where most
arithmetic is done with numbers at least 16 bits in size, we often use numbers
expressed in hexadecimal (base 16). Hexadecimal numbers are easily converted to
either base 10 or base 2 using Table A.1.

Notice that in each system the number of symbols is the same as the base; i.e.,
there are two symbols (0, 1) in the binary system, ten symbols (0 to 9) in the decimal
system and 16 symbols (0 to F) in the hexadecimal system.

The conversion from binary to hexadecimal (or hex, as it is usually called), is
particularly easy when the binary numbers are written in units of 4 bits:
0111 0010 1110 1100 is thus 72EC, which can be converted to decimal as:

$$(7 \times 16^3) + (2 \times 16^2) + (14 \times 16^1) + (12 \times 16^0)$$

which is

$$(15 \times 4096) + (2 \times 256) + (14 \times 16) + (12 \times 1) = 29,420$$

If you are in front of your computer, it is easy to convert hex numbers to
decimal with BASICA; simply type:

> BASICA   (to load BASICA into memory)
> PRINT &H72EC

and the computer will print out the decimal equivalent of hex 72EC. (Notice that
BASICA uses &H as a prefix in front of hex numbers.) To convert decimal numbers
to hex, type:

> PRINT HEX$(29420)

and the hex equivalent of 29,420 will be printed.

We use hexadecimal notation for many numbers in this book, particularly those
that are to be sent out from the computer to another instrument. However, don't
forget that any integer number above &H7FFF (32,767 decimal) is actually a negative number, because the left-most bit is the sign bit. Thus, -1 in decimal is represented as FFFF in hex, because it is 1111 1111 1111 1111 in binary.

If numbers above 7FFF (hex) represent negative numbers, how then do we represent numbers above 32,767, without having them become negative numbers? The
answer is that two methods are generally used: either to use longer "words" to represent the numbers or to use a floating point notation. Many languages support the use

of 32-bit integers but **IBM BASICA** does not. In any case the use of 32-bit integer arithmetic is considerably slower on the IBM PC than 16-bit integer arithmetic.

The usual method for representing numbers other than those integers that fall in the range &H8000 to &H7FFF (-32,768 to 32,767) is to use a **floating point** number where some of the bits are used to represent the exponent and some are used to represent the mantissa. Fortunately, we do not need to know the details of this because almost all of the math used in this book is integer. It shall suffice to say that floating point arithmetic is generally much slower than integer arithmetic (often by one to two orders of magnitude), and that 16-bit integer arithmetic should be used whenever possible. Don't hesitate to use floating point arithmetic for its greater range and greater precision, when that is what is required, however. Languages (including some other dialects of **BASIC**) that use the math coprocessor may give considerably better performance on computations that involve extensive use of floating point arithmetic.

B

SCIENTIFIC ROUTINES DISKETTE

This text contains printed copies of a number of programs designed to help you understand the text material. To make it easier for you, we have included a diskette that has all of the programs in the text. The diskette is double-sided-double-density, which means that it should be able to be read by a standard 5.25 in. floppy diskette drive on a PC, PC/XT, or PC/AT or compatible system using DOS.

B.1 PROGRAMS FROM TEXTBOOK

Listed below are the programs on the diskette that are printed earlier in this text.

B.1.1 Digital to Analog Converter

Name	Page	Title
PROG21.BAS	Page 20	Output One D/A Voltage
PROG22.BAS	Page 23	Output a Range of D/A Voltages
PROG23.BAS	Page 26	Waveform Output with D/A's

B.1.2 Analog to Digital Converter

B.1.3 Noise Detection and Reduction Techniques

B.1.4 Digital Input and Output

B.1.5 IEEE-488

B.1.6 Serial Communications

B.1.7 Timers and Counters

B.1.8 Coordinated Data Collection and Control

B.1.9 User- and Programmer-Friendly Software

B.1.10 Peak Detection

B.1.11 Digital Image Processing

B.1.12 Using Assembly Language for High Speed

PROG141.BAS	Page 250	Calling an Assembly Language Routine from BASIC
PROG142.ASM	Page 250	Assembly Language Program Called from BASIC
PROG143.BAS	Page 253	BASIC Program for Data Collection Using Assembly Language
PROG144.ASM	Page 256	Assembly Language Program for Data Collection
PROG145.BAS	Page 256	Real-Time Signal Averaging Program
PROG146.ASM	Page 256	Assembly Language Program for Signal Averaging

B.1.13 Interrupts and Direct Memory Access

PROG151.BAS	Page 273	Using an Interrupt-Driven Subroutine
PROG152.ASM	Page 273	Assembly-Language Interrupt Routine
PROG153.BAS	Page 288	DMA from Interpreted BASIC
PROG154.BAS	Page 293	DMA from Compiled BASIC
PROG155.ASM	Page 293	Assembly Routine to Get BASIC's Data Segment Register

B.1.14 Corrections Added in Proof

After the diskette was printed, the following changes were made. The programs printed in the book reflect those changes; hence, the user should correct the programs on the diskette to be the same as those printed in the text.

1.   Program 7.2 (page 130). Lines 130-160 have been changed.
2.   Program 7.3 (page 132). Line 90 has been changed.
3.   Program 10.8 (page 183). Line 470 has been deleted.

B.2 ADDITIONAL PROGRAMS

Listed below are programs included in the Scientific Routines Diskette but not
included in the text. These programs all utilize the IBM Data Acquisition and
Control Adapter, and are analogous to programs in the text utilizing the DT2801
card. In addition, programming information used in designing the programs is
included in tabular form.

Name on Disk	Comparison	Title
PROGF1.BAS	PROG21.BAS	Output One D/A Voltage
PROGF2.BAS	PROG22.BAS	Output a Range of D/A Voltages
PROGF3.BAS	PROG23.BAS	Waveform Output with D/A's
PROGF4.BAS	PROG31.BAS	Collect One Datum
PROGF5.BAS	PROG33.BAS	Collect Multiple Data
PROGF6.BAS	PROG51.BAS	Testing Digital I/O
PROGF7.BAS	PROG52.BAS	Digital I/O for LED, Switch and Relay
PROGF8.BAS	PROG91.BAS	Scanning X-Y and Data Collection
PROGF9.BAS	PROG92.BAS	Coordinated Data Collection and Control
PROGF10.BAS	PROG143.BAS	BASIC Program for Data Collection Using Assembly Language
PROGF11.ASM	PROG144.BAS	Assembly Language Program for Data Collection
PROGF12.BAS	PROG145.BAS	Real-Time Signal Averaging Program
PROGF13.ASM	PROG146.ASM	Assembly Language Program for Signal Averaging

B.3 TECHNICAL INFORMATION

Included below is the information needed to understand the programs for the DACA.
Each table shows only those registers and bit meanings used in programming a specific device. The addresses are for Adapter 0. If you have more than one DACA
card in your IBM PC, then each DACA must have a different adapter number. Add
400 (hex) to the address for each higher numbered adapter; for example, the D/A
control register of Adapter 1 is referenced at address 16E2 (hex).

Table B.1.   Registers on the IBM DACA used by the D/A

Register	Name	Address	Meaning of Bits
1	D/A Control	12E2	0-7 are unused 8-15=channel number
3	D/A Datum	32E2	0-11=datum 12-15 are 0's
12	Device Number	C2E2	0-7=device number Set to 9 for D/A

Table B.2.   Registers on the IBM DACA used by the A/D

Register	Name	Address	Meaning of Bits
0	A/D Control	02E2	0=start conversion 8-15=channel number
0	A/D Status	02E2	0=A/D is busy still
2	A/D Datum	22E2	0-11=datum 12-15 are 0's
12	Device Number	C2E2	0-7=device number Set to 9 for A/D

Table   B.3.   Registers on the IBM DACA used for Digital I/O

Register	Name	Address	Meaning of Bits
2	Datum	22E2	0-15=datum
12	Device Number	C2E2	0-7=device number Set to 8 for digital I/O

INDEX